Atomic Absorption, Fluorescence and Flame Emission Spectroscopy
A Practical Approach

Atomic Absorption, Fluorescence and Flame Emission Spectroscopy

A Practical Approach

K. C. THOMPSON
Ph.D., B.Sc., A.R.C.S., D.I.C., M.R.I.C., C. Chem.
Principal Scientist with a U.K. Public Authority
Formerly Chief Chemist, Shandon Southern Instruments Ltd

R. J. REYNOLDS
B.Sc., A.R.C.S.
H.M. Chemical Inspector of Factories
Formerly Chief Chemist, Evans Electroselenium Ltd

Second Edition

CHARLES GRIFFIN & COMPANY LTD
London and High Wycombe

CHARLES GRIFFIN & COMPANY LIMITED
Registered Office:
Charles Griffin House, Crendon Street, High Wycombe
Bucks, HP13 6LE, England

First published 1970
Second Edition 1978

ISBN: 0 85264 246 6

Set and Printed in Great Britain at
The Pitman Press, Bath

Contents

Preface

Since standard commercial equipment became available, atomic absorption spectroscopy has grown more rapidly than any previous instrumental analytical technique. It has passed through the stages of initial development and application in research, and is now regarded as an essential tool in the routine laboratory.

Commercial equipment for the more recently introduced techniques of flameless electrothermal atomic absorption and hydride generation are now available, and as the underlying principles of these new procedures are rapidly being elucidated, the applications of atomic absorption will expand still further.

Atomic absorption has replaced many traditional wet methods for the estimation of metals in solution, and revealed errors in hitherto accepted procedures. Atomic absorption does not demand intricate sample preparation, and is an ideal tool for the non-chemist, e.g. the engineer, biologist or clinician interested only in the significance of the results.

The authors' objective is a practical working guide to a powerful and versatile technique. Because a deep theoretical knowledge of the subject is not necessary to fully utilize the technique, the comprehensive theory section is placed at the end of the book.

The first three chapters deal with the fundamental principles of atomic absorption, procedural consideration (interferences, selection of wavelength, flame system, sensitivities, etc.) and the general techniques of measurement. Chapters IV and V provide reference data on characteristics of the elements and the application of the technique to specific fields of analysis. All the information necessary to develop and perform reliable analyses using conventional atomic absorption spectroscopy is contained in these first five Chapters.

Chapter 6 describes the characteristics of standard equipment and gives some useful hints and advice to a would be purchaser of an atomic absorption instrument. Chapter 7 describes some further techniques including flameless electrothermal atomization which would appear to have almost unlimited potential and is capable of detecting sub-picogram amounts of some elements. Many other very new techniques, most still at the research stage, are described. Mention is also made of various 'historical' techniques that once appeared promising but have since fallen by the wayside.

The principles of the complementary techniques of flame emission and atomic fluorescence spectroscopy are treated in Chapter 8 whilst a comprehensive treatment of the theory of the atomic absorption technique is given in Chapter 9.

Information retrieval from the scientific literature is becoming increasingly expensive, so in order to assist the reader the carefully selected references given at the end of each Chapter are quoted in full to include the title of each paper.

March 1978

K. C. Thompson
R. J. Reynolds

Glossary of Abbreviations

A.A.	atomic absorption
A.A.S.	atomic absorption spectroscopy
A.F.	atomic fluorescence
A.F.S.	atomic fluorescence spectroscopy
A.P.D.C.	ammonium pyrrolidine dithiocarbamate (ammonium tetramethylene dithiocarbamate)
A.R.	analytical grade reagent
D.V.M.	digital voltmeter
E.D.T.	electrodeless discharge tube (microwave excited)
E.D.T.A.	ethylenediaminetetraacetic acid
F.E.S.	flame emission spectroscopy
F.S.D.	full scale deflection
i.d.	internal diameter
M.	Molarity (mol/dm^3)
M.I.B.K.	methyl isobutyl ketone (4-methyl-pentan-2-one)
o.d.	external (outer) diameter
oxine	8-hydroxyquinoline
P.T.F.E.	polytetrafluoroethylene
R.F.E.D.L.	radio frequency excited electrodeless discharge lamp
R.S.D.	relative standard deviation
T.E.L.	tetraethyl lead
T.M.L.	tetramethyl lead
u.v.	ultraviolet

IMPORTANT NOTE ON REAGENT HAZARDS

Some of the digestion procedures described or referenced in this book involve the use of hazardous chemicals[1] and must only be undertaken by or under the direct supervision of an experienced analytical chemist. Face shields and safety aprons must be worn and all digestions must be carried out in fume cupboards.

Procedures involving perchloric acid[2] can be extremely hazardous in unskilled hands and some very serious explosions have occurred with the mis-use of this reagent. **Perchloric acid must never be used in pressure digestion vessels.**

Particular care should be taken with hydrofluoric acid[1] as skin contact with this reagent causes no pain at the actual time of contact, but unless any acid is *immediately* washed off the skin, serious tissue damage will ensue.

The reader is urged to observe all necessary safety precautions when performing digestions. Extra special care is required whenever a new procedure is first attempted.

1. Muir, G. D., Hazards in the Chemical Laboratory. The Chemical Society, London 1977
2. Analytical Methods Committee
Notes on perchloric acid and its handling in analytical work. *Analyst* 84 (1959) 214

1 *Fundamentals*

Introduction

The potential of atomic absorption spectroscopy for the determination of metallic elements in chemical analysis was first realised by Sir Alan Walsh[1] who, during the mid-1950s, developed it into its modern form. Standard commercial equipment became available about 1960. Since that time the use of the technique in routine laboratories has become widespread, where it replaces many traditional wet methods for the estimation of metals in solution.

The outstanding advantage offered by atomic absorption is, that because for all practical purposes it is immune to spectral interferences, sample preparation can generally be confined to dissolution of the sample, or simple extraction of the species to be determined. Handling errors are therefore minimized, and analyses can be performed more rapidly than by procedures requiring more elaborate preparative steps.

The versatility of atomic absorption spectroscopy is demonstrated by the fact that it permits the estimation of between 60 and 70 elements at concentrations that range from trace to macro quantities.[2] Further, it is applicable to estimations of metals in organic and mixed organic–aqueous solvents as well as to those in aqueous solution.

Lastly, repeated analyses performed by atomic absorption are capable of a very high degree of reproducibility.

Basic Principles

If a solution containing a metallic species is aspirated into a suitable flame, an atomic vapour of the metal will generally be formed. Some of the metal atoms may be raised to an energy level sufficiently high to emit the characteristic radiation of that metal, a phenomenon that is exploited in emission flame photometry.

The overwhelming majority of the metal atoms, however, remain in the non-emitting, ground state. If irradiated with light of their own characteristic resonance wavelength, these ground-state atoms will absorb some of the radiation, the absorbance being proportional to the population density of atoms in the flame. This is the principle used in atomic absorption spectroscopy.

The technique thus offers the outstanding intrinsic advantage of specificity, since the atoms of a particular element can only absorb radiation of their own characteristic wavelength. Conversely, light of a particular frequency can only be absorbed by the specific element to which it is characteristic. Spectral interferences, which are so troublesome in emission methods, therefore rarely occur.

Basic Instrument Design

An atomic absorption spectrophotometer consists essentially of the following components, see Fig. 1.1:

1. A stable light source, emitting the sharp resonance line of the element to be determined.
2. A flame system into which the sample solution may be aspirated at a steady rate, and which is of sufficient temperature to produce an atomic vapour of the required species from the compounds present in the solution.
3. A monochromator to isolate the resonance line, and focus it upon a photomultiplier.
4. A photomultiplier that detects the intensity of light energy falling upon it, and which is followed by facilities for amplification and readout of the modulated light output from the hollow cathode lamp.

The light source is usually a lamp having a hollow cathode made of the element to be determined. The emission from this lamp is modulated so that its radiation only, and not that emitted from the flame, will be recorded in the readout signal. The most commonly used flame system is air–acetylene.

Operation

In operation, the readout is adjusted to read zero absorbance when a blank solution is sprayed to the flame, and the unobstructed light of the hollow-cathode lamp passes on to the photomultiplier. When a solution containing the absorbing species is introduced, part of this light is absorbed, resulting in a diminution of light intensity falling upon the photomultiplier and giving rise to a change in signal.

Standard solutions of the element to be determined are employed to construct a calibration curve from which the contents of test solutions can be obtained.

Comparison of Atomic Absorption Spectroscopy with Colorimetry

Colorimetry is perhaps the most commonly used procedure in analytical chemistry. Atomic absorption, however, is undoubtedly revealing errors due to erratic determinations in hitherto trusted colorimetric procedures. The reasons why these errors can, and do, occur in colorimetry, and why atomic absorption spectroscopy is less susceptible to them, become apparent only upon considering how these procedures are practised in the routine laboratory.

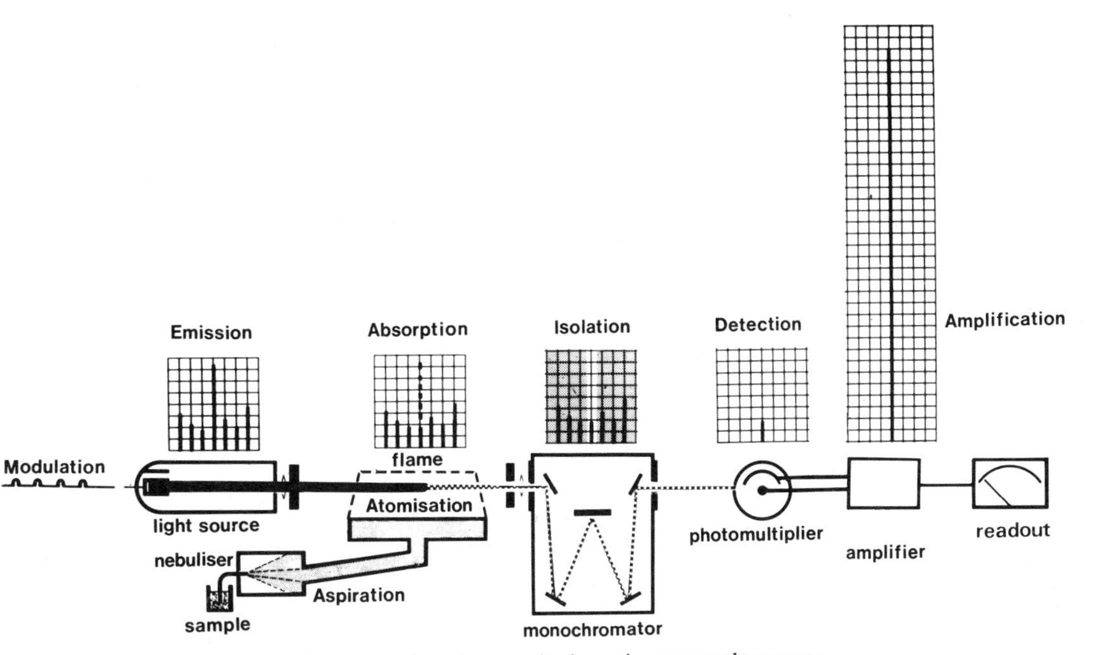

Fig. 1.1 Basic construction of an atomic absorption spectrophotometer

The perceptive and honest analyst will acknowledge that the conditions under which a colorimetric procedure is developed and assessed are usually very different from those under which it will be used in routine practice—particularly after some time has elapsed since its introduction, with perhaps a number of junior staff changes. When carried out to perfection with no mechanical loss of the element to be determined, a colour-change method is usually liable to a positive error. Where final judgement of the change is made with the human eye, this error will be positive and variable. Visual colour matching will also be liable to some error of variation, in this case both positive and negative. Such methods usually entail meticulous attention to detail, extending in some cases even to the time allowed to elapse between addition of reagents and measurement of colour intensity. They frequently involve separations of the species to be determined, with risk of errors through losses brought about by multiple operations.

The errors due to the intrinsic positive bias of a colour-change procedure and those due to mechanical losses tend to balance each other out, and have possibly deluded analysts as to the reliability of some methods.

Because of the simplicity of sample preparation required, atomic absorption is much less liable to these faults. When errors do occur they are usually glaringly large, and cause the analyst to check the instrument's operation and his standard and test solutions—actions that can usually be carried out quickly.

A full theoretical treatment of atomic absorption spectroscopy is given in Chapter 9.

Chap. 1 References and further reading

1. WALSH, A., The application of atomic absorption spectra to chemical analysis. *Spectrochim. Acta* 7 (1955) 108.
2. GATEHOUSE, B. M., and WILLIS, J. B., Performance of a simple atomic absorption spectrophotometer. *Spectrochim. Acta* 17 (1961) 710.

2 *Basic Procedural Considerations*

Interferences

It has been stated that atomic absorption spectroscopy is virtually free from spectral interferences, because a particular element can only absorb light of its own characteristic frequency; and conversely light of a particular frequency can only be absorbed by atoms of a specific element. Interferences, other than non-specific background absorption (see below) are therefore confined mainly to phenomena that affect the number of atoms in the flame and they may be listed under the following headings.

(a) Enhancements due to higher uptake of the test solution medium, as for example when an organic or mixed organic–aqueous solvent is used in place of water.
(b) Depressions due to the formation of refractory compounds that are not dissociated in the flame. The most common example of this condensed phase interference is furnished by the well-known phosphate depression of calcium absorption when the air–acetylene flame is used.
(c) Depressions due to ionization, as for example when a solution containing calcium is aspirated to the nitrous oxide–acetylene flame. At the temperature of this flame calcium compounds are not only dissociated to produce calcium atoms, but some of these atoms are raised to an energy level sufficiently high to produce ionization. The calcium ions so formed will not absorb the characteristic radiation of the ground-state atom and a depression in absorption is noted. Common elements particularly susceptible to this effect are the alkali and alkaline earth metals (Li, Na, K, Rb, Cs, Ca, Sr, Ba).
(d) 'Matrix' interferences can also occur. These are due to the surface tension and viscosity of a test solution being different from that of the standards, usually because of the presence of heavy concentrations of foreign ions in the former. This results in the uptake rate of the sample solution being lower than that for the standards, so that a smaller number of absorbing species per unit time are carried to the flame, and the observed absorption is therefore low.

Interferences of this type present little difficulty to the analyst. Enhancing effects may, of course, be exploited and depressive interferences are either compensated for by the use of matching standards, or overcome by the addition of interference suppressants (releasing agents).

Non-specific Background Absorption (see also p. 239)

This effect manifests itself as an apparent enhancement of the signal. It has been attributed to light being scattered by very small, unvaporized particles of sample matrix constituent(s) reaching the region of the flame through which the light from the hollow-cathode lamp passes. However, this explanation is not now generally accepted.[2,5,14] Current opinion is that it is mainly due to molecular absorption by matrix components that are not completely atomized in the flame. A good example of this interference occurs with the direct estimation of lead in urine at the 217·0 nm line.[12] Light of this wavelength is not only absorbed by the lead in the sample but some is also absorbed or scattered by the high concentration of matrix molecules (mainly sodium chloride) in the sample; erroneously high readings are thereby obtained. For most common sample matrices the degree of background absorption is not very dependent upon the wavelength over short wavelength regions of up to about 2 nm. It is therefore possible to correct for this background absorption by noting the absorbance at a non-resonance line within 2 nm of the analyte absorption line where the analyte does not absorb, and subtracting this value from the absorbance obtained using the analyte absorption line. A simpler procedure, that on many atomic absorption instruments can be performed automatically, is to measure the background absorbance using a hydrogen or deuterium hollow-cathode lamp using the same wavelength setting as the analyte absorption line and to subtract this from the absorbance obtained using the analyte absorption line (see p. 239).

In order to give the approximate magnitude of two typical background absorption signals, a solution containing 4000 μg/ml of calcium and a solution containing 4000 μg/ml of sodium (both present as chlorides) were found by the author (K.C.T.) to give background absorption signals equivalent to 0·38 and 0·09 μg/ml lead respectively at 217·0 nm in the slightly fuel-rich air–acetylene flame.

Where speed is not a major consideration, chelation and extraction of a heavy metal species from extraneous interfering material (e.g., Ca, K, Mg, Na, etc.) can often be employed to avoid the problem.

To summarize, non-specific background absorption can be observed when determining trace levels of elements in media that contain appreciable loadings of foreign ions. The effect can be observed at all wavelengths but generally is more significant at wavelengths below 350 nm. Whenever a new method is developed a check for non-specific background absorption should be made.

Control of Parameters that Influence a Determination

The parameters that lie within the analyst's control are flame conditions, constitution of solution and, in some cases, choice of wavelength.

THE FLAME SYSTEM

Optimum and uniform flame conditions are normally attained by selecting a burner height, setting the oxidizing gas (air or nitrous oxide) to a fixed flow rate, usually specified by the instrument manufacturer, and then regulating the fuel so that peak absorption is obtained when a suitable standard solution is aspirated into the flame. If this procedure is used it is found that for many elements there is a wide range of burner heights over which the maximum sensitivity attainable is sensibly identical.

The most commonly used flame is the air–acetylene system, and the fuel requirements for different elements vary. Some elements, such as calcium and magnesium require a flame that is fairly rich in its fuel content, and with these elements sharply defined conditions for maximum absorption are obtained. Maximum absorption for chromium is obtained with a flame that is just luminous, while molybdenum requires one that is distinctly luminous. It is worth noting that in the air–acetylene flame, calcium can show peak absorptions under two different fuel-support gas ratios. One occurs with a fuel-rich flame and the second with a distinctly leaner flame. The most sensitive conditions with the air–acetylene system are obtained under the latter (more oxidizing) flame conditions.

Maximum absorption for a large number of elements is generally stated to be attained under fuel-lean conditions. With elements exhibiting this feature the absorption can often apparently increase up to the point where the flame eventually lifts off the burner. In many such cases, investigation has shown that the apparent increase in absorption is largely due to the flame gases themselves absorbing more and more light as the fuel supply is diminished, indicated by the 'zero' reading for a blank solution following the absorption for a standard solution. Thus by adjusting to maximum absorption by the general procedure described above a 'false' absorption peak can be obtained. Operating under such conditions a higher level of noise and a lower sensitivity are attained than with a flame having a higher fuel content.

For the elements cobalt, copper, gold, iron, lead (particularly at the 217·0 nm line), nickel (particularly at the 232·0 nm line) and zinc, it is advantageous to use the following procedure when adjusting the flame conditions for optimum absorption:

(a) Light the flame, set the oxidizing gas flow and burner height and adjust to the correct wavelength for the element to be determined.

(b) Aspirate a blank solution and adjust the fuel gas flow so as to obtain maximum throughput of light on to the photomultiplier (i.e. maximum transmission). The flame conditions thus established are then used for the estimation.

In general, it can be said that the hotter the flame used, the less prone will a determination be to depression caused by chemical interferences. For example, the well-known depression caused by phosphate upon the absorption of calcium in the air–acetylene flame can be overcome by using the nitrous oxide–acetylene flame.

The use of a hotter flame, though, can itself give rise to a depressive effect with elements such as calcium, due to their pronounced tendency at high temperatures to form ions which, of course, will not absorb the radiation from the lamp. In the case of calcium, this type of depression can be overcome by dosing the sample and standards with a large excess of potassium chloride. Potassium is even more readily converted to the ionic state than calcium and in doing so gives rise in the flame to a dense population of electrons which annul the tendency of calcium to ionize. The lost absorption is thereby restored, and the benefits of high response and freedom from phosphate interference are obtained.

The absorption of chromium, although lower in the hotter nitrous oxide–acetylene flame than in the air–acetylene flame, is free from the troublesome depressive effect that the presence of iron exerts in the cooler system. Advantage can be taken of this fact to simplify the determination of chromium in steels. Molybdenum exhibits a much more linear response in the nitrous oxide–acetylene flame than with the very fuel-rich air–acetylene system.

Maximum absorption for both silicon and boron is attained with a very fuel-rich (luminous) nitrous oxide–acetylene flame. The noise level, though, also increases as the fuel content of the flame is raised, and depositions of carbon in the burner slot can occur. In practice, therefore, it becomes necessary to select flame conditions that give rise to an acceptable level of noise, and freedom from carbonization, which entails working somewhat below maximum sensitivity.

Tin exhibits a much higher sensitivity and better stability in the air–hydrogen flame than in either the nitrous oxide–acetylene or fuel-rich air–acetylene systems. However, chemical interferences are much more likely to occur in low-temperature flames such as air–hydrogen or air–propane than in higher-temperature flames such as air–acetylene or nitrous oxide–acetylene. For this reason low-temperature flames are seldom used.

A good example of this interference effect is the depression of tin absorption in the air–hydrogen flame in the presence of the sulphate anion.

Wavelength

Choice of wavelength can be an important factor in the success or otherwise of a determination. For most purposes this is limited to mere selection of the most absorbing line, though cases do occur in which the decision requires more consideration.

Both the nitrous oxide-acetylene and air-acetylene flames exhibit a series of bands between 280 and 330 nm from the OH radical. In addition the nitrous oxide-acetylene flame exhibits strong bands around 337 nm from the NH radical and also around 359, 388 and 422 nm from the CN radical.

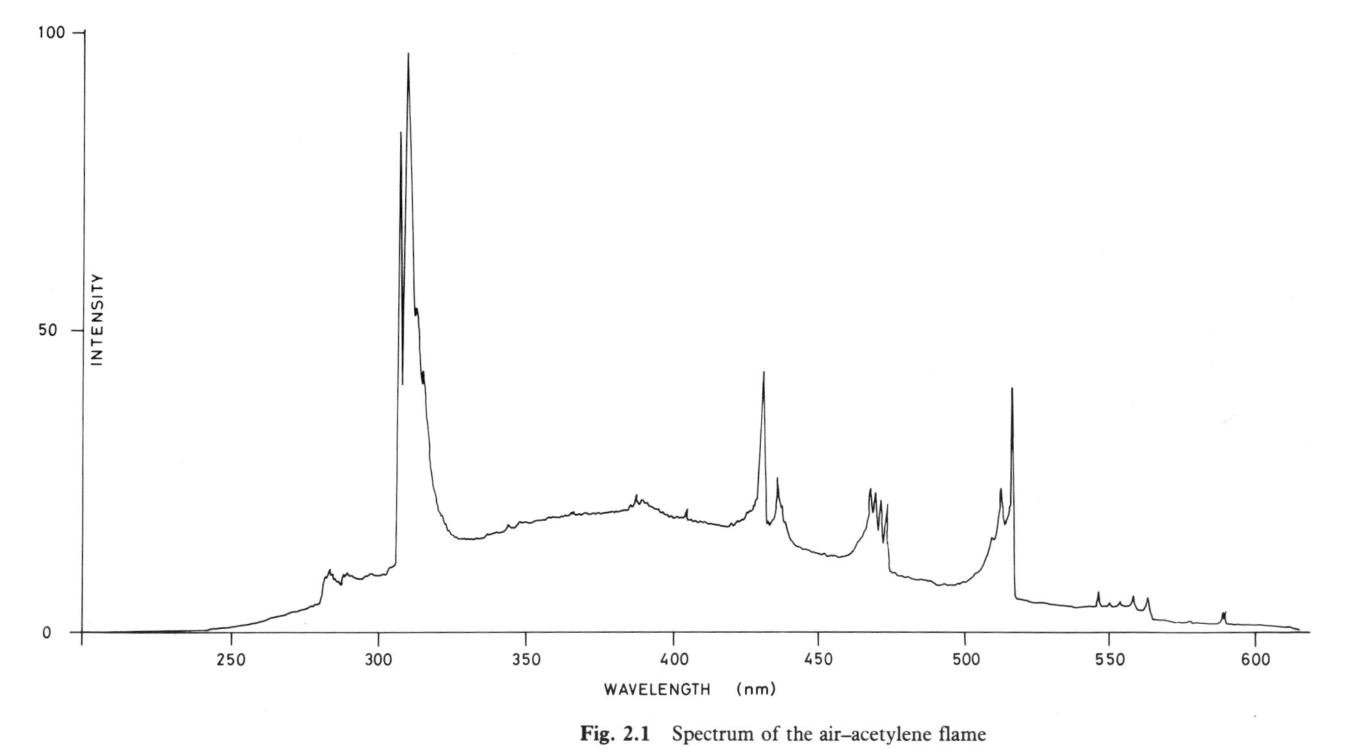

Fig. 2.1 Spectrum of the air–acetylene flame

0·6 nm spectral bandpass
Stoichiometric Flame
Distance from top of burner grid to bottom of monochromator entry slit—10 mm.
Photomultiplier voltage 450 V

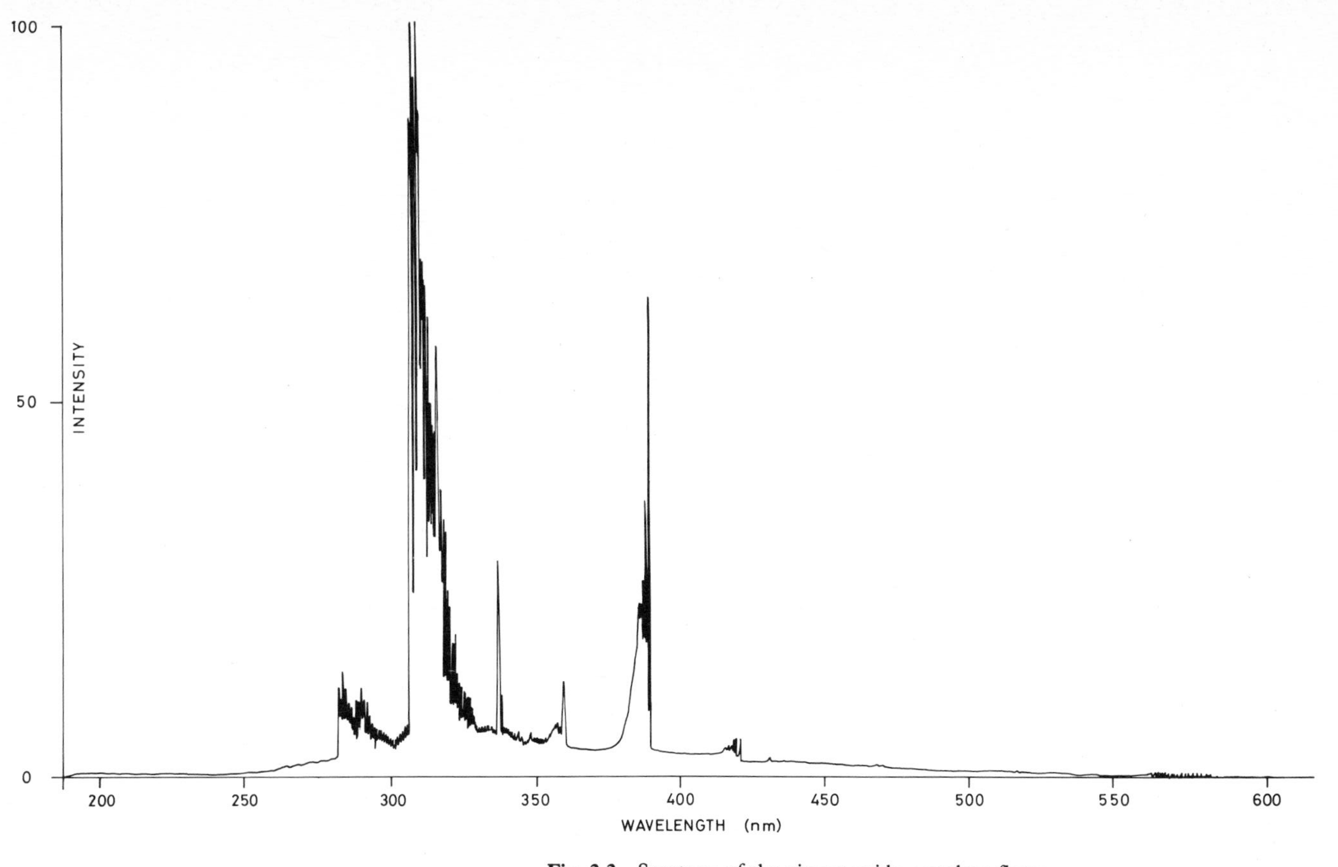

Fig. 2.2 Spectrum of the nitrous oxide–acetylene flame

0·2 nm spectral bandpass
Slightly fuel-rich flame (2 mm 'red feather')
Distance from top of burner grid to bottom
of monochromator entry slit—10 mm.
Photomultiplier voltage 410 V

The emission manifests itself mainly as background noise and can affect the choice of wavelength for a particular determination. This is particularly noticeable if the line being used is of low intensity. Thus the choice of wavelength can be influenced by the flame system selected, which itself frequently depends upon possible interfering substances in the solution to be analysed. For example, it is advantageous to use the nitrous oxide–acetylene flame for the estimation of chromium in steel in order to avoid the troublesome interference from iron that occurs when the cooler air–acetylene flame is used. With the hotter flame, and using a hollow cathode lamp of the original argon-filled type, the background noise at the most sensitive lines at 357·9 and 359·4 nm is excessively high, so that it becomes essential to use the less absorbing but more stable line at 425·4 nm. The modern improved high spectral output lamps for chromium allow the most sensitive lines to be used for determinations with the nitrous oxide–acetylene system.

It may happen that the linearity of the most absorbing line drops off very severely as the concentration of the absorbing species increases. An example of this phenomenon is provided by nickel (see Fig. 2.3). The less sensitive line at 305·1 nm can be used for nickel concentrations of up to 100 μg/ml.

Solution Conditions

The constitution of standard and test solutions, the third parameter within the analyst's direct control, can also be adjusted to enhance sensitivity. The addition of interference suppressants such as lanthanum chloride or E.D.T.A. disodium salt (again to overcome the depressive effect of phosphate upon

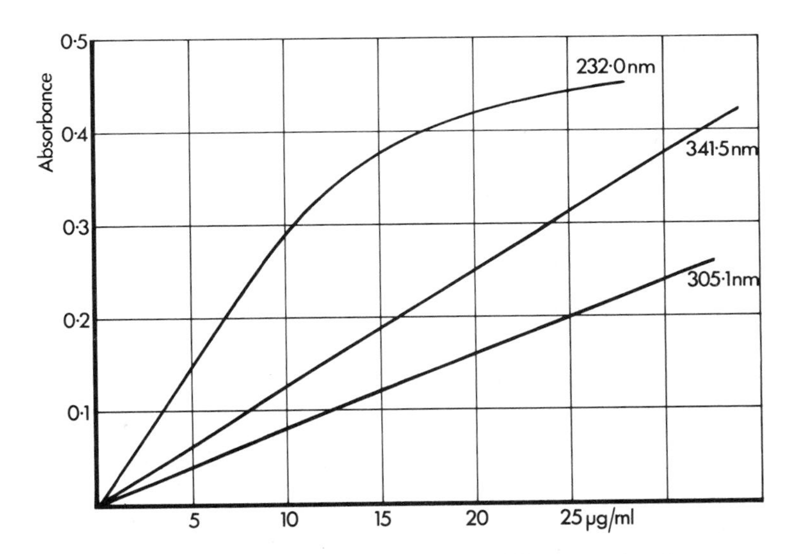

Fig. 2.3 Response curves for nickel at different wavelengths

calcium in the air–acetylene flame) is a well-established procedure. It is of interest to note that the actions of these two reagents, either of which produce the same result, are different. Lanthanum chloride reacts preferentially to form lanthanum phosphate thereby releasing the calcium atoms for absorption in the flame.[13] E.D.T.A. disodium salt, by contrast, exerts a greater affinity than phosphate towards calcium. The calcium–E.D.T.A. compound so formed then breaks up in the flame to yield calcium atoms which are available to absorb light energy of the correct wavelength. Again the absorption of titanium is considerably enhanced in solutions containing 2 per cent HF.

The use of organic solvents that are miscible with water, to enhance an element's response, was at one time a fairly common practice. Industrial methylated spirits, isopropyl alcohol, and acetone were some of the most commonly used. Aqueous mixtures of these materials show little tendency to change composition upon evaporation, have good solvent properties, and give rise to an enhancement that is independent of small changes of solvent composition. The effect of such a solvent is entirely physical. It gives rise to a medium that is more readily and efficiently nebulized so that a greater proportion of small droplets, and hence absorbing species, is carried to the flame.

Determinations that are carried out on extractions made into organic solvents, as, for example, when heavy metals (such as lead) are chelated at dilution in an aqueous medium and then concentrated into methyl isobutyl ketone, gain in sensitivity by both the concentration factor, and the greater aspiration rate of the solvent. This latter phenomenon also occurs in the direct determination of the metallic constituents of oils and petrols.

Sensitivity and Limit of Detection

Conventions for Specifying the Performance Capabilities of Instruments

There are two commonly quoted figures by which instrument manufacturers declare the performance capabilities of their equipment; sensitivity for 1 per cent absorption and limit of detection.

Sensitivity for 1 per cent Absorption (now known as Characteristic Concentration)

This figure was the one originally used to specify the performance of atomic absorption spectrophotometers, and is more appealing to instrument engineers and physicists than to chemists. Sensitivity for 1 per cent absorption (or characteristic concentration) is a theoretical number dependent on the efficiencies of the lamp, atomizer, flame system and monochromator. Its greatest drawback is that it takes no account of noise level, so is of little value as a guide to the least quantity of an element that can be determined.

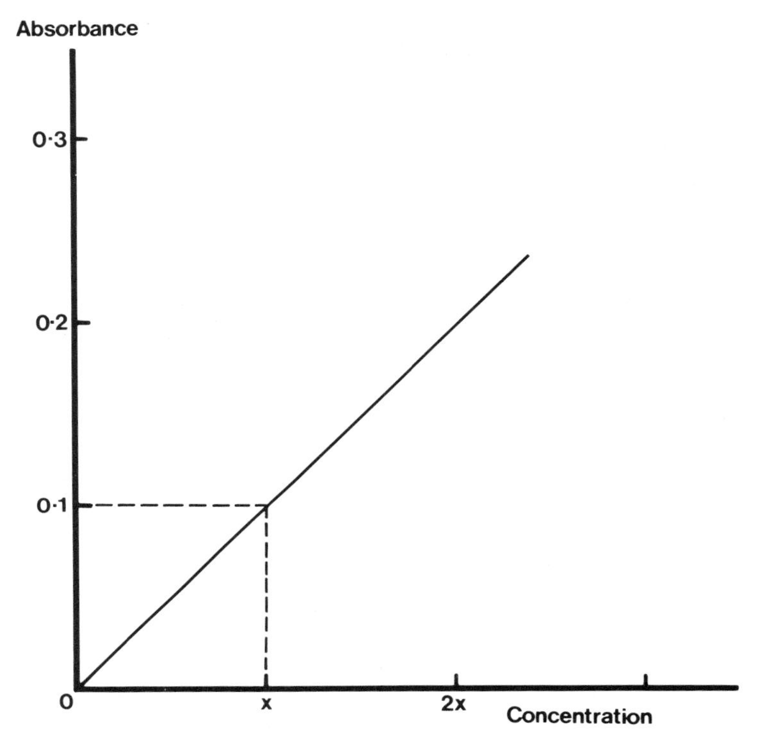

Fig. 2.4 Calculation of sensitivity for 1 per cent absorption

The 1 per cent absorption (or characteristic concentration) figure is obtained in the following manner.

1. The instrument is adjusted so that with a blank solution (normally deionized water) aspirated to it, a reading of zero is registered, and if the light from the lamp were totally obscured, a reading of infinity should be obtained on the absorbance scale. Under these conditions, if Beer's law is obeyed, a direct linear relationship will be obtained between the concentration of the absorbing species in a solution aspirated to the unit, and the absorbance registered.

   At absorbances below about 0·6, this condition is approximately attained for many elements.

2. A concentration response curve is then constructed from aqueous solutions of the species and the concentration ($C_{0 \cdot 1}$) that gives rise to a signal equivalent to an absorb*ance* of 0·1 read from it. If Beer's law is obeyed

$$C_{0 \cdot 1} = K \times 0 \cdot 1 \text{ (where K is a constant.)}$$

3. Optical density or absorbance is defined as $\log_{10} I_0/I$ where I_0 is the intensity of incident light and I is the intensity of transmitted light. If 1 per cent of the

incident light has been absorbed (1 per cent absorp*tion*) then the absorb*ance* of the medium is given by: $\text{Log}_{10}100/99 = 0{\cdot}0044$. Again applying the Beer's Law conditions

$$C_{1\%} = K \times 0{\cdot}0044$$

where $C_{1\%}$ is the concentration that gives rise to 1 per cent absorption (i.e. the sensitivity for 1 per cent absorption),

$$\therefore C_{1\%} = \frac{C_{0\cdot1} \times 0{\cdot}0044}{0{\cdot}1} = C_{0\cdot1} \times 0{\cdot}044$$

Thus the sensitivity for 1 per cent absorption is the concentration of an aqueous solution that gives a signal equivalent to an optical density of 0·1, multiplied by 0·044.

Limit of Detection

For low-level determinations of metals in solution (absolute absorbance less than 0·1), it is usually necessary to resort to the use of scale expansion. In order to assess whether or not the technique is applicable to a particular low-level determination, it is necessary to carry out an actual test run over the range required, using the appropriate scale expansion facility. From the data obtained from such a run the true detection limit, which until recently has usually been taken as twice the noise level, can be obtained.

Fig. 2.5(a) and (b) show recorder traces for two elements of about the same sensitivity, one of which (Cd) exhibits a much higher noise level than the other (Li). The limit of detection for cadmium is therefore less favourable than for lithium in this instance. Errors in interpreting the output trace can be overcome by using the integration mode of operation, which is now standard on all modern atomic absorption units. In this mode the signal is integrated for a preselected fixed time period (typically 3–15 seconds). The integrated readings so obtained are thus absolutely steady, so that there is no uncertainty in taking a single reading.

Obviously the former criterion for the limit of detection will not apply to such readings. For this reason it has been proposed, by the Atomic Absorption Spectroscopy Group of the Society of Analytical Chemistry (U.K.), that the limit of detection shall be defined as: 'The minimum amount of an element which can be detected with a 95 per cent certainty. This is, that quantity of the element that gives a reading equal to twice the standard deviation of a series of at least ten determinations at or near the blank level.' This definition assumes a normal distribution of errors and can, of course, be equally well applied to the mean of readings taken in the 'direct' mode. As the distribution of errors is not always normal, this procedure often results in somewhat optimistic detection limits. For a good discussion on the significance of detection and determination limits ref. 15 should be consulted.

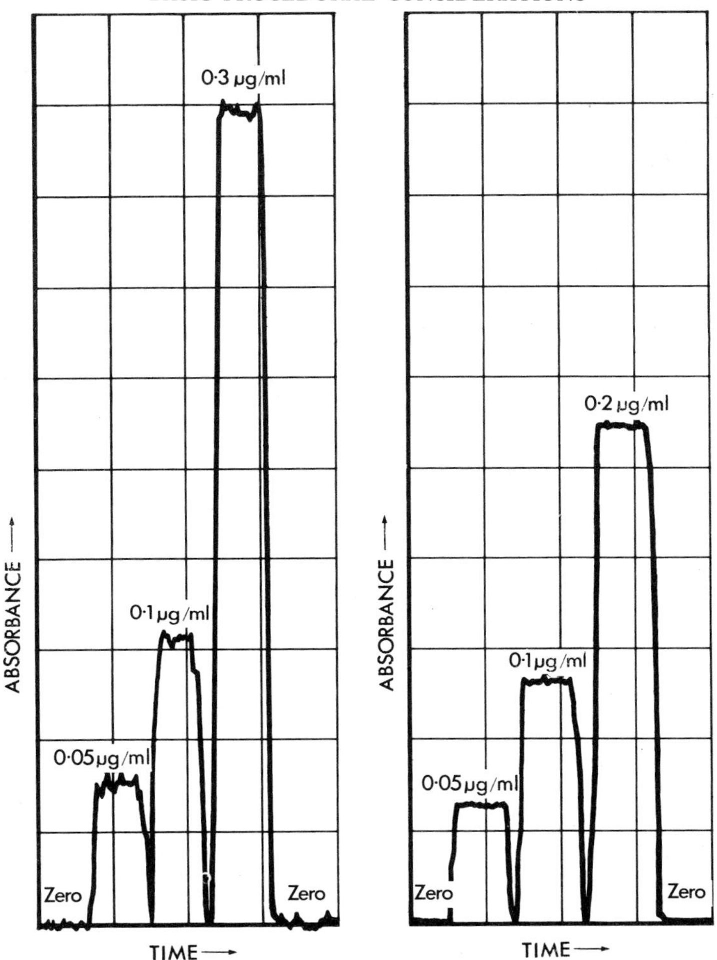

Fig. 2.5 (a) Example of an element showing a fairly high noise level. Cadmium in aqueous solution. Wavelength 228·8 nm

Fig. 2.5 (b) Example of an element showing a low noise level. Lithium in aqueous solution. Wavelength 670·8 nm

Factors Affecting Limit of Detection Values

As mentioned above, 'limit of detection' provides a much more meaningful guide of an instrument's analytical capability, to a chemist, than 'sensitivity for 1 per cent absorption' (or characteristic concentration). However, a straightforward statement of 'limit of detection' by a single value for a particular element can still be misleading, since the performance attainable by an atomic absorption spectrophotometer, when extended to the limit of its capabilities, is dependent upon many operating parameters.

Specifically, the performance of the same instrument can vary from operator to operator, from day to day or even determination to determination, according to the ambient temperature, humidity, etc., and in extreme cases can be affected by fluctuations in the mains supply. Different instruments of nominally identical specifications are likely to vary slightly in performance, the main cause for this variation being unavoidable differences in the characteristics of nebulizer and hollow-cathode lamps.

The stability and light intensity from the lamp, and the stability of the support gas supply, particularly air if delivered from a compressor, will affect the low level performance of an atomic absorption spectrophotometer to a considerable extent. Furthermore, in an analytical estimation the limit of detection will be dependent upon the solution matrix, and will in general be less favourable than that for a pure solution of a salt of the element. The flame system used will, of course, also affect the limit of detection value.

A realistic way of stating limit of detection values might, therefore, be as a range, into which the performance attainable from a particular model ex-works, for pure aqueous solution of a salt of the element, could confidently be expected to lie. This procedure has been adopted for the figures quoted in Chapter 4.

Precision of Analyses

The instrumental precision with which a determination can be made will vary from element to element. For a given element the precision will be dependent upon the concentration of the element as well as the matrix in which the element is being determined (i.e. the concentrations of foreign species, composition of the solvents, etc.).

The most striking feature of atomic absorption methods is the excellent reproducibility that is obtained upon successively aspirating the same sample solution.

For calcium determinations, work carried out on a very basic atomic absorption spectrophotometer, without scale expansion, showed a relative standard deviation of 0·007 at the 5 μg/ml level and 0·01 at the 15 μg/ml level.

Safety Considerations

Safety aspects in atomic absorption are often overlooked but there are various points that should be considered carefully. It is essential to read thoroughly the instrument manufacturers' handbook.

1. Acetylene

Ideally acetylene cylinders should be placed in a secure, weatherproof, adequately ventilated and fire resistant store. The cylinders should be stored upright and securely fastened to a wall or immovable bulkhead. A flashback arrestor should be fitted close to the acetylene cylinder. Approved regulators and hose connectors should always be used and it is important to remember that acetylene cylinders and hose connectors have left-hand thread fittings. A shut-off valve should be positioned close to the atomic absorption spectrophotometer. Ideally seamless mild steel of a quality not less than that required to British Standard 1387:1967, medium or heavy quality not galvanised on internal surfaces, or seamless stainless steel tubing should be used to convey acetylene from an external store. It should be labelled and painted an appropriate colour (maroon in the U.K.). Copper tubing or fittings must never be used because explosive copper acetylide can be formed. Only certain types of brass are permitted. The acetylene manufacturer should always be consulted before installing any pipework as he will be well acquainted with the appropriate legislation and all safety aspects. The acetylene pressure should not exceed 0.621 BAR (9 p.s.i.g.) above that of the atmosphere. Regular tests for leaks using soap solution should be made at all joints. The acetylene should always be turned off at the cylinder at the end of each run.

2. Fume Extraction

It is essential to fit an efficient fume extraction system to vent the exhaust gases. The fan should be located as close to the outlet as possible and the fan blades should be of a material that will withstand the hot exhaust gases from a nitrous oxide-acetylene flame. Most instrument manufacturers give comprehensive details for construction of an efficient fume extraction system. Ideally the extractor should be wired up so that it is automatically switched on when the atomic absorption unit is switched on.

The exhaust gases from a nitrous oxide-acetylene flame contain appreciable quantities of toxic nitrogen oxides, so even if only 'harmless' substances are nebulized it is essential to have an efficient fume extraction system.

3. Flame System

Always check that the water trap is full prior to igniting the flame and closely follow the manufacturers' operating instructions for lighting and extinguishing the flame. Many instruments now have completely automated systems for these procedures. The waste liquid from the spray chamber should never be collected in a glass container; always use a suitable plastic container, which should be placed in a well ventilated area.

If cyanide solutions are to be used, always ensure that the contents of the water trap and waste reservoir are distinctly alkaline and wash them out at the end of each run.

Extra care should be exercised when nebulizing organic solvents. If chlorinated hydrocarbons are nebulized, the exhaust gases will contain appreciable quantities of phosgene.

The nitrous oxide-acetylene flame should *never* be viewed directly with the naked eye as it emits strongly in the ultra-violet region of the spectrum. Flames should be viewed through the protective screen supplied by the manufacturer.

If appreciable carbon build-up occurs along the burner slot during operation of the nitrous oxide-acetylene flame, the flame must be shut down and the burner slot carefully cleaned. (The burner head will be very hot!). Never attempt to scrape off the carbon deposits with the flame alight.

4. LIGHT SOURCES

Ultra-violet radiation has a detrimental effect upon the eyes. This effect is not immediately apparent but comes on some hours after exposure. (It is similar to sunburn).

Never directly view hollow-cathode lamps through the silica end window. Extra special care should be taken with radiofrequency excited and microwave excited electrodeless discharge lamps as these emit very large amounts of ultra-violet radiation, but visually can appear of low intensity. Ordinary crown glass spectacles do *not* give full protection against ultra-violet radiation. Suitably tinted glass is required.

Never directly view a flameless electrothermal atomizer device during the atomization stage as repeated exposure to high levels of infra-red radiation can result in cataracts.

Chapter 2 References and further reading

1. BARNET, W. B. Acid interferences in atomic absorption spectrometry. *Anal. Chem.* 44 (1972) 695.
2. BILLINGS, G. K., Light scattering in trace-element analysis by atomic absorption. *Atomic Abs Newsletter* 4 (1965) 357.
3. CHAKRABARTI, C. L. and SINGHAL, S. P., Effect of complexing agents and organic solvents on the sensitivity of atomic absorption spectroscopic technique. *Spectrochim. Acta.* 24B (1970) 663.
4. DE GALAN, L. and SAMAEY, G. F. Trivial causes for the bending of analytical curves in atomic absorption spectrometry. *Spectrochim. Acta.* 24B (1970) 245.
5. KOIRTYOHANN, S. R. and PICKETT, E. E., Light scattering by particles in atomic absorption spectrometry. *Anal. Chem.* 38 (1966) 1087.
6. MARKS, J. Y. and WELCHER, G. G., Interelement interferences in atomic absorption analyses with the nitrous oxide–acetylene flame. *Anal. Chem.* 42 (1970) 1033.
7. MARUTA, T., SUZUKI, M. and TAKEUCHI, T., Interferences of acids in atomic absorption spectrophotometry. *Anal. Chim. Acta.* 51 (1970) 381.
8. RAMPON, H., Atomic absorption spectrophotometry interference. *Chim. Anal.* 51 (1969) 627.
9. REED, M. F., and REES, E. D., Noise, sensitivity and interference in atomic absorption spectroscopy. *Amer. Lab.* 2 (1970) 45 (Nov).
10. SKOGERBOE, R. K., and GRANT, C. L., Comments on the definitions of the terms sensitivity and detection limit. *Spectry. Lett.* 3 (1970) 215.
11. SLAVIN, S., BARNETT, W. B., and KAHN, H. L., The determination of atomic absorption detection limits by direct measurement. *Atomic Abs. Newsletter* 11 (1972) 37.
12. WILLIS, J. B., Determination of lead and other heavy metals in urine by atomic absorption spectroscopy. *Anal. Chem.* 34 (1962) 614.
13. YOFE, J., and FINKELSTEIN, R., Elimination of anionic interference in flame photometric determination of calcium in the presence of phosphate and sulphate. *Anal. Chim. Acta.* 19 (1958) 166.
14. KOIRTYOHANN, S. R., and PICKETT, E. E., Spectral interference in A.A.S. *Anal. Chem.* 38 (1966) 585.
15. ANNUAL REPORTS ON ANALYTICAL ATOMIC SPECTROSCOPY, 1973, vol. 3, p. 55. Society for Analytical Chemistry (Chemical Society).

3 *The Techniques of Measurement*

Types of Determinations

There is no branch of chemical analysis involving the determination of metals in solution that cannot without advantage use atomic absorption spectroscopy. The technique is already a firmly established procedure in such widely varying fields as clinical chemistry, ceramics, petroleum chemistry, metallurgy, mineralogy, biochemistry, soil analysis, water supplies, and effluents. The general procedures used in performing analyses are conveniently categorized according to the technique required for sample and standard preparation, as shown below:

1. Determinations that can be made against simple standard solutions containing only the element being sought.
2. Trace estimations performed by simple scale expansion.
3. Determinations that require the addition of a chemical interference suppressant to the sample and standard solutions.
4. Determinations where enhancement by the use of mixed organic–aqueous solvents is exploited.
5. Determinations requiring the use of complex standard solutions prepared so as to match approximately the composition of the test solutions.
6. The use of extractive concentration techniques.
7. Determinations carried out directly and entirely in non-aqueous solvents.
8. Determinations of macro constituents.
9. Indirect determinations, usually exploiting chemical amplification reactions.
10. The method of additions.

1. Determinations made against simple standard Solutions containing only the Element (or Elements) being sought

Although interferences of one metal upon another, or of excess acid[8] upon a metal's absorption do occur, atomic absorption spectroscopy is in general remarkably free from such effects. When it is required to develop a procedure for large numbers of similar analyses (particularly when only a few of the elements present in the sample are to be estimated) it is always worthwhile examining the possibility of exploiting this simple approach.

Comparison against simple standards is valid for many analyses and because of its speed is of great value for sighter trials. It is especially likely to be possible when the nitrous oxide–acetylene flame is being used. It is also worth noting that nickel, zinc, iron, copper, and lead are remarkably free of interferences in the air–acetylene flame from the elements with which they are likely to be found in association.

Sometimes, indeed, the analyst is prohibited from using complex standards, an example of this situation occurring in the trace determinations of copper, cadmium, and zinc in tin–lead solders. These determinations are complicated by the fact that even the purest sources of lead and tin salts (readily available to the chemist) can contain substantial quantities of cadmium and zinc. The preparation of matching standards that contain the two major constituents of the alloy is therefore impracticable. An exhaustive series of recovery trials using that well-known analytical gambit, the method of additions (see later), proved that accurate results could be obtained merely by using standards that contained copper, cadmium, and zinc in concentrations that covered the ranges of these elements in the test solutions.

This elementary procedure finds application in a very wide range of analyses, but is of particular value in the analysis of waters, effluents and soil-extracts, where there is little else in the test solutions other than the elements to be determined and even these are present at low levels.

2. Trace Estimations performed by simple Scale Expansion

All atomic absorption spectrophotometers now possess facilities that permit scale expansion. Such facilities have enhanced the capability of the technique for trace estimations performed directly in aqueous media.

In most instruments the expansion is continuous up to a maximum of 20 to 50 times.

Fig. 3.1 shows a typical atomic absorption response curve for an element obtained using unit (X1) scale expansion. It has been marked to indicate the ranges over which scale expansion factors of 5 and 10 times would be used. Over the X1 range it is, of course, possible to improve the precision by resorting to the procedure of zero suppression followed by scale expansion (see below).

3. Determinations that require the Addition of a Chemical Interference Suppressant to the Sample and standard Solutions

One of the most firmly established and expanding applications that atomic absorption has found is for the determination of calcium and magnesium in blood serum. The phosphate ion occurs in this fluid at a sufficiently high concentration level to exert a depressive effect upon the calcium absorption when an air–acetylene flame is used. This depression is conveniently overcome by the addition of a heavy excess of lanthanum chloride or E.D.T.A. disodium salt. Comparison must, of course, be made against a blank and standard solutions containing the suppressant.

Strontium chloride may also be used as a releasing agent to overcome

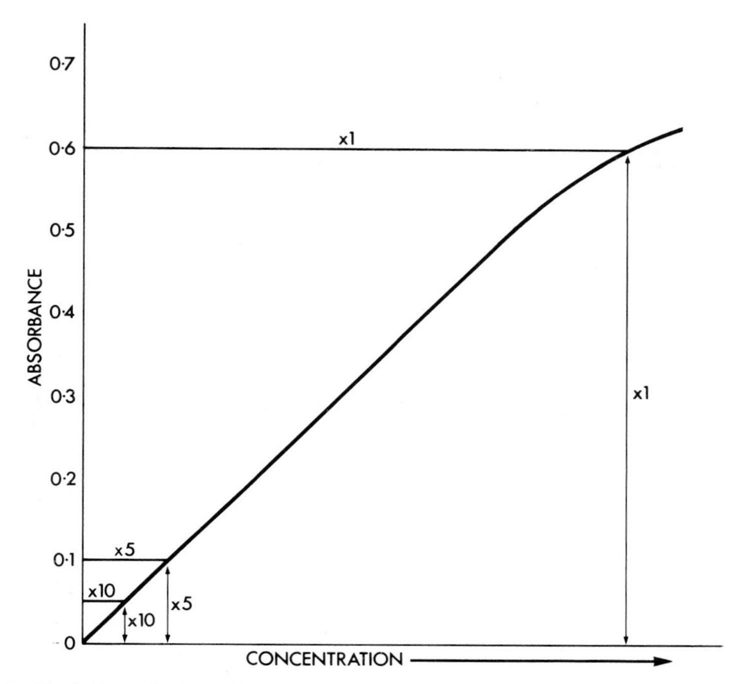

Fig. 3.1 Typical atomic absorption response curve. Showing ranges over which scale expansion could be used

phosphate interference, but is less efficient than lanthanum chloride, and the quantity added must be controlled more stringently in order to avoid non-uniform over-enhancement of the calcium response. Figure 3.2 illustrates this. Lanthanum chloride is undoubtedly the most useful of the releasing agents employed to overcome chemical interference. It effectively overcomes the depressive effect of aluminium upon calcium and magnesium. It may also be used to overcome the depression of iron upon chromium that is so pronounced in the air–acetylene flame. The addition of a heavy excess of an alkali–metal to a solution containing calcium, strontium, or barium, that is to be estimated in the nitrous oxide–acetylene flame may be considered as belonging to this category. Here, though, the interference to be suppressed is a physical one (production of unwanted calcium ions) rather than one of a chemical nature.

4. Determinations where Enhancement by the use of Organic /Aqueous Solvents is Exploited

The use of mixed aqueous solvents was, until 1968, the most common procedure used by the analyst to enhance an element's absorption. It is now much less frequently employed, since the overall capability of atomic

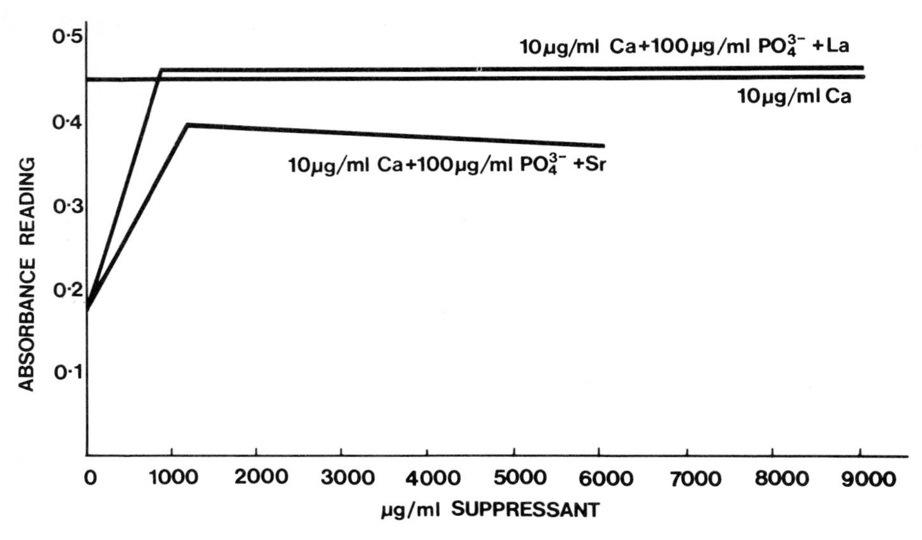

Fig. 3.2 Comparison of lanthanum chloride and strontium chloride as interference suppressants

absorption equipment has so much improved. The greatest disadvantage of this procedure is that although the sensitivity is certainly improved, the noise level also very often increases, so that the final limit of detection is not improved.

As stated previously under 'Solution Conditions' (p. 11), industrial methylated spirits and isopropyl alcohol are the two solvents most commonly used to bring about this type of enhancement.

The direct determination of trace metals in liquors such as whisky is, of course, facilitated by the fact that the matrix itself enhances the absorption of the element being sought. Comparison must, of course, be made against standards and a blank that also contain alcohol at the same level as the whisky.

5. Determinations requiring the use of Complex Standard Solutions, prepared so as to match approximately the Composition of the Test Solutions

Determinations in which measurement is made against complex standards prepared to resemble, in constitution, the solution to be analysed are perhaps the most common carried out by the analyst. Most chemical interferences reach a plateau value, after which further addition of the interfering substance has no effect upon the element to be determined. Examples of the types of analyses tackled by this procedure are multitudinous, but the technique is especially valuable for the analysis of minerals, ores, silicates, ceramics, slags, and alloys.

Apart from the main intention of overcoming chemical interferences the procedure reduces the number of separate standard solutions required. For example, in the analysis of a soil extract containing calcium, magnesium, sodium, and potassium, although none of the elements present interferes with the absorption of any of the others, when the air–acetylene flame is used, complex standards containing all of them will reduce the number of flasks containing solutions from, say, 12 to 3 (if three concentrations for each element are prepared to cover the expected ranges in the test solutions).

Examples of the use of complex standards to counteract interference effects, and also to rationalize the number of solutions to be prepared, are provided by the determination of the minor constituents of a duralumin alloy (p. 142), cement analysis (p. 157) and silicate analysis (p. 150).

6. The use of Extractive Concentration Techniques[1–4]

Procedures that involve prior extraction of the species to be determined into an organic solvent have been widely employed. They are particularly valuable for the determination of trace metals in water, effluents, and extracts from food and biological samples. Extraction also provides a useful means of separating metals such as lead and zinc from solutions that contain heavy loadings of foreign materials, such as Ca, K, Mg and Na, the presence of which can cause errors due to non-specific background absorption.

The most useful method for extractive concentration of heavy metals is that employing chelation with ammonium pyrrolidine dithiocarbamate (A.P.D.C.) followed by extraction into methyl isobutyl ketone (M.I.B.K.). The A.P.D.C. chelation complexes of the common metals are formed over a wide *p*H range, e.g. Mo, *p*H 2–6; Sn, *p*H 2–8; Mn^{II}, Fe, Co, Ni, Cu, Zn, Pb and Cd, *p*H 2–14. Most authorities recommend a *p*H of between 2·2 and 2·8 for the extractions.

100–200 ml of the clear aqueous solution that contains the metal(s) to be determined, and which may have been previously reduced in volume by evaporation, is transferred to a separating funnel. The *p*H is adjusted by the addition of 0·5 M hydrochloric acid or 0·5 M sodium hydroxide solution to lie in the range 2·2 to 2·8. Four ml of a 1 per cent aqueous solution of A.P.D.C. is added, followed by 10 ml of M.I.B.K. The vessel is stoppered and shaken for two minutes, the phases allowed to separate, and the aqueous layer run to waste.

The trace heavy metals are now determined in the organic extract against standards that cover the appropriate range and have also been prepared by chelation and extraction as described above. The absorption obtained upon nebulizing the extracts is enhanced by both the concentration factor and the greater volatility of the solvent. A recorder trace illustrating the sensitivity obtainable for lead extracted into M.I.B.K. with a simple atomic absorption spectrophotometer at $\times 10$ scale expansion is shown in Fig. 3.3.

7. Determinations carried out Directly and Entirely in Non-Aqueous Solvents

One of the major advantages that atomic absorption spectroscopy offers to the chemist is that solutions need not be confined to an aqueous medium. This means that digestions of or extractions from non-aqueous media that are so time-consuming, necessitate the running of reagent blanks, and are always open to the risk of handling losses, are avoided. This advantage is especially convenient in the field of petroleum analysis. The direct determinations of aluminium and vanadium in crude and fuel oils is readily accomplished, as are

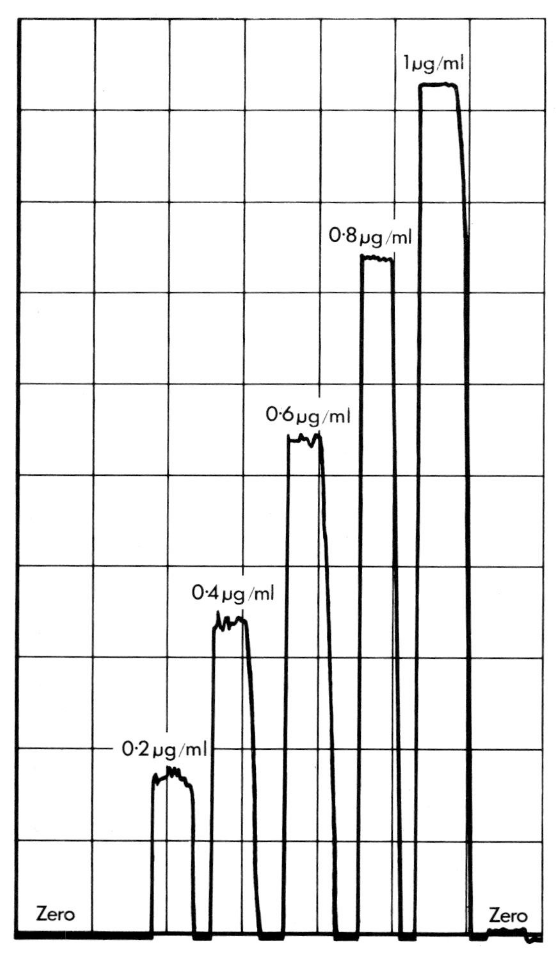

Fig. 3.3 Lead chelated and extracted into isobutyl methyl ketone. Wavelength 217 nm (concentration refer to M.I.B.K. extract)

those of additives (zinc, calcium and barium) and even wear metals in lubricating oils.

It is, of course, necessary to dilute oils and greases with a suitable solvent before they can be aspirated to an atomic absorption spectrophotometer. References will be found in the literature that recommend the use of iso-octane (2,2,4-trimethylpentane), heptane, xylene and other hydrocarbons for this dilution. It is my experience though (R.J.R.), that such solvents are unsuitable, since they give rise to such a rich smoky flame that it is found necessary (in order to establish non-luminous conditions) to restrict the acetylene flow to the point where the flame eventually lifts off the burner. The use of a diluent containing an oxygenated compound, such as 1 + 9 propan-2-ol + white spirit, overcomes this difficulty. With this diluent, solutions may be prepared that allow determinations to be safely carried out with both the air–acetylene and nitrous oxide–acetylene flames.

The standards for such determinations must, of course, be prepared from suitable oil-soluble compounds, or from previously analysed oils, so as to resemble the test solutions.

8. The Determination of Macro Constituents

The estimation of elements present as major constituents of a material is one of the most exacting requirements for any instrumental technique. The combined facilities of zero suppression, scale expansion and signal integration, though, now available with all standard atomic absorption spectrophotometers give them an impressive capability for such analyses. Typical examples are the estimations of copper and zinc in brasses, aluminium in aluminium bronzes, chromium in high chrome steels, lead and tin in solders, etc.

In operation, a series of standards are prepared that contain the element of interest in concentrations that straddle fairly closely the expected contents of that element in the test samples. Complex matching standards would normally be used. The response for the lowest standard is suppressed to read zero, offset on the pen recorder 10 chart divisions, and scale expansion then applied so that the response for the highest standard gives a suitable reading.

Examples are provided by the estimation of calcium in cement (p. 154); SiO_2 and Al_2O_3 in refractories (p. 151) and chromium in steel (p. 144).

The instrumental performance levels attainable for copper and zinc in brass are illustrated in Figs. 3.4 and 3.5. Alternatively, the use of a $3\frac{1}{2}$-digit digital voltmeter (D.V.M.) readout unit allows non-zero suppressed integrated readings to be taken with sufficient precision in most cases. For example, in the determination of calcium, a 4 μg/ml calcium solution may give an integrated absorbance meter reading of 0·35. This would be registered as a D.V.M. reading of 350. By suitable use of the scale expansion this D.V.M. reading could be increased to 400 (or 800) to obtain precise direct (or direct multiple) concentration readout. In this case the readout precision would be 0·25

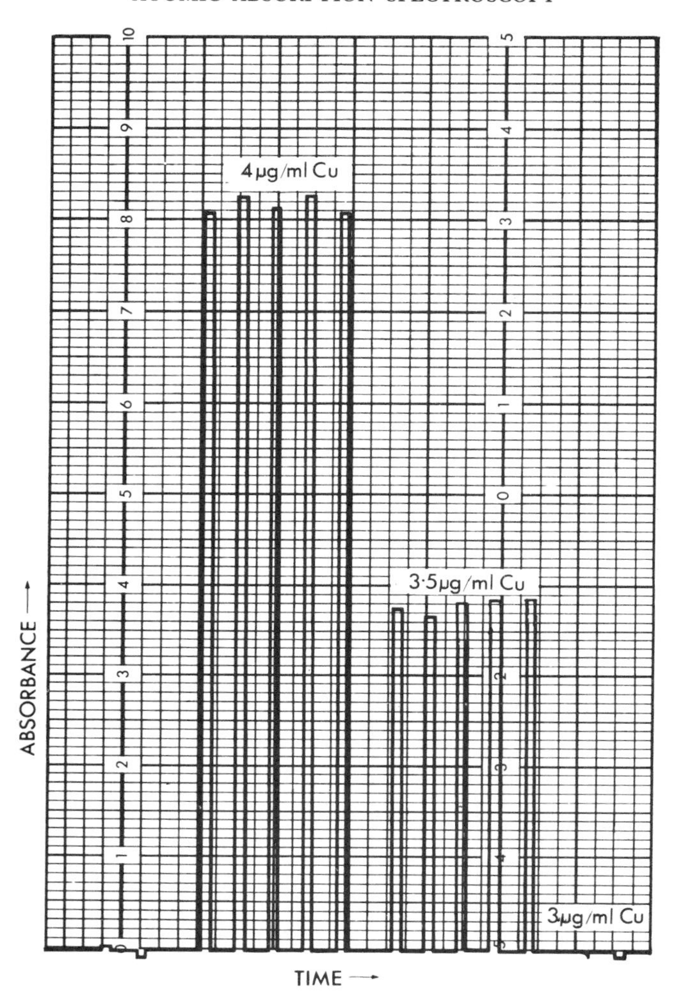

Fig. 3.4 Estimation of copper in brass. Common brass contains about 70% Cu. Effective dilutions for estimation 200 000 times. Standard deviation calculated from response for 4 μg/ml standard is 0·025 μg/ml (Integrated mode).

(0·125)% and in most determinations this would not be the limiting factor in the ultimate precision attainable. For non-linear calibration curves some form of curve correction would be necessary. All modern instruments now have facilities to automatically correct non-linear calibration curves (see p. 197).

9. INDIRECT DETERMINATION USING CHEMICAL AMPLIFICATION PROCEDURES[5]

Interest in amplification (or multiplication) reactions was revived in several branches of analytical chemistry because they allow trace analyses to be performed to a very high degree of precision. In the field of atomic absorption spectroscopy, WEST and co-workers, have applied indirect amplification reactions, in which molybdenum complexes are produced, to the determinations of phosphate, silicate, and niobium.

Taking the determination of phosphate as an example, the principle of the method lies in the formation, and extraction into an organic phase, of phosphomolybdic acid $H_3PO_4(MoO_3)_{12}$. The molybdenum is then determined

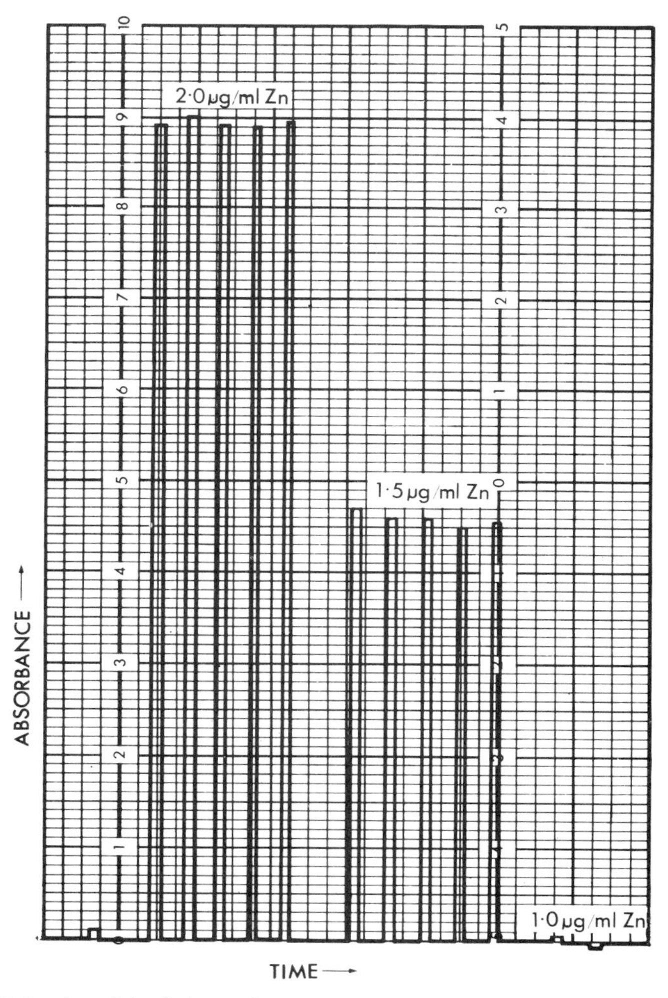

Fig. 3.5 Estimation of zinc in brass. Common brass contains about 30% Zn. Effective dilution for estimation 200 000 times. Standard deviation calculated from responses for 2·0 μg/ml standard is 0·013 μg/ml (Integrated mode).

by atomic absorption, and the estimation benefits from the fact that 12 molybdenum atoms are associated with every atom of phosphorus.

The procedure developed by WEST and his associates for phosphate and silicate allows the two radicles to be determined sequentially from a solution in which they are both present. The test solution is acidified, treated with an excess of molybdate reagent, and the phosphomolybdic acid preferentially extracted away from the excess reagent and silicate into isobutyl acetate. Estimation of the molybdenum content by atomic absorption is made on the organic extract.

The residual aqueous phase, which contains the silicate, is treated with aqueous ammonia to lower the acidity and the silicomolybdic acid $H_4SiO_4(MoO_3)_{12}$ is extracted into butanol. The organic phase is washed free from excess molybdate reagent and its molybdenum content determined by atomic absorption spectroscopy.

The same group of workers[(6)] have developed a similar procedure for the determination of niobium, in which molybdoniobophosphoric acid is formed and extracted into butanol. The phosphomolybdic acid, which is also formed, is previously removed from the solution by selective extraction into isobutyl acetate. In this case eleven molybdenum atoms are associated with every atom of niobium. The procedure is of particular importance, as tantalum does not form a similar complex. Cerium[(9)] has also been determined after the formation of molybdocerophosphoric acid.

10. THE METHOD OF ADDITIONS

The method of additions is a well-known procedure in analytical chemistry and can be useful for checking the accuracy of a determination carried out in the presence of foreign substances, where interference effects upon the element being determined are unknown. For it to be applicable to atomic absorption spectroscopy the output signal must bear a linear relationship to the concentration of the element in the solution. In addition to the necessity for a linear response/concentration relationship, the method suffers from the disadvantage that at least two additional test solutions must be prepared for each sample. However, see p. 204.

The procedure will not correct for errors due to non-specific background absorption. As stated above the method can be useful for checking the validity of certain determinations and for this reason alone it is worthy of brief mention. The procedure consists of dividing the sample into, say, three aliquots A, B and C. Aliquot A is retained untreated, while to aliquots B and C known additions are made of the metal to be determined. The solutions are then aspirated to the atomic absorption spectrophotometer, the meter readings noted and plotted against the concentration increases of the metal, made to the original sample. The curve is extrapolated and the point of intersection with the x-axis is taken to indicate the concentration of the sample. The principle of the method is illustrated in Fig. 3.6.

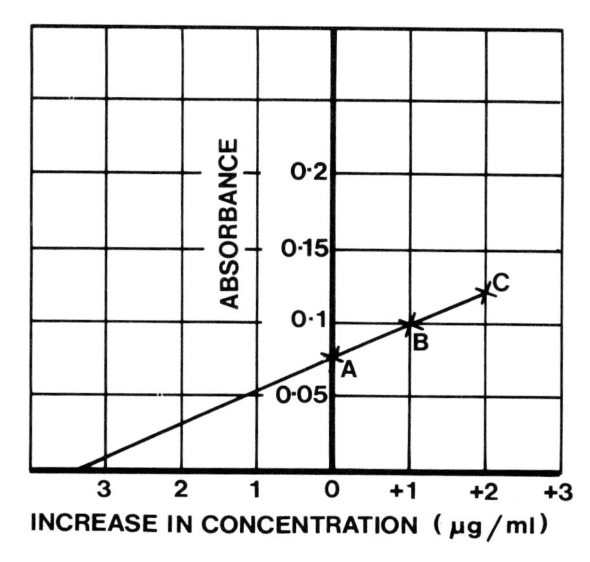

Fig. 3.6 Method of additions

A Simple Direct Concentration Readout Method when using the Standard Addition Technique

The conventional method of standard addition requires a graph of absorbance versus the standard addition concentration in the final solution to be plotted. The graph is then extrapolated back to the concentration axis and the negative intercept gives the unknown concentration. When using bipolar digital readout this procedure can be simplified[9] by nebulizing the unknown solution and setting the output to read zero. Then the sample plus a suitable standard addition is nebulized and using the scale expansion control the readout is set to give a reading corresponding to the magnitude of the standard addition. The blank solution is then nebulized and a negative readout will be obtained, the magnitude of the reading corresponding to the concentration of the element in the sample solution. The following points should be noted:

(i) The sample and sample plus standard addition concentration should fall on the linear portion of the calibration graph. This can be checked by nebulizing a different standard addition and checking that direct concentration readout of the new standard addition is observed.

(ii) There should be negligible non-specific background absorption, unless automatic background correction is applied.

The linearity and maximum negative reading that can be obtained without scale expansion must be ascertained before attempting this technique. This can be rapidly determined by nebulizing a 1 μg/ml copper solution (or any other solution that will give an absolute absorbance of 0·1–0·15) and measuring the non-scale expanded absorbance in the conventional manner. The blank should then be nebulized and the readout set to give a D.V.M. absorbance reading of −0·1. The copper solution should then be nebulized and the indicated absorbance reading plus 0·1 should be the same as that by conventional measurement. This should then be repeated by setting the blank to give an absolute absorbance of −0·2, nebulizing the copper solution and checking the indicated absorbance reading plus 0·2 is still unaltered. This procedure should be repeated until ultimately a reduced overall absorbance is observed. The maximum linear negative reading (without scale expansion) can thus be determined. This value should never be exceeded.

Chap. 3 References and further reading

1. ALLAN, J. E., The use of organic solvents in atomic absorption spectrophotometry. *Spectrochim. Acta* 17 (1961) 467.
2. MALISSA, H., and SCHOFFMANN, E., The use of substituted dithiocarbamates in microanalysis III. *Mikrochim. Acta* 1 (1955) 187.
3. MULFORD, C. E., Solvent extraction techniques for A.A.S. *Atomic Abs. Newsletter* 5 (1966) 88.
4. LAKANEN, E., Separation and concentration of trace metals by means of pyrrolidine dithiocarbamic acid. *Atomic Abs. Newsletter* 5 (1966) 17.
5. KIRKBRIGHT, G. F., SMITH, A. M., and WEST, T. S., The indirect sequential determination of phosphorus and silicon by atomic absorption spectrophotometry. *Analyst* 92 (1967) 411.
6. KIRKBRIGHT, G. F., SMITH, A. M., and WEST, T. S., An indirect amplification procedure for the determination of niobium by atomic absorption spectroscopy. *Analyst* 93 (1968) 292.
7. JOHNSON, H. N., KIRKBRIGHT, G. F. and WHITEHOUSE, R. J., Molecular and A.A.S. methods for the determination of cerium utilising the formation of molybdocerophosphoric acid. *Anal. Chem.* 45 (1973) 1603.
8. BARNETT, W. B., Acid interferences in atomic absorption spectrometry. *Anal. Chem.* 44 (1972) 695.
9. FULLER, C. W., A simple standard additions technique using the model 306 atomic absorption spectrophotometer. *Atomic Abs. Newsletter* 11 (1972) 65.

4 *Characteristics of the Elements*

Information on the atomic absorption characteristics of the elements is given in this chapter.

The values quoted within the sensitivity sections for each element refer to pure aqueous solutions of the element. It is important to appreciate that in the presence of appreciable quantities of foreign ions the limits of detection will usually be degraded compared to those obtained using pure solutions.

Abbreviated data are tabulated for convenient reference in Tables 4.1, 4.2 and 4.3, and fuller information is contained in the text.

It should be stressed that the actual concentration figures given for 1 per cent absorption (characteristic concentration) and detection limit will depend on the actual instrument and the care exercised in the setting up of that instrument (e.g., burner alignment, burner height setting, gas flow settings, etc.). The values quoted are considered to be typical values for a modern atomic absorption spectrophotometer set-up according to the manufacturer's handbook.

A Note on Interference Effects

It is important to stress that interelement effects in atomic absorption spectrophotometry are not simply dependent upon the presence of other cationic species in the test solution but are also dependent upon a large number of other parameters as well.

Viz.—Acetylene flow rate; e.g., the depression of the chromium absorption signal caused by the presence of iron in the air–acetylene flame is considerably less in a fuel-lean than in a fuel-rich flame.

Burner height setting; e.g., the depression of the calcium absorption signal caused by the presence of phosphate in the air–acetylene flame is less when measured higher up in the flame.

Burner head temperature; it is essential to allow the burner head to attain thermal equilibrium whilst nebulizing the blank solution prior to commencing measurements.

The actual area of the light beam passing through the flame; a hollow-cathode lamp with a narrow-bore cathode will give a light beam with a smaller cross-sectional area at a given point in the flame than a lamp with a wide-bore cathode. This could result in a difference in the magnitude of the interference observed for some analyses.

The design of the nebulizer, spray chamber and burner grid; a badly adjusted or corroded nebulizer can affect the incidence and magnitude of interference effects.

The anions present in the test solution; e.g., the cationic interference effects observed for tin and iron in the presence of chloride in the air–acetylene flame are different from those observed in the presence of sulphate.

Occasionally an interference effect observed using one manufacturer's instrument is substantially different from that observed using another instrument. Many interference tests, reported in the literature, are carried out by nebulizing a solution containing a fixed amount of the analyte and a 100- or 1000-fold excess of the interfering element (usually added as the chloride or nitrate) and comparing the response with that of a solution containing the same amount of the analyte only. Unfortunately for most analyses a number of other elements are present in the test solution and although these might not cause interference when each is present alone, it cannot always be assumed that the combination of these elements will not result in an interference effect. It is generally impossible to state dogmatically that no interference will be observed for a proposed determination. The comments given under the interference sections in this Chapter should only be taken as a general indication and each time a new method is developed it should be checked for potential interferences. It is *not* advisable simply to assume that interelement effects for a proposed new method are negligible, therefore aqueous standards or standards prepared in the dissolution media used will be suitable for a particular analysis.

A good example of an unexpected interference effect has been reported by CRESSER and MACLEOD.* The interference was observed when burner rotation or an absorption line of poorer sensitivity than the main absorption line was used to determine high levels of magnesium, cobalt and nickel (15, 50 and 50 μg/ml respectively). A considerable difference in response was observed depending upon whether the elements were present as sulphates or chlorides. This effect was not significant at the lower concentration levels normally used to determine these elements. The degree of interference in this case is markedly dependent upon the analyte concentration.

For a comprehensive discussion on the parameters influencing the incidence and magnitude of interferences in flame atomic absorption spectrophotometry the reader is specifically referred to the references below.*†

Aluminium

WAVELENGTHS

The most absorbing resonance lines in the aluminium spectrum is at 309·3 nm. The line at 396·2 nm is slightly less absorbing but is often used. Other less absorbing lines are given in Table 4.2.

* CRESSER, M. S., and MACLEOD, D. A., Observations on the limitations imposed by interferences in flame A.A.S. at high analyte concentrations. *Analyst* 101 (1976) 86.

† CRESSER, M. S., Literature interpolation: A possible source of error in flame A.A.S. *Laboratory Practice* 26 (1977) 171.

Table 4.1 Common elements that can be determined by A.A.S.

Main Resonance Lines

Element	Wave-length nm	Flame System	Characteristic concentration (Sensitivity for 1% absorption) μg/ml	Limit of detection μg/ml	Useful range for determination
			AQUEOUS SOLUTION		
Aluminium	309·3	N_2O–acetylene	1·0	0·1–0·2	2–200
Antimony	217·6	Air–acetylene	0·4	0·05–0·1	1–60
Arsenic	193·7	Argon–hydrogen	0·5	0·1–0·2	1–50
Arsenic	193·7	Air–acetylene	1–2	1–2	10–100
Barium*	553·6	N_2O–acetylene	0·3	0·05–0·2	0·6–60
Beryllium	234·9	N_2O–acetylene	0·02	0·002–0·01	0·04–4
Bismuth	223·1	Air–acetylene	0·3	0·1–0·2	1–50
Boron	249·7 (doublet)	N_2O–acetylene	12	3–10	25–1 500
Cadmium	228·8	Air–acetylene	0·01	0·001–0·003	0·02–2·0
Caesium*	852·1	Air–acetylene	0·1	0·01–0·03	0·2–10
Calcium	422·7	Air–acetylene	0·05	0·003–0·006	0·1–8
Calcium*	422·7	N_2O–acetylene	0·02	0·002–0·008	0·05–4
Chromium	357·9	Air–acetylene	0·08	0·01–0·02	0·2–15
Chromium	357·9	N_2O–acetylene	0·15	0·03–0·15	0·4–25
Cobalt	240·7	Air–acetylene	0·1	0·01–0·02	0·2–20
Copper	324·8	Air–acetylene	0·05	0·005–0·01	0·1–10
Gallium	287·4	Air–acetylene	1·0	0·05–0·1	2–150
Germanium	265·2	N_2O–acetylene	1·5	0·1–0·4	3–200
Gold	242·8	Air–acetylene	0·2	0·02–0·05	0·5–20
Hafnium	307·3	N_2O–acetylene	15	2–4	30–2 000
Indium	303·9	Air–acetylene	0·5	0·1–0·2	1–100
Iridium	208·9	Air–acetylene	1·0	0·3–0·6	2·5–200
Iridium	208·9	N_2O–acetylene	8·0	2·5–3·0	20–1 500
Iron	248·3	Air–acetylene	0·1	0·005–0·02	0·2–20
Lead	217·0	Air–acetylene	0·15	0·02–0·05	0·3–25
Lead	283·3	Air–acetylene	0·3	0·03–0·07	0·75–50
Lithium	670·8	Air–acetylene	0·02	0·002–0·005	0·05–4
Magnesium	285·2	Air–acetylene	0·005	<0·001	0·01–1
Manganese	279·5	Air–acetylene	0·03	0·004–0·01	0·05–5
Mercury (Hg^{2+})	253·7	Air–acetylene	4·0	0·2–0·5	10–500
Mercury Cold Vapour	253·7	—	—	ca. 0·005 μg	—
Molybdenum	313·3	Air–acetylene	1·9	0·05–0·15	4–200
Molybdenum	313·3	N_2O–acetylene	0·5	0·04–0·2	1–100
Nickel	232·0	Air–acetylene	0·07	0·005–0·02	0·2–15
Nickel	341·5	Air–acetylene	0·4	0·03–0·1	1–80
Niobium	334·9	N_2O–acetylene	25	25–50	100–4 000
Osmium	290·9	N_2O–acetylene	1·5	0·15–0·5	5–250
Palladium	247·6	Air–acetylene	0·25	0·04–0·1	1–40
Palladium	247·6	N_2O–acetylene	2·2	0·6–1·0	5–200
Platinum	265·9	Air–acetylene	1·5	0·1–0·3	5–300
Platinum	265·9	N_2O–acetylene	10	1–3	25–2 000
Potassium†	766·5	Air–acetylene	0·02	0·004–0·01	0·04–4
Rhenium	346·1	N_2O–acetylene	10	1–3	20–1 500

Table 4.1—*Contd.*

Element	Wave-length nm	Flame System	Characteristic concentration (Sensitivity for 1% absorption) μg/ml	Limit of detection μg/ml	Useful range for determination
			AQUEOUS SOLUTION		
Rhodium	343·5	Air–acetylene	0·5	0·05–0·1	1–100
Rhodium	343·5	N_2O–acetylene	0·5	0·1–0·2	1–100
Rubidium*	780·0	Air–acetylene	0·05	0·01–0·03	0·1–10
Ruthenium	349·9	Air–acetylene	1·0	0·15–0·3	2–150
Selenium	196·0	Argon–hydrogen	0·2	0·05–0·2	1–20
Selenium	196·0	Air–acetylene	0·5	0·5–2	2–50
Silicon	251·6	N_2O–acetylene	2·0	0·3–0·7	4–400
Silver	328·1	Air–acetylene	0·05	0·003–0·006	0·1–10
Sodium	589·0	Air–acetylene	0·01	0·002–0·004	0·02–2
Strontium*	460·7	N_2O–acetylene	0·1	0·01–0·02	0·2–20
Tantalum	271·5	N_2O–acetylene	12	2–4	30–2 000
Tellurium	214·3	Air–acetylene	0·5	0·1–0·2	1–80
Thallium	276·8	Air–acetylene	0·4	0·03–0·06	1–80
Tin	224·6	Air–hydrogen	0·4	0·05–0·1	1–60
Tin	224·6	N_2O–acetylene	2·4	0·3–0·6	5–300
Tin	286·3	Air–hydrogen	0·8	0·1–0·2	2–100
Tin	286·3	N_2O–acetylene	4·5	0·5–1·0	10–600
Titanium	364·3	N_2O–acetylene	2·0	0·2–0·4	4–400
Tungsten	255·1	N_2O–acetylene	10	1–3	20–1 500
Uranium	358·5	N_2O–acetylene	150	40–80	300–20 000
Vanadium	318·4 (triplet)	N_2O–acetylene	1·0	0·1–0·2	2–150
Zinc	213·9	Air–acetylene	0·01	0·001–0·003	0·02–2
Zirconium	360·1	N_2O–acetylene	15	1·5–5	50–2 000

* In the presence of 1 000 μg/ml potassium in order to suppress ionization.
† In the presence of 1 000 μg/ml sodium in order to suppress ionization.

Table 4.2 Table of less-sensitive absorption lines

Note: The most sensitive absorption line is listed first followed by increasingly less-sensitive lines. The number within the brackets following each line gives the approximate decrease in sensitivity relative to the most sensitive line.

(All wavelengths in nm)

Aluminium	309·3 (×1)	396·2 (×1·2)	237·3 (×5)	257·4 (×10)
Antimony	217·6 (×1)	206·8 (×1·8)	231·1 (×2·5)	212·7 (×8)
Arsenic	193·7 (×1)	197·2 (×1·7)	189·0[1] (×1)	
Barium	553·6 (×1)	455·4 (×1) (ionic line)	350·1 (×600)	
Beryllium	234·9 (×1)			
Bismuth	223·1 (×1)	222·8 (×3·5)	306·8 (×3·5)	227·7 (×35)
Boron	249·7 (×1)	208·9 (×2·5)		
Cadmium	228·8 (×1)	326·1 (×400)		
Caesium	852·1 (×1)	894·4 (×1·8)	455·5 (×5)	

Table 4.2—*Contd.*

Calcium	422·7 (×1)	239·9[2] (×200)		
Chromium	357·9 (×1)	359·4 (×1·3)	425·4 (×3)	429·0 (×8)
Cobalt	240·7 (×1)	252·1 (×3)	352·7 (×15)	346·6 (×40)
Copper	324·8 (×1)	327·4 (×2)	217·9 (×7)	222·6 (×30)
			249·2 (×120)	244·2 (×350)
Gallium	287·4 (×1)	294·4 (×1·1)	417·2 (×1·6)	403·3 (×3)
Germanium	265·2 (×1)	269·1 (×5)	303·9 (×20)	
Gold	242·8 (×1)	267·6 (×3)		
Indium	303·9 (×1)	410·5 (×3)	271·0 (×20)	
Iridium	208·9 (×1)	264·0 (×2·5)	285·0 (×3)	254·4 (×5)
				292·5 (×5)
Iron	248·3 (×1)	252·3 (×2·5)	271·9 (×4)	372·0 (×10)
			344·1 (×30)	392·0 (×270)
Lead	217·0 (×1)	283·3 (×2)	261·4 (×50)	405·7 (×5000)
Lithium	670·8 (×1)	323·3 (×400)		
Magnesium	285·2 (×1)	202·5[3] (×30)		
Manganese	279·5 (×1)	279·8 (×1·3)	280·1 (×2)	403·1 (×12)
Mercury	253·7[4] (×1)			
Molybdenum	313·3 (×1)	317·0 (×1·3)	386·4 (×2)	320·9 (×8)
Nickel	232·0 (×1)	231·1 (×2)	234·6 (×4)	341·5 (×6)
	352·5 (×7)	351·5 (×12)	247·7 (×250)	362·5 (×500)
Palladium	247·6 (×1)	244·8 (×1)	276·3 (×3)	340·5 (×4·5)
Platinum	265·9 (×1)	299·8 (×6)	264·7 (×6)	
Potassium	766·5 (×1)	769·9 (×2)	404·4 (×400)	
Rhodium	343·5 (×1)	369·2 (×3)	365·8 (×5)	370·1 (×8)
				328·1 (×50)
Rubidium	780·0 (×1)	794·8 (×2)	420·2 (×100)	
Ruthenium	349·9 (×1)	392·6 (×11)		
Selenium	196·0 (×1)	204·0 (×15)		
Silicon	251·6 (×1)	250·7 (×3)	288·2 (×18)	
Silver	328·1 (×1)	338·3 (×2)		
Sodium	589·0 (×1)	589·6 (×2)	330·3 (×500)	
Strontium	460·7[5] (×1)			
Tellurium	214·3 (×1)	225·9 (×16)	238·6 (×20)	
Tin	224·6 (×1)	235·5 (×1·5)	286·3 (×2)	254·7 (×5)
				266·1 (×50)
Titanium	364·3 (×1)	365·4 (×1·1)	399·0 (×2·5)	394·9 (×6)
Tungsten	255·1 (×1)	272·4 (×2)	400·8 (×3)	407·4 (×8)
Vanadium	318·4 (×1)	306·6 (×3)	439·0 (×8)	
Zinc	213·9 (×1)	307·6 (×8000)		

[1] Very noisy.
[2] Most calcium lamps have a pyrex window and do *not* transmit the 239·9 nm line.
[3] Certain magnesium lamp windows will not transmit the 202·5 nm line.
[4] No alternative lines above 200 nm.
[5] Other ground state neutral atomic lines at 293·2 and 689·3, no actual results reported.

Flame System

It is necessary to use the nitrous oxide–acetylene flame.

Sensitivity

A sensitivity of 1·0 μg/ml for 1 per cent absorption can be expected with standard instruments.

The limit of detection is between 0·1 and 0·2 μg/ml.

The optimum range for determination is from about 2 to 200 μg/ml.

Interferences and Special Characteristics

Solutions containing equal quantities of aluminium as the chloride or nitrate give rise to identical responses; but a solution of potash–alum containing the same aluminium loading produces a distinctly higher absorption.

Excess hydrochloric acid when present at the 1·0 M level does not affect the absorption figure, but a slight depression is produced when the acid concentration is raised to 5 M.

The presence of heavy excesses of Ca, Mg, Cu, Fe, Zn, Si and PO_4^{3-} have negligible influence upon the absorption of aluminium. Distinct enhancement occurs in the presence of heavy excesses of K, Na and La, probably due to suppression of ionization. For this reason it is good practice when determining aluminium by atomic absorption, to add about 1000 μg/ml sodium or potassium to test and standard solutions.

Table 4.3 Less common elements that have been determined by AAS

Element	Wavelength	Flame system	Characteristic concentration (Sensitivity for 1% absorption) (μg/ml)
Dysprosium	421·2	Nitrous oxide–acetylene	1·0
Erbium	400·8	Nitrous oxide–acetylene	1·0
Europium	459·4	Nitrous oxide–acetylene	0·4
Gadolinium	368·4	Nitrous oxide–acetylene	40·0
Holmium	410·4	Nitrous oxide–acetylene	1·0
Lanthanum	550·1	Nitrous oxide–acetylene	50·0
Lutecium	336·0	Nitrous oxide–acetylene	10·0
Neodymium	492·5	Nitrous oxide–acetylene	10·0
Praseodymium	495·1	Nitrous oxide–acetylene	50·0
Samarium	429·7	Nitrous oxide–acetylene	10·0
Scandium	391·2	Nitrous oxide–acetylene	0·5
Terbium	432·7	Nitrous oxide–acetylene	10·0
Thulium	371·8	Nitrous oxide–acetylene	0·5
Ytterbium	398·8	Nitrous oxide–acetylene	0·15
Yttrium	410·2	Nitrous oxide–acetylene	5·0

Antimony

Wavelengths

The resonance line at 206·8 nm is quoted in the earlier atomic absorption literature as giving rise to the most sensitive conditions. Determinations made at 231·1 nm are slightly less sensitive than those made at 206·8 nm, but the signal is much more stable. More recent work has shown that the line at 217·6 nm is in fact the most absorbing wavelength, but in order to use it effectively the instrument must be capable of resolving it from the non-absorbing line at 217·9 nm.

Flame Systems

Either the air–acetylene or air–propane flames can be used with equal advantage.

Sensitivity

The essential data for the three wavelengths is tabulated below.

	Wavelength		
	217·6 nm	206·8 nm	231·1 nm
Useful range for determination at ×1 scale expansion (μg/ml)	1–60	1·5–100	2–150
Sensitivity for 1% absorption (μg/ml)	0·4	0·7	1·0
Limit of detection (μg/ml)	0·05–0·1	0·1–0·2	0·1–0·2

Interferences and Special Characteristics

With the air–acetylene flame system, solutions of the chloride, or potassium antimony tartrate, containing equal concentrations of antimony give rise to similar responses.

The presence of the following elements exerts very little effect upon the absorption for antimony: Al, Ba, Bi, Ca, Mg, Cr, Pb, Mn, Ni, K, Na, Sn, Sr, Zinc, when present as the nitrate, has very little effect upon the antimony absorption, but the presence of zinc chloride exerts a distinct depression. The presence of heavy excesses of Cd, Cu, and Fe cause slight depression of the antimony absorption. Excess hydrochloric acid gives rise to a slight enhancement, but this levels out as the acid concentration increases.

With the air–propane flame system, interference effects occur in the presence of high concentrations of Bi, which gives rise to a pronounced enhancement, and zinc (as zinc chloride) which causes slight depression.

To ensure the reliable determination of antimony by atomic absorption spectroscopy the analyst must approximately match the acidity and metal composition of the standard solutions to those of the test samples. Antimony can be determined by the hydride generation technique to give a detection limit better than 0·001 μg/ml for a 1 ml sample (see p. 79).

Arsenic

Wavelengths

The resonance line at 193·7 nm is the one most likely to be of practical value to the analyst. There is a slightly less sensitive line at 197·2 nm. All absorbing lines are listed in Table 4.2.

Flame Systems and Sensitivity

All premixed air–fuel gas flame systems (e.g., air–acetylene, air–propane, air–hydrogen) absorb radiation strongly at wavelengths below 200 nm (80–90% at 193·7 nm). The sensitivity for 1% absorption in these flames is 1–2 μg/ml with a similar detection limit. The fuel-rich nitrous oxide–acetylene flame gives much less flame absorption but gives a poorer 1% absorption figure, and the use of this flame results in no improvement in detection limit. However, the main advantage of this hotter flame is the elimination of almost all interelement effects. The argon–hydrogen–entrained air flame (nebulizing gas–argon, fuel gas–hydrogen) exhibits very little flame absorption at 193·7 nm. The sensitivity for 1% absorption is approximately 0·5 μg/ml at 193·7 nm, with a limit of detection of 0·1–0·2 μg/ml. The main drawback of this flame is that owing to its relatively low temperature interelement effects are very pronounced.

Interferences and Special Characteristics

Hollow-cathode lamps capable of providing a sufficiently intense light output for estimations to be made with standard commercial equipment are now available. Unfortunately, though, the determination of this element is frequently required at very low levels in solutions containing heavy loadings of foreign ions, e.g., sodium. Under these conditions severe background absorption is observed, rendering simple direct estimations of arsenic at low levels by atomic absorption impossible for many applications.

This difficulty is not likely to be alleviated by the use of intense microwave or radio frequency excited electrodeless discharge tubes, or by the replacement of atomic absorption by atomic fluorescence as the measuring technique. Arsenic can be determined by the hydride generation technique to give a detection limit better than 0·001 μg/ml for a 1 ml sample (see p. 79).

Barium

Wavelength

The atomic resonance line at 553·6 nm is normally used. A less absorbing line exists at 350·1 nm; and the ionic line at 455·4 nm can be used for some applications.

Flame System

For realistic determinations it is necessary to use the nitrous oxide–acetylene flame.

Sensitivity

The sensitivity for 1 per cent absorption for standard instruments is about 0·3 μg/ml at both 455·4 and 553·6 nm.

The limit of detection is between 0·05 and 0·2 μg/ml.

The optimum range for determination is from about 0·6 to 60 μg/ml.

Interferences and Special Characteristics

Barium exhibits a much lower sensitivity to atomic absorption than either calcium or strontium and the use of the nitrous oxide–acetylene flame is essential for realistic determinations.

It is necessary, in order to attain optimum absorption, to dose the sample and standard solutions with about 1000 μg/ml K. At the 455·4 nm ionic line the presence of an alkali metal severely depresses the barium absorption. In the *absence* of an alkali metal, or other ionization suppressor, a very stable response is obtained at this wavelength.

Equal concentrations of barium present in solution as either the chloride or nitrate give rise to identical absorptions.

The presence of heavy loadings of Sr, Mg, Fe, and Zn hardly affect the barium response.

The presence of even moderate amounts of calcium results in an increased noise level on the barium signal. This is caused by the intense molecular emission at 553·6 nm from CaOH species in the flame. At sufficiently high calcium levels (500–2000 μg/ml) complete overload of the demodulator occurs, and erroneous results are obtained.[10] The demodulator is the part of the electronics where the modulated hollow-cathode lamp signal is separated from the unwanted flame emission signals. In order to minimise this overload effect the hollow-cathode lamp should be run at a high current (80–90%) of maximum-rated current, and a narrow spectral bandpass should be selected.

Beryllium

Wavelength

234·9 nm.

Flame System

For practical purposes the only system of value is nitrous oxide–acetylene.

SENSITIVITY

With standard atomic absorption equipment a sensitivity for 1 per cent absorption of about 0·02 μg/ml can be attained. The limit of detection with standard equipment is between 0·002 and 0·01 μg/ml.

Without scale expansion the element can be determined over a range from 0·04 to 4 μg/ml.

INTERFERENCES AND SPECIAL CHARACTERISTICS

The absorption of beryllium is almost unaffected by the presence of heavy loadings of Ca, Mg, and Cu.

Aluminium in heavy excess slightly depresses the beryllium response.

Bismuth

WAVELENGTHS

The line at 223·1 nm provides the highest sensitivity for solutions containing up to 50 μg/ml Bi. The 306·8 nm line is useful for the determination of bismuth in solutions containing up to about 200 μg/ml.

Other less-absorbing lines are given in Table 4.2.

FLAME SYSTEMS

Either the air–acetylene or air–propane flame may be used. At 306·8 nm using the air–acetylene flame, strong hydroxyl band emission occurs and if a low intensity lamp is used, a poor signal to noise ratio will be observed.

SENSITIVITY

The sensitivity for 1 per cent absorption at 223·1 nm is about 0·3 μg/ml. At 306·8 nm it is about 1 μg/ml. At 223·1 nm the limit of detection is about 0·1–0·2 μg/ml. At this wavelength, without scale expansion the element can be determined over a range from 1 to 50 μg/ml.

At 306·8 nm determinations can be made without scale expansion over a range from 4 to 200 μg/ml.

INTERFERENCES AND SPECIAL CHARACTERISTICS

Of the common elements likely to be found in association with bismuth the following have negligible effect, even when present in heavy excess, upon that element's absorption in either the air–acetylene or air–propane flames: Sb, Ba, Cd, Cu, Fe, Pb, Sn, Zn.

Excess hydrochloric acid gives rise to a small but noticeable increase in absorption, and it is necessary to match the acidity of standard solutions to that of test solutions.

Bismuth can be determined by the hydride generation technique to give a detection limit better than 0·0005 μg/ml for a 1 ml sample (see p. 79).

Boron

Wavelength

The line at 249·77 nm is stated to be more sensitive than that at 249·68 nm. With standard commercial instruments, though, the two lines cannot be resolved, so that the mean sensitivity has to be accepted for determinations.

Flame System

Nitrous oxide–acetylene, very fuel rich.

Sensitivity

The sensitivity for 1 per cent absorption attainable with standard atomic absorption equipment is at best about 12 μg/ml.

A realistic value for the limit of detection is between 3 and 10 μg/ml.

Flame emission[14, 15] is more sensitive but less specific using the 518 nm BO_2 band. Boron cannot satisfactorily be determined by flameless electrothermal atomization.

Cadmium

Wavelength

228·8 nm. There is a less absorbing line at 326·1 nm.

Flame System

Air–acetylene. Air–propane and air–hydrogen are also suitable, but have no advantages over air–acetylene.

Optimum flame conditions are best established by adjusting to maximum transparency (see p. 7).

Sensitivity

The sensitivity for 1 per cent absorption is about 0·01 μg/ml on standard commercial instruments. The limit of detection is 0·001–0·003 μg/ml. Without the use of scale expansion the optimum range for determination is from about 0·02 to 2·0 μg/ml.

Interferences and Special Characteristics

Cadmium is one of the most sensitive of all metals to determination by atomic absorption spectroscopy.

Solutions of the chloride, sulphate, or nitrate, containing the same concentrations of cadmium all give rise to identical responses.

The following elements, even when present in heavy excess, have very little effect upon the absorption of cadmium: Sb, Cu, Fe, Pb, Zn, Sn, Ag. The presence of excess hydrochloric acid slightly enhances the cadmium response, but this effect levels out and the same enhancement is obtained in solutions that are 1 M and 5 M.

The absorption for this element is strongly dependent upon lamp operating current. Maximum absorption is attained at low currents; 3 to 5 mA.

Caesium

Hollow-cathode lamps for this element are now reasonably satisfactory. Vapour discharge lamps provide an alternative radiation source for caesium.

Wavelength

852·1 nm. In order to make measurements at this wavelength the instrument must be fitted with a suitable extended wavelength range photomultiplier. An improved signal-to-noise ratio can often be obtained by inserting a simple cut-off filter (opaque to all wavelengths below 650 nm) between the flame and the monochromator. This filter transmits at 852 nm, but absorbs the wavelengths that correspond to the second (426 nm) and third (284 nm) order radiation that would be diffracted from the grating on to the photomultiplier.

Less absorbing lines are listed in Table 4.2.

Flame System

Either the air–acetylene or air–hydrogen flame systems can be used. Very pronounced ionization depression occurs with both flame systems, and it is essential to dose standard and test solutions with a heavy excess of an easily ionized species (2000 μg/ml K).

Sensitivity

Under the most favourable conditions at ×1 scale expansion the element can be estimated over the range 0·2–10 μg/ml. The sensitivity for 1 per cent absorption is 0·1 μg/ml.

The limit of detection is 0·01–0·03 μg/ml.

Interferences and Special Characteristics

Of the elements commonly found in association with caesium, potassium, sodium, lithium and calcium do not affect the absorption provided that the test and standard solutions are adjusted to contain a heavy excess of an alkali metal.

Calcium

Wavelength

422·7 nm. (A less sensitive line is given in Table 4.2.)

Flame Systems

Calcium is readily determined by atomic absorption spectroscopy, with both the air–acetylene and nitrous oxide–acetylene flame systems. At the temperature of the nitrous oxide–acetylene flame the ground-state calcium atoms tend to lose electrons to form calcium ions which will not absorb radiation of the atomic resonance frequency. This ionization interference is overcome by loading the solution with potassium or sodium chloride. Under these conditions calcium exhibits greater sensitivity in the nitrous oxide–acetylene flame than in the air–acetylene system.

In the air–acetylene flame, calcium can show peak absorptions under two different fuel-support gas ratios. One occurs with a distinctly fuel-rich and the second with a much leaner flame. The most sensitive conditions are obtained under the latter (more oxidizing) flame conditions.

Sensitivity

The essential data for the two flame systems are tabulated below.

	Flame system	
	Air–acetylene	N_2O–acetylene
Useful range for determination at ×1 scale expansion (μg/ml)	0·1–8	0·05–4
Sensitivity for 1% absorption (μg/ml)	0·05	0·02
Limit of detection (μg/ml)	0·003–0·006	0·002–0·008

Interferences and Special Characteristics

Air–acetylene Flame System. With this flame the same concentration of calcium present as the nitrate will exhibit a slightly lower response than if present as the chloride. Of the foreign ions commonly found in association with calcium, Ba, La, Sr, and E.D.T.A. disodium salt enhance the calcium response, but this enhancement levels out at high concentrations of the foreign ion.

Heavy excesses of Fe, Mg, Na and K have negligible effect on calcium in the air–acetylene flame.

Depression of the calcium response occurs in the presence of Al, Mn and PO_4^{3-}. The interference is overcome (in the cases of Mn and PO_4^{3-}) by the addition of lanthanum chloride or E.D.T.A. disodium salt to the solution. In the fuel-rich air–acetylene flame the extent to which phosphate depression occurs depends on the height at which the light path traverses the flame and at a point high in the flame the interference entirely disappears. At this region of the flame the noise level is rather high, and the fact is, therefore, of little more than academic interest. It is much more convincing, reliable, and practical to overcome phosphate interference by the addition of a suppressant such as lanthanum chloride. Bicarbonate alkalinity (in natural waters) will depress the calcium response.[20] This effect, though, is easily overcome by the addition of lanthanum chloride, or by adjusting the *p*H of the solution to lie in the range 1·8 to 3·8. In solutions containing hydrochloric acid, at concentrations above 1 M, the calcium response is seriously depressed.

Nitrous Oxide–Acetylene Flame System. At the higher temperature of the nitrous oxide–acetylene flame some of the calcium atoms are further dissociated into calcium ions, a depression of the response thereby occurs.

This ionization interference can be overcome by adding 1000–2000 μg/ml potassium to both test and standard solutions.

Of the common elements likely to be found in association with calcium those which will enhance the response in the absence of an ionization suppressant are Ba, Sr, Na, and K, due to their ability to supply electrons to the flame system and thereby inhibit the tendency for calcium to ionize.

Equal concentrations of calcium as either the nitrate or chloride exhibit the same responses, while the presence of heavy loadings of PO_4^{3-}, Fe, La, and Mn have negligible influence on the calcium absorption. An excess of Al depresses the calcium response, but this depression is overcome by the addition of lanthanum chloride.

It is possible to determine calcium with greatly reduced sensitivity by utilising the overlap of the germanium 422·657 nm non-resonance line profile with the calcium 422·673 nm resonance line profile.[23] An intense germanium source is essential in order to minimise the effect of the intense calcium thermal emission emitted by the flame. The sensitivities for 1% absorption in the air and nitrous oxide-acetylene flames were 8·3 and 6 μg/ml respectively and the calibration graphs were linear up to 1500 and 1000 μg/ml respectively. (See also reference 186, chapter 5.)

REFERENCES

A very large volume of literature has appeared on the determination of calcium by atomic absorption spectroscopy, and it is felt that to give an exhaustive list of references would be more confusing than helpful. As a matter of courtesy, though, it is right that recognition should be given to two workers whose activities did so much to establish atomic absorption as a tool in the analytical laboratory. WILLIS pioneered the application of the technique to calcium determination in clinical chemistry and DAVID worked in the area of agricultural analysis.[16,17,18]

Chromium

Wavelengths

Improvements in lamp technology have drastically enhanced the capabilities of simple commercial atomic absorption spectrophotometers for the estimation of chromium. The two most absorbing lines in the chromium spectrum are located at 357·9 and 359·4 nm.

With the older argon-filled lamps there is a gas line very close to the most sensitive resonance line at 357·9 nm and using an instrument with a low resolution monochromator, a grossly bent response curve is obtained at this wavelength. Indeed in practice it is found that with such an instrument the 359·4 nm line is apparently more sensitive and gives superior results.

With the newer lamps which also have superior outputs, an instrument having a fairly low resolution monochromator possesses a capability equal to one having high resolution optics. With these lamps a low resolution instrument can take advantage of the more absorbing nature of the 357·9 nm line.

Furthermore, because the argon filled lamps, previously supplied, required the use of fairly wide monochromator slit settings it was impossible to make reliable measurements at either 357·9 or 359·4 nm with the nitrous oxide–acetylene flame, due to severe flame 'noise' at these wavelengths. (See Fig. 2.2.) Determinations with nitrous oxide–acetylene were therefore restricted to the less sensitive 425·4 nm line.

The fact that the spectrum from the newer lamps is cleaner, and of higher intensity permits the use of narrow monochromator slit widths, so that measurements can be made with the nitrous oxide–acetylene flame at 357·9 nm as well as at the less sensitive 425·4 nm line.

Spectral traces from the two types of lamps are shown in Fig. 4.1.

Flame Systems

Either the very fuel-rich air–acetylene or nitrous oxide–acetylene flame can be used.

Sensitivity

The essential data for the two flame systems obtained at the 357·9 nm line with a neon-filled lamp are tabulated below.

Flame system	Sensitivity for 1% absorption (μg/ml)	Limit of detection (μg/ml)	Useful range for determination without scale expansion (μg/ml)
Air–Acetylene	0·08	0·01–0·02	0·2–15
N_2O–Acetylene	0·15	0·03–0·15	0·4–25

INTERFERENCES AND SPECIAL CHARACTERISTICS

Chromium is reasonably sensitive to atomic absorption spectroscopy, but the fact that the presence of iron and vanadium can both cause serious chemical interferences, necessitate its determination to be carried out with particular care.

Air–Acetylene Flame System. The chromium response with respect to acetylene flow rate exhibits a very sharp peak in the very fuel-rich air–acetylene flame. OTTAWAY *et al.*[(26)] have reported that by making the flame more oxidizing in character an increase in CrO molecular emission occurs, which corresponds to the decrease in chromium absorption observed under the same conditions. These workers, therefore, reasoned that a shift in the Cr/CrO equilibrium is produced by changing the oxygen content of the flame.

The most notable characteristic of the behaviour of chromium in the air–acetylene flame is the chemical interference of iron. This troublesome

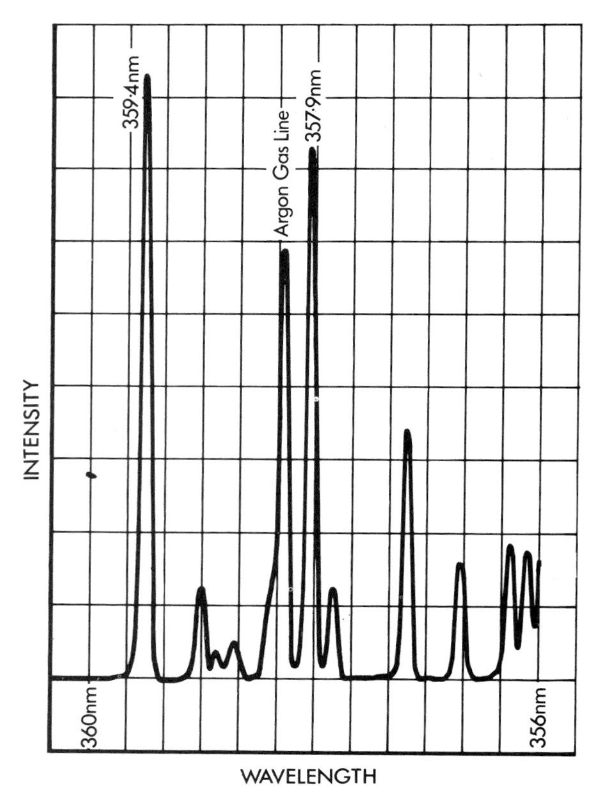

Fig. 4.1a Spectral trace from argon-filled Cr-lamp of older design. Note prominent gas line next to the sensitive 357·9 nm resonance line.

interference is overcome by the addition of lanthanum chloride, 8-hydroxyquinoline, or a heavy excess of ammonium chloride. All of these reagents cause some enhancement of the chromium response and care must be taken to match standard with test solutions when they are used to suppress the effect of iron.

In the air–acetylene flame the same concentrations of chromium present as the alum or chloride give rise to the same absorption. The response for a similar concentration of chromium as the dichromate is higher especially in very fuel rich flames.

The presence of vanadium and nickel, like that of iron, depresses the absorption of chromium. Aluminium and titanium enhance chromium absorption. These interferences disappear in a fuel-lean flame, but this is of academic interest because the chromium absorption is itself severely reduced under these conditions.

The following elements have little effect upon the absorption of chromium:

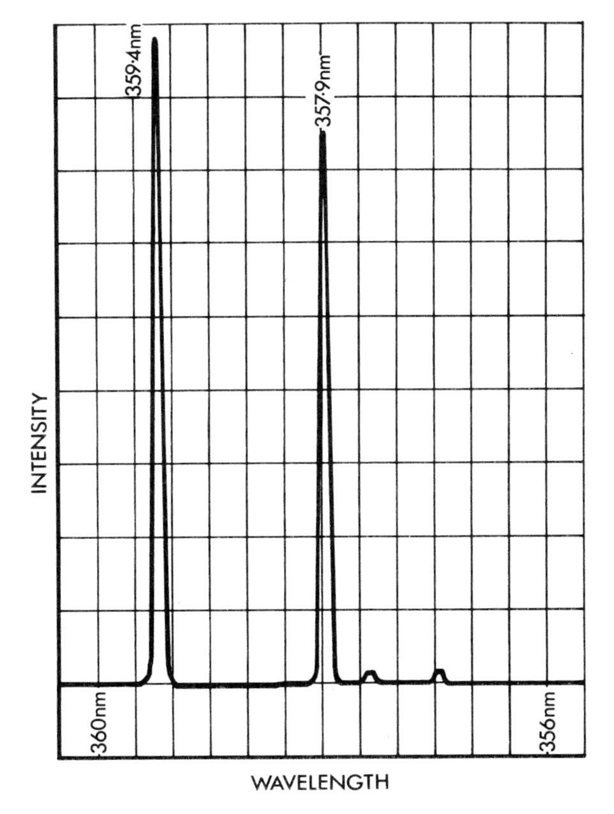

Fig. 4.1b Spectral trace from newer neon-filled Cr-lamp. Note cleanliness of spectrum.

Ca, Co, Cu, Mn, Na. The presence of excess hydrochloric acid does not influence the absorption but excesses of sulphuric and nitric acid do.

Nitrous oxide–Acetylene Flame System. With the nitrous oxide–acetylene flame system similar concentrations of chromium present as the chloride, alum, or dichromate give rise to the same response. 1000 μg/ml potassium should be added to both test and standard solutions to suppress ionization.

The presence of the following metals has very little influence upon the absorption of chromium: Al, Ca, Co, Cu, Fe, Mn, Mo, Ni, V, Na. Once again the presence of excess hydrochloric acid does not affect the absorption, whereas excess of sulphuric or nitric acids does, but to a lesser extent than with the air–acetylene flame.

Cobalt

Wavelengths

240·7 nm affords maximum sensitivity, and is most commonly used. It is essential to distinguish this line from the less absorbing one at 241·2 nm. Other less sensitive lines are given in Table 4.2.

Flame System

The air–acetylene flame is most commonly used. Air–propane and air–hydrogen have also been found suitable, but offer no outstanding advantages. With the air–acetylene flame, cobalt can be determined without scale expansions over a range from 0·2 to 20 μg/ml. Optimum flame conditions are best established by adjusting to maximum transparency (see p. 7).

Sensitivity

The sensitivity for 1 per cent absorption for standard commercial instruments is about 0·1 μg/ml. The limit of detection is about 0·01–0·02 μg/ml.

Interferences and Special Characteristics

Figure 4.2 compares typical response curves, obtained for aqueous solutions, with two types of lamp, the older-type wide-bore argon-filled lamp and the neon-filled high spectral output lamp. These indicate the distinct improvements in the modern lamp, which does much to overcome the effect of the complex cobalt spectrum and gives rise to a more sensitive and linear response.

Solutions of the chloride, nitrate or sulphate that contain an equal concentration of cobalt give rise to identical responses.

Of the common metals and radicals likely to be found in association with cobalt, the following exert negligible effect upon that element's absorption, even when present in heavy excess: Cr, Cu, Fe, Mg, Mn, Mo, Ni, Na, W, V, Si, Ti, and PO_4^{3-}.

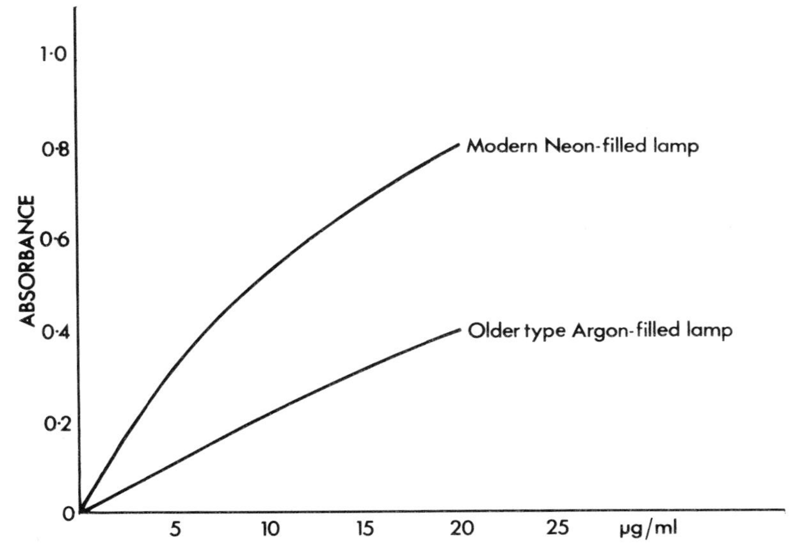

Fig. 4.2 Response curve for cobalt with two types of lamp.

Copper

Wavelength

324·8 nm is the line most commonly used. Less absorbing lines are listed in Table 4.2.

Flame System

The air–acetylene flame is the most commonly used, but the air–propane flame is also suitable. With the air–acetylene flame optimum conditions are best selected by adjusting to maximum transparency (see p. 7).

Sensitivity

The sensitivity for 1 per cent absorption for standard commercial instruments is around 0·05 μg/ml and the limit of detection is about 0·005–0·01 μg/ml.

Without scale expansion the element can be determined at 324·8 nm, over a range from about 0·1 to 10 μg/ml.

Interferences and Special Characteristics

The absorption for copper is remarkably free from interference effects. Solutions of the chloride, sulphate, or nitrate containing the same concentrations of copper exhibit identical responses. The following elements commonly found in association with copper have very little effect upon its absorption, even when present in heavy excess: Al, Ca, Cr, Co, Fe, Pb, Mg, Mn, Ni, Ag, Na, Sn, V, Zn.

With a tubular steel air–acetylene burner, gross non-uniform enhancement occurs in the presence of excess acids. This effect does not occur if a flat-sided laminar flow burner is used.

Gallium

Wavelength

287·4 nm is the optimum wavelength for atomic absorption measurements. Other less absorbing lines are given in Table 4.2.

Flame System

Air–acetylene.

Sensitivity

The sensitivity for 1 per cent absorption is about 1·0 μg/ml. The useful range for determination without scale expansion is 2–150 μg/ml. The limit of detection is 0·05–0·1 μg/ml.

Interferences and Special Characteristics

Heavy loading of Cu, Mg, Zn, Cl, NO_3, PO_4^{3-}, and SO_4^{2-} have little effect upon the absorption of gallium, but very heavy loadings of Al produce a slight depression.

Germanium

Wavelength

265·2 nm.

Other less absorbing lines are given in Table 4.2.

Flame System

Nitrous oxide–acetylene.

Sensitivity

The sensitivity for 1 per cent absorption is about 1·5 μg/ml, and without scale expansion the element can be estimated over a range from 3 to 200 μg/ml. The limit of detection is between 0·1 and 0·4 μg/ml.

Germanium can also be determined by hydride generation techniques (see p. 79).

Gold

Wavelength

242·8 nm. There is a less absorbing line at 267·6 nm.

Flame System

Air–acetylene. Optimum flame conditions are best established by adjusting to maximum transparency (see p. 7).

Sensitivity

The sensitivity for 1 per cent absorption is about 0·2 μg/ml.

The limit of detection is about 0·02–0·05 μg/ml.

Without scale expansion the element can be determined over a range from 0·5 to 20 μg/ml.

Interference and Special Characteristics

The noble metals might reasonably be expected to exhibit freedom from chemical interference effects, but in fact very pronounced interferences do occur,[37] especially with cooler flames. The absorption for gold is distinctly higher in the air–acetylene flame than in the nitrous oxide acetylene system; and in this case interference effects are not greatly reduced at the higher temperature, so that in practice the air–acetylene system is most commonly used.

The same concentration of gold gives a slightly higher absorbance reading when present as the cyanide than if present as the chloride, and interelement effects appear to be less pronounced in cyanide solution.[38]

The presence of iron, nickel, palladium and reducing agents such as hydroxylamine hydrochloride exert a considerable depression upon gold absorption in chloride solution. Palladium continues to depress gold response in cyanide medium, while the presence of copper produces an enhancement.

Hafnium

Wavelength

307·3 nm.

Flame System

A very fuel-rich nitrous oxide–acetylene flame is required.

Sensitivity

The sensitivity for 1 per cent absorption is 15 μg/ml. The limit of detection is 2–4 μg/ml. Without scale expansion the element can be determined over the range 30–2000 μg/ml.

Interferences and Special Characteristics

Hafnium is subject to complex interference effects.

Indium

Wavelength

The line at 303·9 nm exhibits maximum sensitivity. Other less absorbing lines are listed in Table 4.2.

Flame System

Air–acetylene.

Sensitivity

The sensitivity for 1 per cent absorption is around 0·5 μg/ml.

The limit of detection is about 0·1–0·2 μg/ml. Without the use of scale expansion the useful working range is from about 1–100 μg/ml.

Interferences and Special Characteristics

Heavy excess of chloride, nitrate, phosphate, sulphate, Al, Cu, Mg, and Zn have little influence upon the absorption of indium.

Iodine (see p. 212)

Iridium

Wavelength

The iridium spectrum contains a large number of resonance lines. The one at 208·9 nm is the most absorbing but is of low intensity, and until recently the lamps available did not possess sufficient output to allow it to be used. The newer lamps give rise to much improved light outputs and do allow estimations to be made at 208·9 nm. There is a fairly broad peak at this wavelength consisting presumably of the 208·9 nm atomic resonance line and the 209·6 nm ionic line plus background radiation. On some older instruments distinctly better wavelength setting, leading to optimum sensitivity, can be obtained by approaching the line from the lower end of the spectrum.

The unresolved doublet at 263·94 and 263·97 nm (usually shown as 264·0 nm) has been widely used for iridium estimations and possesses a good linear concentration/response characteristic. The lines at 285·0 nm and 292·5 nm are also of value (see Table 4.2).

Flame System

Air–acetylene or nitrous oxide–acetylene.

Sensitivity

The essential data for various conditions are tabulated below:

Air—Acetylene

	Wavelength	
	208·9 nm	264·0 nm
Useful range for determination at × 1 scale expansion (μg/ml)	2·5–200	7·5–400
Sensitivity for 1% absorption (μg/ml)	1·0	3·0
Limit of detection (μg/ml)	0·3–0·6	0·7–1·0

N_2O—Acetylene

	Wavelength	
	208·9 nm	264·0 nm
Useful range for determination at × 1 scale expansion (μg/ml)	20–1500	40–2500
Sensitivity for 1% absorption	8	19
Limit of detection (μg/ml)	2·5–3·0	5–6

INTERFERENCES AND SPECIAL CHARACTERISTICS

The most notable feature of iridium absorption is that a solution of an iridite produces a distinctly lower response than that for an iridate. For this reason it is essential to ensure that the metal is present in the higher oxidation state, by acidification and addition of potassium dichromate. Test solutions should be prepared so that they contain up to about 200 μg/ml Ir (for estimation at 208·9 nm), hydrogen ion concentration adjusted to approximately 0·5 mol dm^{-3} with hydrochloric or sulphuric acid, and treated with potassium dichromate solution so that the final solution contains about 500 μg/ml Cr.

The absorption of iridium, surprisingly for a noble metal, is noticeably depressed in the presence of iron, zinc, palladium, rhodium and nickel in both the air–acetylene and nitrous oxide–acetylene flames.

The presence of platinum and gold has little effect upon iridium absorption with either flame, and copper causes no interference when the nitrous oxide–acetylene system is used, but does give rise to some depression with the air–acetylene system.

Iron

WAVELENGTHS

The spectrum of iron possesses a doublet with peaks at 248·3 and 248·8 nm.

Hollow-cathode lamps for iron are now commonly filled with neon in contrast to the earlier models that contained argon. These modern lamps (containing also an improved electrode configuration) have sufficiently high output to permit the use of narrow monochromator slit widths. It therefore becomes possible to separate the most absorbing line at 248·3 nm from the less absorbing one at 248·8 nm. The adjustment is best accomplished by approaching from the lower end of the spectrum.

Other less absorbing lines are given in Table 4.2.

Flame System

The air–acetylene flame gives optimum response. The nitrous oxide–acetylene flame can also be used with much reduced sensitivity. Iron does not exhibit a sharp absorption peak with respect to acetylene flow in the air–acetylene flame (unlike chromium). Optimum flame conditions are best established by adjusting to maximum transparency (see p. 7).

Sensitivity

The sensitivity for 1 per cent absorption for standard instruments is about 0·1 μg/ml at the 248·3 nm line, in the air–acetylene flame.

The limit of detection is between 0·005 and 0·02 μg/ml. Without the use of scale expansion the element can be determined over a range from 0·2 to 20 μg/ml.

Interferences and Special Characteristics

In chloride solution, iron absorption is remarkably free from chemical interferences. Of the common substances likely to be found in association with the element Al, Co, Cu, Mg, Ni, K, Na, and PO_4^{3-} cause negligible alteration of the absorption, even when present in heavy excess over the iron concentration. Excesses of hydrochloric, nitric and sulphuric acid enhance the absorption, but this effect plateaus out above a hydrogen ion concentration of 0·5 mol dm^{-3}. The presence of silicate depresses the absorption of iron, but this is overcome by the addition of calcium or lanthanum chloride.

Although aqueous solutions of ferric chloride and ferrous ammonium sulphate containing the same concentrations of iron exhibit the same absorption, cationic interference occur in sulphate media that do not occur in the presence of chloride. For example, cobalt, nickel and copper depress iron absorption in sulphate media, while titanium, aluminium, calcium, chromium, vanadium and zinc enhance the iron response.

Lead

Wavelength

The most absorbing line occurs at 217·0 nm. The line at 283·3 nm is also useful and indeed was the one initially used for lead determinations before the lamps that are now available, possessing improved output characteristics, were produced.

The line at 261·4 nm is also of value for determinations upon solutions containing heavy loadings of lead.

Flame System

The air–acetylene flame is the most useful. Determinations can also be made with air–propane but this system offers no advantages in sensitivity over air–acetylene. Optimum flame conditions are best established by adjusting to maximum transparency (see p. 7).

Sensitivity

The essential data for the two wavelengths is tabulated below.

	Wavelength	
	217·0 nm	283·3 nm
Useful range for determination at ×1 scale expansion (μg/ml)	0·3–25	0·75–60
Sensitivity for 1% absorption (μg/ml)	0·15	0·35
Limit of detection (μg/ml)	0·02–0·05	0·03–0·07

Interferences and Special Characteristics

Measurements for lead are remarkably free from interference effects in both the air–acetylene and air–propane flame systems. Heavy excess of Cu, Fe, Zn, and Sn cause negligible alteration of the response.

Enhancement attributed to non-specific background absorption can be troublesome and when lead is determined at trace levels in solutions (e.g., urine) containing heavy excess of foreign ions it is essential that background correction (see p. 239) should be used.

Lithium

Wavelength

The unresolved doublet at 670·8 nm is normally used.

Flame System

Either the air–acetylene or air–propane flame is suitable for the determination of lithium.

Sensitivity

The sensitivity for 1 per cent absorption is about 0·02 μg/ml with standard instruments. The limit of detection is about 0·002–0·005 μg/ml.

Without the use of scale expansion the element can be determined over a range from 0·05 to 4 μg/ml.

Interferences and Special Characteristics

At low levels, using the air–acetylene flame, appreciable ionization occurs. Under these conditions, the standard and test solutions should be buffered with potassium.

Solutions of the chloride, sulphate, and nitrate that contain the same concentrations of lithium give rise to identical absorptions. The response for lithium is unaffected by the presence of heavy loadings of Al, Ba, Ca, Fe, Mg, K, Na, Zn, and PO_4^{3-}.

Magnesium

Wavelengths

For practical purposes determinations are always carried out at the intense resonance line at 285·2 nm. Weak absorption is also obtained from the 285·2 nm iron line, and using an iron lamp at this wavelength, magnesium can be estimated over the range 20–1000 μg/ml, with a sensitivity for 1 per cent absorption of 10 μg/ml. It is essential that the cathode of the iron lamp should not contain traces of magnesium.

Flame Systems

The air–acetylene flame is most commonly used, and with this the element can be determined in aqueous solution without scale expansion over a range from 0·01–1 μg/ml.

Magnesium can also be determined with the nitrous oxide–acetylene flame over a range from about 0·05 to 5 μg/ml in aqueous solution. 1000 μg/ml potassium should be added to both test and standard solutions to suppress ionization.

Sensitivity

Magnesium is the most sensitive of the elements to determination by atomic absorption.

The sensitivity for 1 per cent absorption in the air–acetylene flame is generally stated to be 0·005 μg/ml and the limit of detection is less than 0·001 μg/ml. Using the nitrous oxide–acetylene flame the sensitivity for 1 per cent absorption is about 0·03 μg/ml.

Interferences and Special Characteristics

Air–Acetylene Flame. Equal quantities of magnesium, present either as the nitrate or chloride, exhibit the same response, but when present as the sulphate, the sensitivity is slightly diminished. The addition of hydrochloric acid to a solution of magnesium sulphate raises the response to the same level as the response of the chloride solution.

Copper and calcium each exert a small enhancement, while phosphate and aluminium seriously depress the absorption. Phosphate interference can be overcome by the addition of lanthanum chloride.

The presence of Fe, Cr, Mn, Ni, Mo, Pb, Zn, Na, and K have negligible effect upon the absorption of magnesium in the air–acetylene flame.

Nitrous Oxide–Acetylene Flame. In this flame system solutions of the nitrate, chloride, and sulphate all exhibit identical responses for the same magnesium concentration, and the presence of phosphate does not depress the magnesium absorption.

In the absence of an ionization suppressant, calcium, sodium and potassium cause a slight enhancement of the response.

The presence of Al, Cu, Cr, Fe, Pb, Mn, Mo, Ni, and Zn has negligible effect upon the absorption of magnesium in the nitrous oxide–acetylene flame.

References

A very large number of references to the determination of magnesium in a wide variety of materials will be found in the literature. Allan, Willis, and David contributed largely to the initial development of atomic absorption by their work upon the element.

Manganese

Wavelengths

There is a triplet in the manganese spectrum with peaks at 279·5, 279·8 and 280·1 nm. It is possible to separate these lines with most modern instruments.

A less absorbing line at 403·1 nm is of value for determinations of solutions containing manganese up to a level of about 60 μg/ml (see Table 4.2).

Flame System

Air–acetylene. Optimum flame conditions are best established by adjusting to maximum transparency (see p. 7).

Sensitivity

At 279·5 nm a sensitivity for 1 per cent absorption of 0·03 μg/ml is attainable with standard instruments.

The limit of detection is between 0·004 and 0·01 μg/ml.

The optimum range for determinations, without the use of scale expansion, is from about 0·05 to 5 μg/ml.

INTERFERENCES AND SPECIAL CHARACTERISTICS

Solutions containing the same concentrations of manganese as the chloride, sulphate, nitrate, or permanganate give rise to identical absorptions.

The following cations have very little effect upon the manganese response, even when present in heavy excess: Ca, Cr, Co, Cu, Fe, Pb, Mg, Mo, Ni, K, Na, W, Zn, PO_4^{3-}. In massive excess (e.g. a 1 to 5 per cent solution of the salt) zinc sulphate depresses the response for manganese, whereas zinc chloride at the same level has little effect.

The presence of heavy excesses of Al and Si slightly depresses the manganese absorption, but these interferences can be overcome by adding lanthanum or calcium chloride to the test solution.

It should be noted that iron and nickel salts commonly contain manganese. In order to ascertain reliably the interferences of these two elements upon manganese absorption, therefore, it is necessary to make comparison against a relevant blank.

In the presence of perchloric acid, as for example when manganese has been extracted by wet digestion from plants or soils, the absorption characteristics depend on the oxidation state of the element. A solution of the permanganate exhibits a depressed absorption.

For determinations that are carried out in the presence of perchloric acid, therefore, it is essential to reduce the manganese to Mn(II) by the action of sodium nitrite or hydroxylamine hydrochloride.

Mercury

WAVELENGTH

The line at 253·7 nm is that most commonly used for atomic absorption measurements.

FLAME SYSTEM

Air-acetylene.

SENSITIVITY

The most notable characteristic is that a solution containing the metal in the mercury(I) form exhibits a higher absorption than one containing the same concentration of mercury in the mercury(II) state. This phenomenon is explained by the unique characteristic of the mercury(I) ion to dissociate readily into the mercury(II) ion and elemental mercury which has significant vapour pressure at room temperature:

$$Hg_2^{2+} \rightarrow Hg + Hg^{2+}$$

Typical response curves obtained with the air–acetylene flame for solutions of mercury(I) and mercury(II) salts are shown in Fig. 4.3. It is seen that in the mercury(II) state the element can be determined without scale expansion over the range 10 to 500 μg/ml. The sensitivity for 1 per cent absorption is about 4·0 μg/ml. The practical limit of detection for aqueous mercuric solutions is 0·2–0·5 μg/ml. Mercury(I), it is seen, may be determined over the range 4·0 to 200 μg/ml. The sensitivity for 1 per cent absorption is 1 μg/ml, and the practical limit of detection is 0·1–0·3 μg/ml in aqueous solution. Above about 100 μg/ml the response curve for mercury(I) salts shows very pronounced curvature and readings above this concentration are less reliable than those below 100 μg/ml.

By resorting to flameless atomic absorption it is possible to estimate mercury in solution down to very low levels. This procedure is described on p. 102.

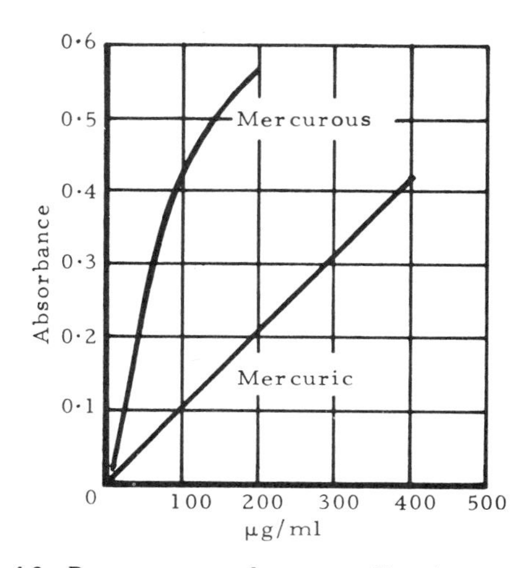

Fig. 4.3 Response curves for mercury(I) and mercury(II).

INTERFERENCES AND SPECIAL CHARACTERISTICS

As stated above, solutions of mercury(I) salts exhibit a higher absorbance than solutions of mercury(II) salts containing the same concentration of mercury. Further enhancement to a peak value may be obtained by treating solutions of either mercury(I) or mercury(II) salts with an excess of ascorbic acid[60] or tin(II) chloride. This phenomenon is due to the complete reduction of the mercury to the volatile colloidal elemental form. Due to the rapid aggregation of the colloidal mercury, this procedure can only be utilized to enhance the response for very dilute solutions, and even then the absorption measurement must be made immediately after the addition of the reducing agent.

Heavy excess of the following cations do not affect the absorption response for mercury: Sb, Bi, Cd, Cu, and Zn.

Molybdenum

Wavelength

The 313·3 nm line is most frequently used. The line at 379·8 nm has also been used by Mostyn and Cunningham.[62] Other less absorbing lines are listed in Table 4.2.

Flame Systems

Either the luminous air–acetylene or nitrous oxide–acetylene flame may be used. The higher temperature flame gives rise to the most sensitive and linear absorption characteristics, and interferences are also less pronounced.

Sensitivity

The essential data for the two flame systems is tabulated below.

	Flame System	
	Air-Acetylene	N_2O–Acetylene
Useful range for determination at ×1 scale expansion (μg/ml)	4–200	1–100
Sensitivity for 1% absorption (μg/ml)	1·9	0·5
Limit of detection (μg/ml)	0·05–0·15	0·04–0·2

Interferences and Special Characteristics

Heavy excess of Cd, Cu, Pb, and Sn have little effect upon the absorption of molybdenum when either of the above flame systems is used.

Interference from Cr, Fe, Mn, and Ni has been reported, but this could be overcome by dosing standards and test solutions with ammonium chloride (2% m/v).

The determination of molybdenum by atomic absorption spectroscopy is, of course, the basis of the indirect amplification procedures for phosphorus, silicon, niobium, and titanium that have been developed by West and co-workers.[61,65,103,119]

The effect of various acids has been thoroughly investigated.[64]

Nickel

Wavelengths

The most sensitive wavelength is at 232·0 nm. This is very close to an ionic line at 231·6 nm.

In setting to the 232·0 nm wavelength, it is essential to distinguish it from the less absorbing line at 231·1 nm, which is of only slightly lower intensity. This is accomplished by using narrow monochromator slit widths and approaching from the higher end of the spectrum. 341·5 nm and 352·5 nm were originally the most commonly used nickel lines and are still valuable for many determinations. Other less absorbing lines are given in Table 4.2.

FLAME SYSTEM

Air–acetylene. Optimum flame conditions are best established by adjusting to maximum transparency (see p. 7).

SENSITIVITY

	Wavelength		
	232·0 nm	341·5 nm	352·5 nm
Useful range for determination at ×1 scale expansion (μg/ml)	0·2–15	1–80	1–75
Sensitivity for 1% absorption (μg/ml)	0·07	0·4	0·5
Limit of detection (μg/ml)	0·005–0·02	0·03–0·1	0·1–0·2

INTERFERENCES AND SPECIAL CHARACTERISTICS

Nickel is remarkably free from interference effects from foreign ions, and solutions of the nitrate, sulphate, and chloride that contain equal concentrations of the element give rise to identical responses. The following elements, even when present in heavy excess, hardly affect the absorption of nickel: Al, Ca, Cd, Cr, Co, Cu, Fe, Pb, Mg, Mn, Mo, Ag, Na, Sn, W, V, and Zn.

Niobium

Niobium is not an easy metal to determine by classical analytical procedures and it would indeed be cause for elation in the laboratories responsible for its routine estimation if it could be reported that atomic absorption provided a rapid and reliable means for the analysis. Unfortunately this is not so.

The most convincing procedure utilizing atomic absorption is the indirect amplification method developed by WEST and co-workers.[65] In principle the method depends upon the formation of molybdoniobophosphoric acid which is extracted into butanol. The phosphomolybdic acid, which is also formed during the reaction, is previously removed by selective extraction into isobutyl acetate.

The molybdenum content of the molybdoniobophosphoric complex is estimated by atomic absorption spectroscopy, and amplification is obtained because eleven molybdenum atoms are associated with every niobium atom. The procedure is of particular value because tantalum does not form a similar complex. The procedure allows niobium to be determined down to a concentration level of about 0·03 μg/ml Nb in the original solution. Heavy excess of Al, Ag, Bi, Be, Ca, Cd, Cr, Co, Cu, Fe, Ni, Mg, Mn, Sb, Ta, V, W, Zn, Zr, SO_4^{2-}, NO^-_3, F^-, and E.D.T.A. do not interfere with the analysis. The presence of titanium causes interference.

WAVELENGTHS (For direct estimation of niobium.)

334·9 and 358·0 nm.

FLAME SYSTEM

A very fuel-rich nitrous oxide–acetylene flame is required.

SENSITIVITY

The sensitivity for 1 per cent absorption is about 25 μg/ml. The limit of detection is 25–50 μg/ml. Without scale expansion the element can be determined over the range 100–4000 μg/ml. Potassium should be added to all solutions including the blank to suppress ionization.

INTERFERENCES AND SPECIAL CHARACTERISTICS

Niobium is subject to complex interferences.

Osmium

WAVELENGTHS

290·9 nm, also 305·9 nm (less sensitive).

FLAME SYSTEM

Nitrous oxide–acetylene.

SENSITIVITY

The sensitivity for 1 per cent absorption is 1·5 μg/ml. The limit of detection is 0·15–0·5 μg/ml. Without scale expansion the element can be estimated over the range 5–250 μg/ml.

INTERFERENCES AND SPECIAL CHARACTERISTICS

Very little work on interference effects observed in the determination of osmium has been reported.

Palladium

WAVELENGTHS

Three useful resonance lines exist. Those at 247·6 and 244·8 nm are the most sensitive, but the responses obtained at them become grossly non-linear at concentrations above 50 and 20 μg/ml respectively. A much higher light intensity is obtained at the 340·5 nm line, and the response at this wavelength is sensibly linear up to at least 120 μg/ml (see Table 4.2).

FLAME SYSTEM

Early work suggested that the use of a cool air–propane flame would be advantageous for the estimation of palladium. It is now established, though, that the air–acetylene and nitrous oxide–acetylene systems are preferable.

SENSITIVITY

The essential data for the various conditions are tabulated below:

Air–Acetylene

	Wavelength		
	244·8 nm	247·6 nm	340·5 nm
Useful range for determination at ×1 scale expansion (μg/ml)	1–20	1–40	2–120
Sensitivity for 1% absorption (μg/ml)	0·18	0·25	0·9
Limit of detection (μg/ml)	0·04–0·1	0·04–0·1	0·3–0·5

N_2O—Acetylene

	Wavelength		
	244·8 nm	247·6 nm	340·5 nm
Useful range for determination at ×1 scale expansion (μg/ml)	2·5–100	5–200	5–200
Sensitivity for 1% absorption (μg/ml)	1·5	2·2	2·2
Limit of detection (μg/ml)	0·4–0·8	0·6–1·0	1·0–1·2

INTERFERENCES AND SPECIAL CHARACTERISTICS

The noble metals might reasonably be expected to exhibit freedom from chemical interferences. This expectation is more nearly realized for palladium than the other metals of the group, but chemical interferences do nevertheless occur.

With the air–acetylene system severe depression occurs in the presence of iridium and nickel, and smaller but noticeable depression in the presence of iron, chromium, platinum and copper. Rhodium and gold produce a slight enhancement.

With the nitrous oxide–acetylene system only iridium, rhodium and gold affect the absorption of palladium, producing slight enhancement.

The absorption of palladium is affected by the presence of excess acid, but this effect levels out as the acid concentration rises and can be sensibly overcome by working in solutions that are made 10% v/v with concentrated hydrochloric acid. (The investigation of chemical interferences detailed above was carried out in the presence of 10% v/v HCl.)

Phosphorus

This element can be determined indirectly by atomic absorption spectroscopy by using an indirect amplification procedure.[75]

The principle of the method depends upon the formation and extraction into an organic phase of phosphomolybdic acid $H_3PO_4(MoO_3)_{12}$. The molybdenum is then determined by atomic absorption. Amplification is obtained because 12 molybdenum atoms are associated with every atom of phosphorus. The method allows as little as 0·08 μg/ml of phosphorus in the original test solution to be determined.

The presence of heavy loadings of Al, Sb, As, Au, Bi, Ca, Cd, Co, Cr, Cu, Fe, Ge, Ni, Pb, Mg, Mn, Se, Te, Ti, V, Zn, F^-, E.D.T.A., NO_3^-, and SO_4^{2-}, do not interfere. Tungsten, present in heavy excess in the original test solution, gives rise to low results for phosphorus.

Direct determination of phosphorus by flame A.A.S. (see p. 212).

Direct determination of phosphorus by flameless electrothermal atomization using the 213·6 nm non-resonance line. (See chapter 7, refs 140, 201 and 202.)

Platinum

WAVELENGTH

The only line of real value to the analyst is that at 265·9 nm. When setting to this wavelength, it is advisable to approach from the higher end of the spectrum to ensure that it, and not the less absorbing line at 264·7 nm, is selected. Another less absorbing line is given in Table 4.2.

Flame System

Platinum can be determined with the air–propane, air–acetylene or nitrous oxide–acetylene flame systems.

Sensitivity

	Flame System	
	Air–Acetylene	N_2O–Acetylene
Useful range for determinations at ×1 scale expansion (μg/ml)	5–300	25–2000
Sensitivity for 1% absorption (μg/ml)	1·5	10
Limit of detection (μg/ml)	0·1–0·3	1–3

Interferences and Special Characteristics

The absorption for the noble metals might reasonably be expected to be free from chemical interference effects, but in fact very pronounced interferences have been reported in the literature, especially for cooler flame systems.

With the air–acetylene flame severe depression of the platinum absorption is brought about by the presence of iron, chromium, palladium, iridium, rhodium, silver, copper or nickel in the test solutions. Of the common metals likely to be found in association with platinum only gold and zinc exert negligible effect upon that element's absorption.

The depressions observed with the air–acetylene system can be sensibly overcome by the addition of a mixture of cadmium and copper sulphates so that the treated solutions contain 5000 μg/ml of Cd and Cu, or of lanthanum chloride (at a level of about 5000 μg/ml La). The addition of this latter reagent causes a noticeable enhancement of the platinum absorbance, so that it must be added to both test and standard solutions if reliable results are to be obtained.

It has been reported that the use of the nitrous oxide–acetylene flame system completely eliminates chemical depressive interferences of most metals upon platinum absorption. This experience has not been confirmed by our own assessment trials (R.J.R.). Interference from the metals listed above, we find, still occurs, but to a lesser degree, and the addition of lanthanum chloride to test and standard solutions is still necessary to overcome its effect.

The absorption in the nitrous oxide–acetylene flame is considerably lower than with the air–acetylene system.

Potassium

The determination of the alkali metals by flame emission, is of course, a well-established procedure, and this fact, together with the difficulties initially encountered with the production of hollow-cathode lamps for these elements, tended to retard the development of atomic absorption procedures for them, but reliable lamps are now available.

Wavelength

766·5 nm is the main line of practical value. In order to make determinations at this wavelength, the instrument must be fitted with a suitable red-sensitive photomultiplier tube. Less absorbing lines are given in Table 4.2.

Flame System

Air–acetylene. Air–propane can also be used. The sensitivity obtained with this flame is similar to that with air–acetylene, but the response curve is distinctly more linear at concentrations above 2 μg/ml.

Sensitivity

The sensitivity for 1 per cent absorption is in the region of 0·02 μg/ml. The limit of detection is about 0·004–0·01 μg/ml.

Without scale expansion the element can be determined over a range from 0·04 to 4 μg/ml.

Interferences and Special Characteristics

Potassium is surprisingly free from interference effects from other anions and cations that may be present. Solutions of the chloride, nitrate, and sulphate that contain the same concentrations of potassium give rise to identical responses. Ionization interference can be suppressed by the addition of 1000 μg/ml caesium or sodium to both test and standard solutions. Under these conditions the presence of heavy excess of Ca, Li, Mg, and Na do not affect the absorption of potassium.

The Rare Earths and Rarer Metals (see Table 4.3)

Of this group of elements the following have been determined by atomic absorption spectroscopy: dysprosium, erbium, europium, gadolinium, holmium, lanthanum, lutetium, neodymium, praseodymium, samarium, scandium, terbium, thulium, ytterbium, yttrium. Most of these elements undergo significant ionization in the nitrous oxide–acetylene flame and potassium should be added to both test and standard solutions in order to minimize this effect.

The need to determine these elements though, is rather infrequent in most laboratories.

Hollow-cathode lamps are expensive for the rare earths, and unfortunately for the rare earth chemist, he is likely to require a different lamp for almost every element of the group.

It so happens that these elements are amenable to determination by high-temperature flame emission spectroscopy.[87,88] The technique of separated flame spectroscopy holds out further possibility of reliable, rapid, and inexpensive determinations in this field.[82]

Rhenium

Wavelength

346·1 nm.

Flame System

A fuel-rich nitrous oxide–acetylene flame is required.

Sensitivity

The sensitivity for 1 per cent absorption is 10 μg/ml. The limit of detection is 1–3 μg/ml. Without scale expansion, the element can be determined over the range 20–1500 μg/ml.

Interferences and Special Characteristics

The rhenium signal is enhanced in the presence of sulphuric acid and depressed in the presence of calcium, barium or magnesium, or most transition metals. The effect is minimized, with loss of sensitivity, by working with a slightly less fuel-rich flame.

Rhodium

Wavelengths

The most absorbing line is at 343·5 nm. Other less absorbing lines are given in Table 4.2.

Flame System

Rhodium is readily determined by atomic absorption spectroscopy with either the air–acetylene or nitrous oxide–acetylene flame systems. At lower concentrations both systems show a similar sensitivity, but above about 50 μg/ml a higher and more linear absorption is obtained with the nitrous oxide–acetylene flame.

Sensitivity

The essential data for the two flame systems is tabulated below:

	Flame System	
	Air–Acetylene	N_2O–Acetylene
Useful range for determination at ×1 scale expansion (μg/ml)	1–100	1–100
Sensitivity for 1% absorption (μg/ml)	0·5	0·5
Limit of detection (μg/ml)	0·05–0·1	0·1–0·2

INTERFERENCES AND SPECIAL CHARACTERISTICS

Rhodium exhibits greater freedom from chemical interferences than most other noble metals. Heavy excesses of iron, platinum, palladium, gold, zinc and copper have negligible effect upon the absorption of rhodium in the nitrous oxide–acetylene system, but chromium, iridium and nickel produce noticeable depressions.

Interferences are much more pronounced with the air–acetylene flame, iridium and zinc being the only metals commonly found in association with rhodium that do not affect its absorption. The presence of iron, chromium, and nickel depress rhodium absorption while platinum, palladium, gold and to a lesser extent copper enhance the readings.

Rubidium

WAVELENGTH

780·0 nm. A cut-off filter positioned between the flame and the monochromator can sometimes prove useful (see under Caesium).

FLAME SYSTEM

Air–acetylene. Pronounced ionization depression occurs with the air–acetylene system and it is essential to dose standard and test solutions with a heavy excess of potassium (say 2000 μg/ml K).

SENSITIVITY

The sensitivity for 1 per cent absorption is 0·05 μg/ml. The limit of detection is 0·01–0·03 μg/ml. Without scale expansion the element can be determined over the range 0·1–10 μg/ml.

INTERFERENCES AND SPECIAL CHARACTERISTICS

Of the elements commonly associated with rubidium; potassium, sodium, lithium and calcium do not affect the absorption provided the test and standard solutions are adjusted to contain a heavy excess of an alkali metal.

Less absorbing lines are given in Table 4.2.

Ruthenium

Wavelength

349·9 nm. A less absorbing line is given in Table 4.2.

Flame System

Air–acetylene.

Sensitivity

The sensitivity for 1 per cent absorption is 1·0 μg/ml.

The limit of detection is 0·15–0·3 μg/ml. Without scale expansion the element can be estimated over the range 2–150 μg/ml.

Interferences and Special Characteristics

Ruthenium is subject to complex interference effects. The addition of uranyl nitrate as a releasing agent (0·4% m/v U) has been found to minimize interelement effects.

Selenium

Wavelength

196·0 nm. Less absorbing lines are given in Table 4.2.

Flame System

The remarks made for arsenic apply also to selenium, namely that all flame systems themselves absorb strongly at wavelengths below 200·0 nm. The argon–hydrogen entrained air flame is much more transparent than air–acetylene at 196·1 nm and so is likely to be advantageous for the determination of selenium in relatively pure solutions. Interference due to non-specific background absorption is again likely to be troublesome when the solutions contain heavy loadings of foreign materials.

Sensitivity

The sensitivity for 1 per cent absorption is 0·5 μg/ml in the air–acetylene flame. The limit of detection is about 0·5–2 μg/ml. With the argon–hydrogen-entrained air flame the corresponding figures are 0·2 and 0·05–0·2 μg/ml.

Other Possibilities

West *et al.*[101] have employed atomic fluorescence, using microwave discharge tubes as spectral sources, and have obtained a limit of detection of 0·25 μg/ml at the 204·0 nm line in the air–propane flame.

General improvements in hardware now make atomic absorption measurements possible at the 196·0 nm line. The fact that atomic fluorescence is much more seriously affected by 'light scatter' from refractory oxides (e.g., AlO species) than is atomic absorption, does not suggest that fluorescence will provide a more reliable procedure for trace estimations of selenium than absorption.

Selenium can be determined by the hydride generation technique to give a detection limit better than 0·002 μg/ml for 1 ml sample (see p. 79). It can also be determined using flameless electrothermal atomization. (See chapter 7, refs 120 and 164.)

Silicon

Wavelength

251·6 nm. Less absorbing lines are given in Table 4.2.

Flame System

Nitrous oxide–acetylene. Very fuel-rich.

Sensitivity

The sensitivity for 1 per cent absorption is about 2 μg/ml.

The limit of detection is about 0·3–0·7 μg/ml.

Without scale expansion the element can be determined over a range from 4 to 400 μg/ml.

Interferences and Special Characteristics

Solutions of sodium metasilicate and hydrofluosilicic acid that contain the same concentrations of silicon give rise to identical responses. The presence of heavy loadings of the following cations have little effect upon the absorption of silicon: Al, Ca, Fe.

The determination of silicon by an indirect amplification procedure, utilizing the silicomolybdic acid, and allowing silicon to be estimated over a concentration range from 0·08 to 1·2 μg/ml has been described by West[103] and co-workers. This procedure is reported to be remarkably free of interference from foreign ions.

Silver

Wavelength

The line at 328·1 nm offers maximum sensitivity. The less absorbing line at 338·3 nm is of value for determinations in solutions containing up to about 20 μg/ml Ag.

FLAME SYSTEM

Air–acetylene.

SENSITIVITY

The sensitivity for 1 per cent absorption can be expected to be about 0·05 μg/ml.

The limit of detection is about 0·003–0·006 μg/ml.

Without the use of scale expansion, the element can be determined at the 328·1 nm line over the range 0·1 to 10 μg/ml.

INTERFERENCES AND SPECIAL CHARACTERISTICS

Atomic absorption determinations for silver are remarkably free from interference effects and the following anions and cations, commonly found in association with the element, do not affect its absorption, even when present in heavy excess: Al, Ba, Be, Bi, Cd, Ca, Cr, Co, Cu, Fe, La, Pb, Li, Mg, Mn, Hg, Mo, Ni, K, Na, Sr, Sn, Ti, Zn, NH_4^+, CN^-.

Sodium

The determination of sodium in low concentrations is easily carried out by either atomic absorption or emission flame photometry. The latter technique is, of course, a firmly established procedure in most laboratories and this fact has tended to retard the development of absorption procedures for sodium determinations.

WAVELENGTHS

The doublet with peaks at 589·0 and 589·6 nm provides the only wavelengths of practical value. Hollow-cathode lamps capable of providing sufficiently intense emission for the use of narrow monochromator slit settings are now available, and under these conditions it is possible to resolve the two lines. It is found that the absorption at 589·0 nm is twice as sensitive as that at 589·6 nm.

A less absorbing line is listed in Table 4.2.

FLAME SYSTEMS

Either the air–acetylene or air–propane flame may be used.

SENSITIVITY

The sensitivity for 1 per cent absorption is about 0·01 μg/ml. The limit of detection is 0·002–0·004 μg/ml, frequently limited by the purity of the blank solution.

Without scale expansion the element can be estimated over a range from 0·02 to 2 μg/ml.

INTERFERENCES AND SPECIAL CHARACTERISTICS

Ionization can be suppressed by working in the presence of a large excess of potassium (2000 μg/ml).

Solutions of sodium chloride, nitrate, or sulphate containing equal concentrations of sodium exhibit identical responses.

Heavy loadings of the following cations and anions do not affect the absorption: Ca, Li, Mg, K, PO_4^{3-}.

Strontium

WAVELENGTHS

460·7 nm. (There is an ionic line at 407·7 nm that has been used for determinations in the range 25–1000 μg/ml strontium, in the absence of other alkaline earth or alkali metals.)

FLAME SYSTEMS

The air–acetylene or nitrous oxide–acetylene flames can be used. The latter flame gives slightly better sensitivity and is the preferred flame because of lack of interference effects. Strontium is partially ionized in both flames (especially in the nitrous oxide–acetylene flame) and 2000 μg/ml of potassium should be added to all solutions including the blank in order to suppress the ionization.

SENSITIVITY

The sensitivity for 1 per cent absorption is 0·15 μg/ml in the air–acetylene flame and 0·1 μg/ml in the nitrous oxide–acetylene flame. The detection limit is about 0·01–0·02 μg/ml for both flames.

INTERFERENCES AND SPECIAL CHARACTERISTICS

Solutions of the nitrate or chloride that contain equal concentrations of strontium give rise to identical absorptions. In the air–acetylene flame Al, Si, Ti, Zr, PO_4^{3-} and SO_4^{2-} depress the absorption of strontium, this interference can be overcome by the addition of lanthanum chloride.

In the nitrous oxide–acetylene flame these interference effects, except for aluminium, are overcome. However, aluminium still exerts a slight depressive effect.

Sulphur: see p. 212.

Tantalum

Like niobium, this element is insensitive to determination by atomic absorption, although one will find lamps advertised, and limits of detection specified, in most instrument makers' catalogues.

WAVELENGTH

271·5 nm.

FLAME SYSTEM

A very fuel-rich nitrous oxide–acetylene flame is required.

SENSITIVITY

The sensitivity for 1 per cent absorption is about 12 μg/ml. The limit of detection is 2–4 μg/ml. Without scale expansion the element can be determined over the range 30–2000 μg/ml.

INTERFERENCES AND SPECIAL CHARACTERISTICS

Tantalum is subject to complex interference effects. Tantalum can be extracted into M.I.B.K. from 10 M HF/6 M H_2SO_4/2·2 M NH_4F media.

Tellurium

WAVELENGTH

214·3 nm. Other less absorbing lines are given in Table 4.2.

FLAME SYSTEM

Air–acetylene. Optimum flame conditions are best established by adjusting to maximum transparency (see p. 7).

SENSITIVITY

The sensitivity for 1 per cent absorption is 0·5 μg/ml. The limit of detection is 0·1–0·2 μg/ml. Without scale expansion the element can be determined over the range 1–80 μg/ml.

INTERFERENCES AND SPECIAL CHARACTERISTICS

No significant interferences have been observed in the air–acetylene flame.

Tellurium can be determined by the hydride generation technique to give a limit of detection better than 0·002 μg/ml for a 1 ml sample (see p. 79).

Thallium

WAVELENGTH

276·8 nm.

FLAME SYSTEM

Air–acetylene.

SENSITIVITY

The sensitivity for 1 per cent absorption is 0·4 μg/ml. The limit of detection is 0·03–0·06 μg/ml. Without scale expansion, the element can be estimated over the range 1–80 μg/ml.

INTERFERENCES AND SPECIAL CHARACTERISTICS

Thallium is remarkably free from interference effects in the air–acetylene flame. Thallium can be sensitively determined using flame emission in the nitrous oxide–acetylene flame using the 377·6 or 535·0 nm lines. Potassium should be added to suppress ionization.

Tin

WAVELENGTH

224·6 nm. (There is a much more intense line at 235·5 nm that is only slightly less absorbing.) Other less absorbing lines are given in Table 4.2.

FLAME SYSTEMS

Tin possesses a refractory oxide and its absorption in the air–acetylene flame is therefore low. An improved response can be obtained by using the much hotter nitrous oxide–acetylene system, but by far the most sensitive conditions are attained with the relatively cool but strongly reducing air–hydrogen flame. Not only is the response enhanced with this latter system, but it is also more stable. Unfortunately, though, interferences are more likely to occur when this cooler flame is used.

SENSITIVITY

The essential data for the two lines 224·6 and 286·3 nm, using the two flame systems are tabulated below.

Wave-length nm	Flame system	Sensitivity for 1% absorption (μg/ml)	Limit of detection (μg/ml)	Useful range for determination without scale expansion (μg/ml)
224·6	Air–hydrogen	0·4	0·05–0·1	1–60
224·6	N_2O–acetylene	2·4	0·3–0·6	5–300
286·3	Air–hydrogen	0·8	0·1–0·2	2–100
286·3	N_2O–acetylene	4·5	0·5–1·0	10–600

INTERFERENCES AND SPECIAL CHARACTERISTICS

The absorption of tin in the air–hydrogen flame, although more stable and pronounced than with the nitrous oxide–acetylene flame, is more prone to interference effects. It is, therefore, considered to be good practice for the determination of tin in alloys, ores, etc., to utilize the hotter flame system.

Air–Hydrogen Flame. The presence of Al and Sb in heavy excess noticeably depresses the absorption response for tin, as does SO_4^{2-}.

The following elements produce very slight depressive effects in chloride media: Cd, Cu, Fe, Pb, Ni, Na, and Zn. The following elements produce an elevated response:[117] Co, Mg, K, and Sr.

The presence of an excess of hydrochloric acid also depresses the absorption of tin.

Argon–Hydrogen–Entrained Air Flame. This cool flame gives a good sensitivity for tin, but interelement effects are very pronounced. It has been found[118] that in the presence of 2000 μg/ml^{-1} of iron(III), many interelement effects are eliminated and the tin sensitivity is considerably enhanced.

Nitrous Oxide–Acetylene Flame. With this hotter flame system, only antimony, when present in heavy excess, exerts a depressive interference upon tin.

Tin can be determined by the hydride generation technique to give a detection limit better than 0·001 μg/ml for 1 ml sample (see p. 79).

Titanium

Wavelength

The titanium spectrum is rather complex. The line most frequently used is at 364·3 nm. Other less absorbing lines are listed in Table 4.2.

Flame System

A fuel-rich nitrous oxide–acetylene flame is used.

Sensitivity

The sensitivity for 1 per cent absorption is about 2 μg/ml. The limit of detection is about 0·2–0·4 μg/ml.

In an aqueous solution titanium can be estimated without scale expansion over the range 4 to 400 μg/ml.

Interferences and Special Characteristics

Titanium is not only insensitive to atomic absorption, but is prone to complex interferences. The presence of potassium dichromate depresses the titanium absorption, whereas the addition of chromic chloride or hydrofluoric acid enhances the response. A solution of the potassium titanium oxalate gives rise to an absorption reading identical with that of a solution of titanium chloride containing the same concentration of titanium.

The presence of vanadium also depresses the titanium absorption. Heavy loadings of nickel and iron have little effect on the absorption of titanium.

It is essential to match the acidity of standard and test solutions when the titanium content of alloys is being determined by atomic absorption spectroscopy.

An indirect amplification procedure, capable of a sensitivity for 1 per cent absorption corresponding to 0·013 μg/ml Ti in the aqueous solution, before extraction, has been developed by WEST *et al.*[119] In principle the method depends upon the formation of molybdotitanophosphoric acid, in which eleven atoms of molybdenum are combined with two of titanium.

In brief the details of the procedure are as follows. Molybdotitanophosphoric acid is formed in aqueous solution, made 0·5 M with hydrochloric acid, by the addition of molybdate, phosphate, and potassium aluminium sulphate (to act as a masking agent when fluoride is present in the solution). The excess molybdophosphoric acid is extracted away from the titanium complex into a mixture of chloroform and butanol (4:1 by volume). The titanium complex is then extracted into butanol, and the organic phase washed with 0·1 M hydrochloric acid. The concentration of molybdenum in the butanol extract is finally measured by atomic absorption spectroscopy using the nitrous oxide–acetylene flame.

The presence of the following ions, in heavy excess in the aqueous phase from which the titanium complex is extracted is reported to have no effect upon the determination: Ba, Be, Bi, Ca, Cr, Co, Cu, Fe, Pb, Mg, Mn, Ni, Se, Sr, Te, Zn, B, Cl, NO_3^-, SO_4^{2-}. Of the elements that also form heteropoly acids under similar conditions, As and Ge can be removed by volatilization of the chlorides, and Si is eliminated by volatilization of fluosilicic acid during sample preparation.

Tantalum, niobium and zirconium produce positive interferences.

Tungsten

WAVELENGTHS

The line at 255·1 nm is most commonly used. Other less absorbing lines are given in Table 4.2.

FLAME SYSTEM

A very fuel-rich nitrous oxide–acetylene flame is required.

SENSITIVITY

The sensitivity for 1 per cent absorption is about 10 μg/ml at 255·1 nm. The limit of detection is about 1–3 μg/ml.

The element can be determined at 255·1 nm without scale expansion over a range from 20 to 1500 μg/ml.

INTERFERENCES AND SPECIAL CHARACTERISTICS

The presence of heavy loadings of vanadium slightly depresses the absorption of tungsten. Little interference is caused by the presence of Cr, Ni, or Mo.

Uranium

WAVELENGTH

358·5 nm.

FLAME SYSTEM

A very fuel-rich nitrous oxide–acetylene flame is required for this element.

SENSITIVITY

The sensitivity for 1 per cent absorption is about 150 μg/ml. The limit of detection is about 40–80 μg/ml. Without scale expansion, the element can be determined over the range 300–20 000 μg/ml.

INTERFERENCES AND SPECIAL CHARACTERISTICS

Uranium is subject to complex interferences and undergoes appreciable ionization in the nitrous oxide–acetylene flame. Excess potassium should be added to both samples and standards to suppress ionization. An indirect method capable of determining 5–500 μg of uranium based on the reaction $U(IV) + 2\,Cu(II) \rightarrow U(VI) + 2\,Cu(I)$ and subsequent extraction of the cuprous ion formed has been reported.[122]

Vanadium

WAVELENGTH

The unresolved triplet with peaks at 318·3, 318·4 and 318·5 nm gives rise to the most sensitive conditions. Other less absorbing lines occur at 306·6 nm and at 439·0 nm. See table 4.2.

FLAME SYSTEM

Nitrous oxide–acetylene.

SENSITIVITY

Sensitivity for 1 per cent absorption is about 1·0 μg/ml. The limit of detection is about 0·1–0·2 μg/ml.

Without scale expansion the element can be estimated over a range from 2 to 150 μg/ml.

INTERFERENCES AND SPECIAL CHARACTERISTICS

The absorption of vanadium is virtually unaffected by heavy excesses of Cr, Co, Fe, Mn, Mo, Ni, or W. 1000 μg/ml potassium should be added to both test and standard solutions to suppress ionization.

Zinc

WAVELENGTH

213·9 nm. The only other zinc absorption line is at 307·6 nm and is about 8000 times less sensitive than the 213·9 nm line.

FLAME SYSTEM

The air–acetylene flame is most commonly used. With this flame optimum conditions are best established by adjusting to maximum transparency (see p. 7). The air–propane flame gives a slightly enhanced, and the air–hydrogen flame a slightly lower, response. Zinc may also be determined with the nitrous oxide–acetylene flame, which gives rise to a lower sensitivity and allows the element to be estimated over a range from 1 to 20 μg/ml. The sensitivity with this flame system is not improved by the addition of a heavy excess of potassium ions.

SENSITIVITY

The sensitivity for 1 per cent absorption for the air–acetylene system is 0·01 μg/ml and without scale expansion the element can be estimated over the range 0·02 to 2 μg/ml. The limit of detection is between 0·001 and 0·003 μg/ml.

INTERFERENCES AND SPECIAL CHARACTERISTICS

Using the air–acetylene flame, which is the one most commonly employed, solutions of the chloride, nitrate, or sulphate that contain equal concentrations of zinc give rise to identical absorptions. The following substances have a negligible effect upon the zinc response, even when present in heavy excess: Al, Cd, Ca, Cu, Fe, Pb, Mn, Mg, Ni, Ag, Na, Sn, and PO_4^{3-}.

Zirconium

WAVELENGTH

360·1 nm.

FLAME SYSTEM

A very fuel-rich nitrous oxide–acetylene flame is required for this element.

SENSITIVITY

The sensitivity for 1 per cent absorption is about 15 μg/ml. The limit of detection is 1·5–5 μg/ml. Without scale expansion, the element can be determined over the range 50–2000 μg/ml.

INTERFERENCES AND SPECIAL CHARACTERISTICS

Zirconium is subject to complex interferences.

Indirect Methods of Atomic Absorption

Many non-metallic elements, radicals or metallic elements with poor detection limits can be determined by indirect atomic absorption techniques. For example, chloride can be estimated by adding an excess of silver nitrate, filtering off the silver chloride, dissolving it in ammonia and then determining the silver content. Amplification techniques based on the formation of molybdoheteropoly compounds (see p. 26) can be used for the indirect determination of As, Ce, Ge, Nb, P, Si, Th, Ti and V.

KIRKBRIGHT and JOHNSON[128] have produced a comprehensive review of indirect methods of atomic absorption analysis. Methods for the following species were reviewed: Al, As,Cl^-, ClO_4^-, CN^-, SCN^-, F^-, Ge, Hg, I^-, IO_3^-, Nb, NH_3, NO_3^-, PO_4^{3-}, Se, Si, SO_4^{2-}, Th, Ti, Tl, V, certain organic compounds and non-ionic surfactants. Table 4.4 gives a brief outline of the methods used for some non-metallic species. It should be stressed that most methods suffer from various interelement effects.

Hydride Generation Techniques

It is now possible to determine a number of elements by generation of the corresponding hydride and then passing the liberated hydride into an argon–hydrogen–entrained air flame. Arsenic and selenium hydrides[131,132] can be generated in the presence of zinc and hydrochloric acid, collected in a suitable reservoir and then passed directly to the flame. The technique has been extended to Bi, Sb and Te[134,135] by using a Mg–$TiCl_3$ reduction. Unfortunately, hydride generation is relatively slow, thus the hydride and hydrogen have to be collected for 3–10 minutes and stored in a suitable reservoir (usually a latex balloon) prior to injection into the flame.[132] The detection limit is limited by the large dilution factor incurred when the liberated hydrides and hydrogen are passed into the argon–hydrogen-entrained air flame and the pulse of the hydrides and hydrogen entering the flame can cause background (non-specific) absorption especially for arsenic and selenium determinations.

A better choice of reducing agent is sodium borohydride;[135–138] this reagent can be used to generate the hydrides of As, Bi, Ge, Pb, Sb, Se, Sn and Te. The hydrides are generated, almost instantaneously, by adding 1 ml of the suitably acidified sample to 2 ml of 1% m/v aqueous sodium borohydride contained in a simple glass cell (Fig. 4.4). A P.T.F.E. coated magnetic stirring bar, placed

Table 4.4 Indirect methods of atomic absorption

Species	Reagent(s) added	Principle	Element actually measured by A.A.
F^-	Mg	Change in Mg absorbance in air–coal-gas flame	Mg
F^-	Zr	Change in Zr absorbance	Zr
Cl^-	$AgNO_3$	Precipitation of AgCl	Ag
ClO_4^-	Cu(I) + neocuproin (nc)	Extraction of $Cu(nc)_2ClO_4$ into ethylacetate	Cu
I^-	Cd + 1,10-phenanthroline	Tris(1,10 P)CdI_2 extracted into nitrobenzene	Cd
IO_3^-	Fe(II)	$Fe(II) \xrightarrow{IO_3^-} Fe(III)$ Extraction of Fe(III) into diethyl ether (9 M HCl)	Fe
NH_3	Zr	Enhancement of Zr absorbance	Zr
CN^-	Ag	Add metallic silver and shake for 1 h	Ag
SCN^-	Cu + pyridine	Extraction of $[Cu_2(Py)_2](SCN)_2$ into chloroform	Cu
NO_3^-	Neocuproin + Cu(I)	Extraction of $[Cu(nc)_2]NO_3$ into M.I.B.K.	Cu
PO_4^{3-}	See p. 64		
SO_4^{2-}	Ba	Filter ppt and either redissolve in Na_2 E.D.T.A. or measure residual Ba	Ba

in the cell, helps to promote better mixing of the acidified sample and the sodium borohydride. It is also possible to mix the neutral sample solution and the sodium borohydride solution in the cell and then add the acid to this. The hydrides and hydrogen gas are liberated almost instantaneously and carried by a stream of nitrogen directly into a narrow bore silica tube (Fig. 4.5a) which is mounted over a wide-path air–acetylene burner. Alternatively, the silica tube can be heated electrically by winding a suitable resistance coil around it and surrounding it with some form of insulation. The use of a narrow bore silica-atomizing tube rather than a flame to atomize the hydrides results in much improved limits of detection because of the limited dilution of the hydrides. The hydride entry port (A) is kept cool by an air-stream directed into annulus (B).

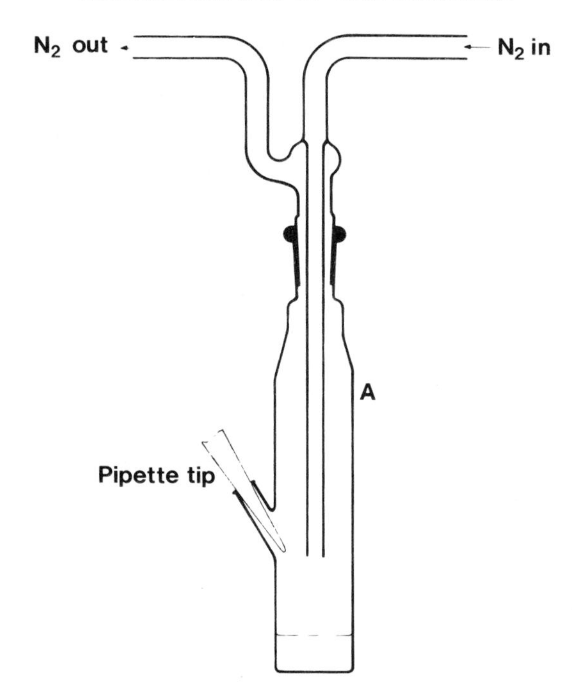

Fig. 4.4 Hydride generator cell

A transverse nitrogen flow through tubes (C) and (C^1) ensures that the liberated hydrogen does not ignite at the ends of the main silica atomization tube. If ignition were to occur on the optical axis a small background (non-specific) absorption signal would be observed, especially when determining arsenic or selenium. For As, Bi, Ge, Sb, Se and Te, the acid concentration is not very critical and it is possible to work with a final hydrogen ion concentration of 1–5 M using hydrochloric, nitric or sulphuric acids. With tin and lead the acid concentration is much more critical and a hydrochloric acid concentration of 0·7 M for tin and 0·2 M for lead has been found to be optimum.

Table 4.5 gives some typical detection limits obtained using a Shandon Southern Instruments (Baird Atomic) A3490 hydride generator (Fig. 4.5b) and an A3400 atomic absorption spectrophotometer. Figure 4.6 shows a typical trace for 1 ml of a 0·01 μg/ml arsenic solution. The blank signal corresponding to 0·0025 μg/ml arsenic was mainly due to arsenic present in the sodium borohydride. The signal observed using a hydrogen hollow-cathode lamp corresponded to less than 0·0008 μg/ml arsenic. Figures 4.7a and 4.7b show some typical calibration curves.

One problem experienced with this technique is that the response depends on the oxidation state of the hydride-forming element. Arsenic and antimony

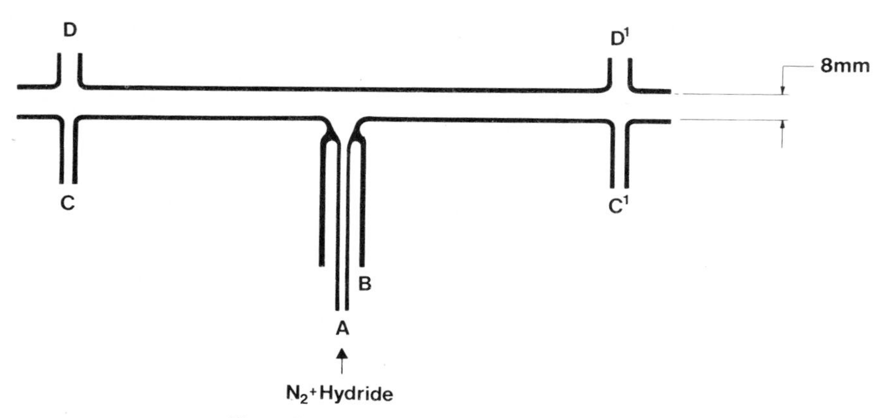

Fig. 4.5a Silica-atomizing tube
A Hydride inlet
B Cooling annulus
CC¹ Transverse nitrogen flow entry
DD¹ Transverse nitrogen flow exit
Length of tube 170 mm.

in the pentavalent(V) state give a lower response than when present in the trivalent(III) state. One method of overcoming this problem[139,140] is to make the samples and blank solution 20% v/v with respect to hydrochloric acid (36% m/m), 1% m/v with respect to potassium iodide and 0·2% m/v with respect to ascorbic acid. The solutions should be allowed to stand for fifteen minutes at 60–70°C. This will then ensure that complete reduction to the trivalent(III) state occurs. The ascorbic acid is added to prevent oxidation of

Table 4.5 Detection limits for hydride generation technique using a flame-heated silica atomizing tube and a sodium borohydride reduction[137]

1 ml Sample volume		
Element	Wavelength nm	Detection limit (2 × standard deviation) μg/ml
As	193·7	0·0008
Bi	223·1	0·0002
Ge	265·1	0·5
Pb	283·3	0·1
Sb	217·6	0·0005
Se	196·2	0·0018
Sn	224·6	0·0005
Te	214·3	0·0015

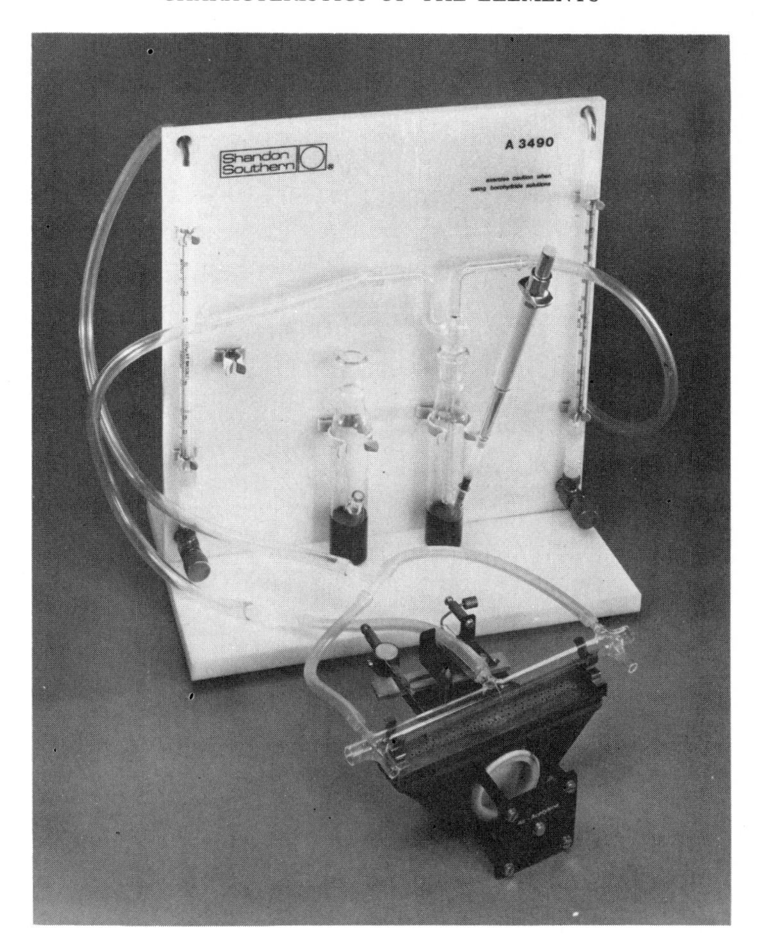

Fig. 4.5b A 3490 Hydride generation apparatus

iodide to iodine (e.g., by air or ferric ions). Selenium and tellurium in the hexavalent(VI) state give a negligible response compared to the tetravalent(IV) state. Solid samples should be dissolved in the minimum volume of boiling aqua regia (3 : 1 HCl : HNO_3) and then diluted with at least an equal volume of water. This will ensure that all the selenium and tellurium ends up in the tetravalent(IV) state. For liquid samples the sample should be diluted with an equal volume of hydrochloric acid (36% m/m) and simmered for about thirty minutes in order to effect reduction to the tetravalent(IV) state. For tin solutions containing less than 0·1 μg/ml it can usually be assumed that the tin will be present in tetravalent(IV) state.

The relatively poor detection limit for lead can be considerably improved

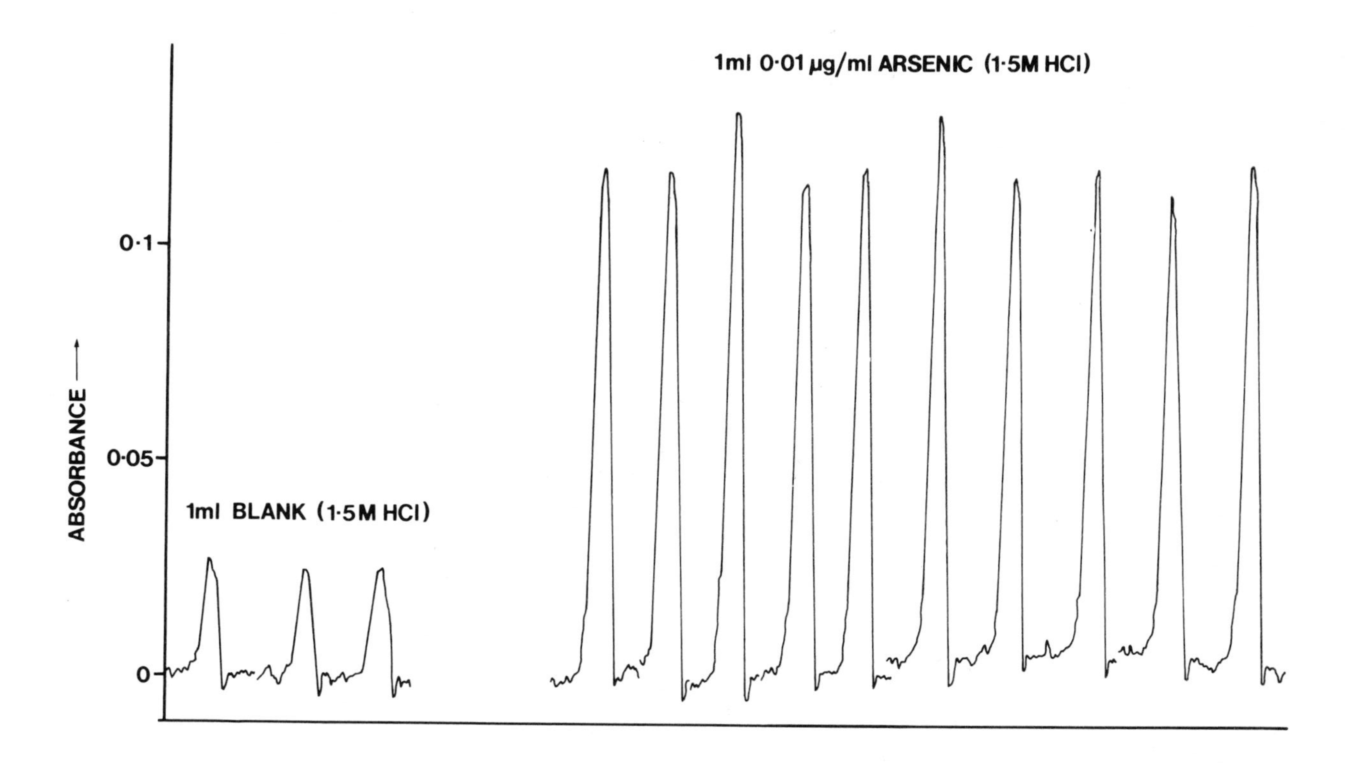

Fig. 4.6 Typical Arsenic Trace (2 ml 1% m/v $NaBH_4$ in cell A)

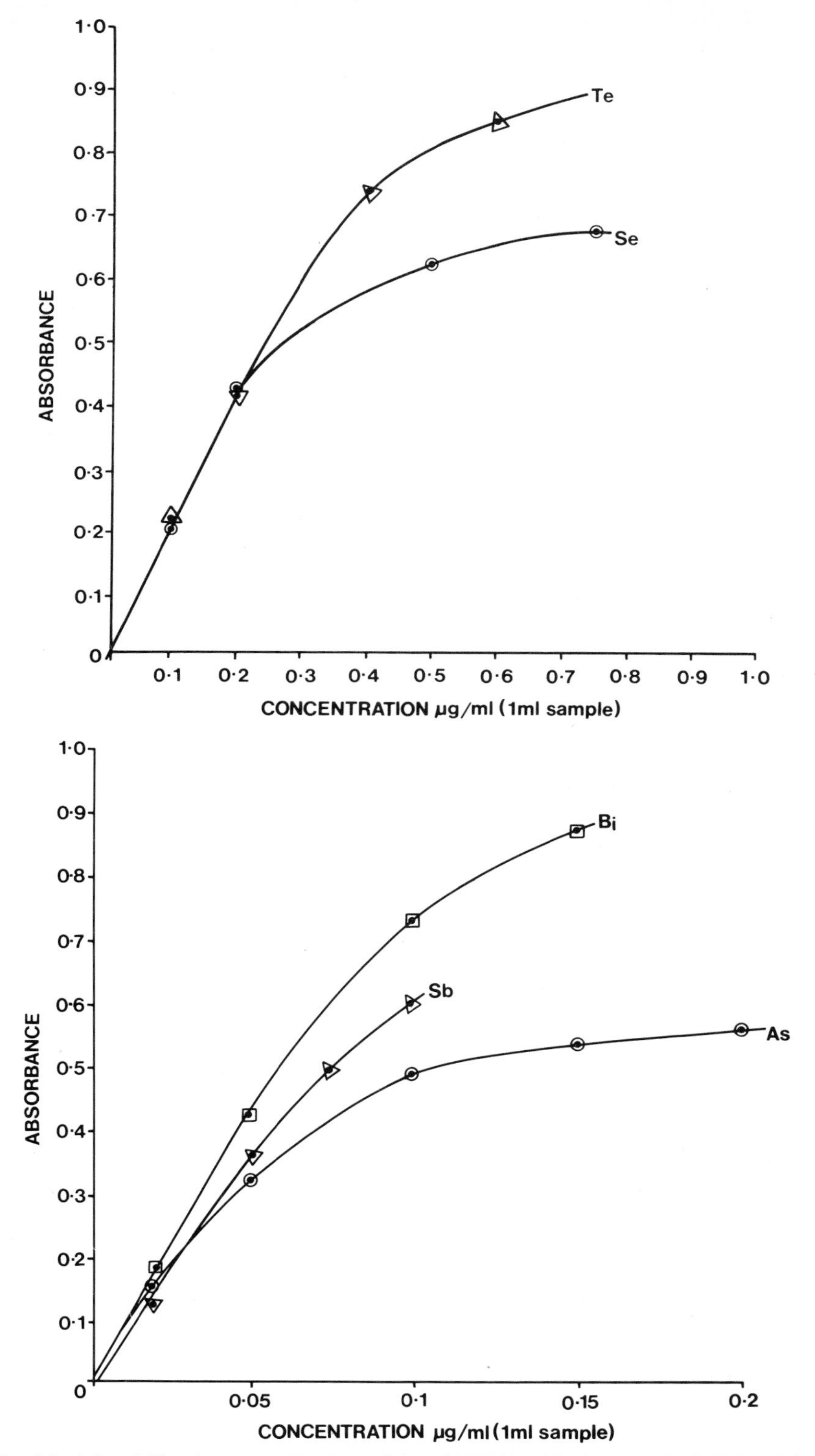

Fig. 4.7a & b Calibration curves (1 ml sample in 1·5 M HCl and 2 ml 1% m/v $NaBH_4$ in cell A)

(up to 500 times) by adding oxidizing agents (hydrogen peroxide[143] or potassium dichromate in conjunction with tartaric acid[144]) to the suitably acidified sample. It would appear that for efficient generation of lead hydride an intermediate, unstable, lead(IV) state is involved.

For most hydride generation technique methods it is essential to use the standard addition technique of calibration as interelement effects can be severe. The effects are manifested as suppressions of the signal and are almost certainly caused by failure to generate the hydride species in the presence of the sample matrix. In general it can be stated that elements that are not readily reduced by sodium borohydride do not interfere significantly (e.g., Al, B, Ba, Ca, K, La, Li, Mg, Na, Si, Sr, Zn, etc.) whilst elements that are easily reduced by sodium borohydride (e.g., Ag, Au, Cu, Ni, etc.) often result in severe interference.[138] Fortunately iron, even at 10 000 μg/ml, does not result in a very severe suppression of signal when determining arsenic or bismuth, but even small amounts of copper (>5 μg/ml) severely depress the selenium response. DRINKWATER[145] has determined bismuth in complex nickel base alloys by adding E.D.T.A. (buffered at pH 9) to the sodium borohydride contained in the glass cell (Fig. 4.4) prior to adding the acidified sample. This procedure effectively overcame the very severe depression of the bismuth signal caused by the complex sample matrix.

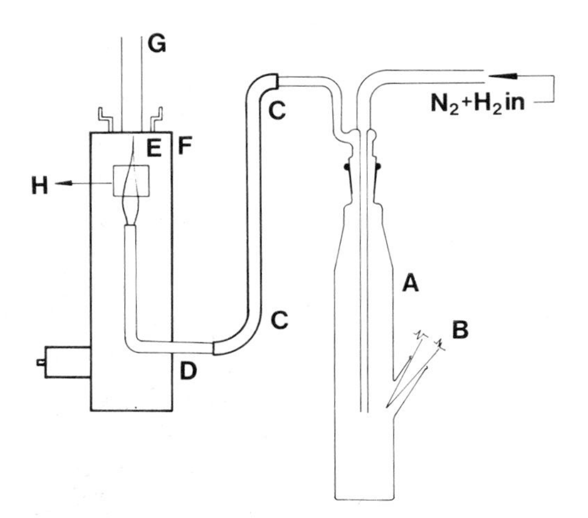

Fig. 4.8 Arrangement of atomic fluorescence spectrophotometer and hydride generator
A. Glass cell
B. Disposable pipette tip
C. P.V.C. tube
D. 10 mm i.d. Pyrex tube
E. Microwave cavity window
F. $\frac{3}{4}$-wave Broida type microwave cavity
G. Microwave lamp
H. Optical axis. (Centre of flame was 8·5 cm from entry slit of monochromator.)

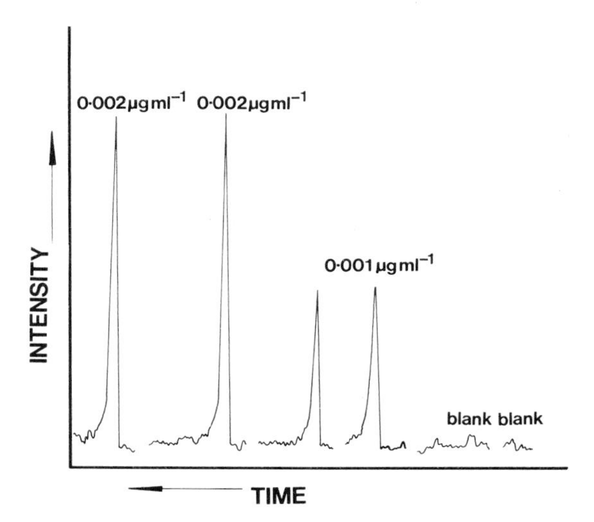

Fig. 4.9 Typical traces for selenium using atomic fluorescence detection (1 ml sample volume) in conjunction with the hydride generation technique

The hydride generation method has been applied to numerous applications and has been successfully automated.[141–143,146–149] Some useful references have been included in the reference section at the end of this Chapter.

The limits of detection for As, Se, Sb and Te can be improved considerably by utilizing the atomic fluorescence technique[153] (see Chapter 8). The hydrides are generated as previously described and then atomized in a nitrogen–hydrogen–entrained air flame maintained on a 10-mm internal diameter Pyrex tube. The experimental set-up is depicted in Fig. 4.8. The fluorescence signal is excited using microwave-powered electrodeless-discharge lamps. The detection limits for a 1 ml sample volume for arsenic and antimony were 0·0001 μg/ml, whilst those for selenium and tellurium were 0·00006 and 0·00008 μg/ml. A typical trace is shown in Fig. 4.9. The calibration curves were linear up to a concentration of 0·1 μg/ml for all four elements.

Chap. 4 References and further reading

ALUMINIUM

1. AMOS, M. D., and WILLIS, J. B., Use of high temperature premixed flames in atomic absorption spectroscopy. *Spectrochim. Acta* 22 (1966) 1325.
2. FERRIS, A. P., JEPSON, W. B., and SHAPLAND, R. C., Evaluation and correction of interference between aluminium, silicon and iron in atomic absorption spectrophotometry. *Analyst* 95 (1970) 574.

ANTIMONY

3. MOSTYN, R. A., and CUNNINGHAM, A. F., The determination of antimony by atomic absorption spectrometry. *Anal. Chem.* 39 (1967) 433.

4. MERANGER, J. C., and SOMERS, E., Determination of antimony in titanium dioxide by atomic-absorption spectrophotometry. *Analyst* 93 (1968) 799.
5. QUARRELL, T. M., POWELL, R. J. W., and CLULEY, H. J., Determination of tin and antimony in lead alloys for cable sheathing by atomic-absorption spectrophotometry. *Analyst* 98 (1973) 443.

ARSENIC
6. JOHNS, P., The atomic absorption of arsenic and selenium. Spectrovision 24 (1970) 6.
7. KASZERMAN, R., and THEURER, K., Effect of valence state on the determination of arsenic by flame A.A. *Atomic Abs. Newsletter*, 15 (1976) 129.
8. KIRKBRIGHT, G. F., and RANSON, L., Use of the nitrous oxide–acetylene flame for determination of arsenic and selenium by atomic absorption spectrometry. *Anal. Chem.* 43 (1971) 1238.

BARIUM
9. KOIRTYOHANN, S. R., and PICKETT, E. E., Spectral interferences in atomic absorption spectrometry. *Anal. Chem.* 38 (1966) 585.
10. RUBESKA, I., The determination of barium in gas shielded flames. *Atomic Abs. Newsletter* 12 (1973) 33.
11. MARUTA, T., TAKEUCHI, T., and SUZUKI, M., A.A. and emission interferences on barium *Anal. Chim. Acta* 58 (1972) 452.
12. MAGILL, W. A., and SVEHLA, G., Study of the elimination of interferences in the determination of barium by A.A.S. *Z. Anal. Chem.* 268 (1974) 180.
13. CIONI, R., MAZZUCOTELLI, A., and OTTONELLO, G., Interference effects in the determination of barium in silicates by flame A.A.S. *Analyst* 101 (1976) 956.

BORON
14. MORROW, R. W., Determination of Boron in Glass by F.E.S. *Analyt. Letters* 5 (1972) 371.
15. PICKETT, E. E., and PAU, J. C. M., Emission Photometric Determination of Boron in Unboronated Fertilisers using the Nitrous Oxide–Hydrogen Flame. *J. Ass. Off. Analyt. Chem.* 56 (1973) 151.

CALCIUM
16. DAVID, D. J., The determination of calcium in plant material by atomic absorption spectrophotometry. *Analyst* 84 (1959) 536.
17. WILLIS, J. B., The determination of calcium and magnesium in urine by atomic absorption spectroscopy. *Anal. Chem.* 33 (1961) 556.
18. WILLIS, J. B., The determination of metals in blood serum by atomic absorption spectroscopy—I. Calcium. *Spectrochim. Acta* 16 (1960) 259.
19. DICKSON, R. E., and JOHNSON, C. M., Interferences associated with the determination of calcium by atomic absorption. *Appl. Spectroscopy* 20 (1966) 214.
20. BENTLEY, E. M., and LEE, G. F., The determination of calcium in natural water by atomic absorption spectroscopy. *Environmental Science and Tech.* 1 (1967) 721.
21. MILLER, T. H., and EDWARDS, W. H., Direct calcium and magnesium determination by a modified atomic absorption spectrophotometry aspiration system. *Atomic Abs. Newsletter* 15 (1976) 75.
22. WARD, D. A., and BIECHLER, D. G., Rapid direct determination of calcium in natural waters by atomic absorption spectrometry. *Atomic Abs. Newsletter* 14 (1975) 29.
23. THOMPSON, K. C., A method for decreasing the A.A. sensitivity of calcium by using the germanium 422·657 nm line. *Analyst* 95 (1970) 1043 (see also ref. 186 chapter 5).

CHROMIUM
24. FELDMAN, F. J., and PURDY, W. C., The atomic absorption spectroscopy of chromium. *Anal. Chim Acta.* 33 (1965) 273.
25. GIAMMARISE, A., The use of ammonium chloride in analyses of chromium samples containing iron. *Atomic Abs. Newsletter* 5 (1966) 113.
26. COKER, D. T., and OTTAWAY, J. M., Overexcitation interferences in atomic absorption spectrophotometry with an air–acetylene flame. *Nature* 227 (1970) 831.

27. YANAGISAWA, M., SUZUKI, M., and TAKEUCHI, T., Cationic interferences in the atomic absorption spectrophotometry of chromium. *Anal. Chim. Acta.* 52 (1970) 386.
27a. RAWA, J. A. and HENN, E. L., Interference effects in the determination of Chromium by A.A. *Amer. Lab.* 9 (Aug. 1977) 31.

COBALT
28. HARRISON, W. W., Factors affecting the selection of a cobalt analysis line for atomic absorption spectrometry. *Anal. Chem.* 37 (1965) 1168.
29. FLEMING, H. D., Chemical and Spectral Interferences in the Determination of Cobalt in Iron and Steel by Atomic Absorption Spectroscopy. *Anal. Chim. Acta.* 59 (1972) 197.

COPPER
30. ALLAN, J. E., The determination of copper by atomic absorption spectrophotometry. *Spectrochim. Acta* 17 (1961) 459.
31. KHALIFA, H., SVEHLA, G., and ERDEY, L., Precision of the determination of copper and gold by atomic absorption spectrophotometry. *Talanta* 12 (1965) 703.
See also reference 44.

GALLIUM
32. POLLOCK, E. N., The determination of gallium and germanium in limonite by A.A. *Atomic Abs. Newsletter* 10 (1971) 77.
33. CHOW, A., and LIPINSKY, W., Determination of gallium by atomic absorption spectrometry. *Anal. Chim. Acta.* 75 (1975) 87.
See also 43.

GOLD
34. TINDALL, F. M., Silver and gold assay by atomic absorption spectrophotometry. *Atomic Abs. Newsletter* 4 (1965) 339.
35. SIMMONS, E. C., Gold assay by atomic absorption spectrophotometry. *Atomic Abs. Newsletter* 4 (1965) 281.
36. TINDALL, F. M., Notes on silver and gold assay by atomic absorption. *Atomic Abs. Newsletter* 5 (1966) 140.
37. ECKELMANS, V., GRAAUWMANS, E., and DE JAEGERE, S., Mutual interference of gold, platinum and palladium in atomic absorption spectroscopy. *Talanta* 21 (1974) 715.
38. ADRIAENSSENS, E., and VERBEEK, F., Atomic absorption spectroscopy of silver, gold, palladium and platinum in a potassium cyanide medium. *Atomic Abs. Newsletter* 12 (1973) 57.
39. CARLSON, G. G., and VAN LOON, J. C., A study on the determination of gold in solutions containing high concentrations of other salts. *Atomic Abs. Newsletter* 9 (1970) 90.
40. MALLETT, R. C., PEARTON, D. C. G., RING, E. J., and STEELE, T. W., Interferences and their elimination in the determination of the noble metals by atomic absorption spectrophotometry. *Talanta* 19 (1972) 181.
41. RUBESKA, I., KORECKOVA, J., and WEISS, D., The determination of gold and palladium in geological materials by A.A. after extraction with dibutyl sulphide. *Atomic Abs. Newsletter* 16 (1977) 1.
42. HEINEMANN, W., Optimizing the measuring conditions for the determination of noble metals by A.A.S. *Z. Anal. Chem.* 280 (1976) 127.
See also 31.

HAFNIUM (See ZIRCONIUM)

INDIUM
43. MULFORD, C. E., Gallium and indium determinations by atomic absorpion. *Atomic Abs. Newsletter* 5 (1966) 28.
44. FUJIWARA, K., HARAGUCHI, H., and FUWA, K., Response surface and atomization mechanisms in air–acetylene flames. Acid interferences in atomic absorption of copper and indium. *Anal. Chem.* 47 (1975) 1670.

IRIDIUM

45. MULFORD, C. E., Iridium absorption. *Atomic Abs. Newsletter* 5 (1966) 63.
46. MANNING, D. C., and FERNADEZ, F., Iridium determination by atomic absorption. *Atomic Abs. Newsletter* 6 (1967) 15.
47. ASHY, M. A., and HEADRIDGE, J. B., The determination of iridium and ruthenium in rhodium sponge by solvent extraction followed by atomic absorption spectrophotometry. *Analyst* 99 (1974) 285.

See also 99.

IRON

48. CURTIS, K. E., Interferences in the determination of Iron by Atomic absorption Spectroscopy in an air–acetylene flame. *Analyst* 94 (1969) 1068.
49. OTTAWAY, J. M., COKER, D. T., ROWSTON, W. B., and BHATTARAI, D. R., The interference of Cobalt, Nickel and Copper in the Determination of Iron by Atomic-absorption spectrophotometry in an Air–Acetylene flame. *Analyst* 95 (1970) 567.
50. MARTIN, M. J., Lanthanum as a releasing agent for iron in atomic absorption spectrophotometry. *Chem. and Ind.* 19 (1971) 514.

See also 2.

LEAD

51. DAGNALL, R. M., and WEST, T. S., Observations on the atomic absorption spectroscopy of lead in aqueous solution, in organic extracts and in gasoline. *Talanta* 11 (1964) 1553.
52. BERMAN, E., The determination of lead in blood and urine by atomic absorption spectrophotometry. *Atomic Abs. Newsletter* 3 (1964) 111.
53. CHAKRABARTI, C. L., ROBINSON, J. W., and WEST, P. W., The atomic absorption spectroscopy of lead. *Anal. Chim. Acta* 34 (1966) 269.

MAGNESIUM

54. ALLAN, J. E., Atomic absorption spectrophotometry with special reference to the determination of magnesium. *Analyst* 83 (1958) 466.
55. DAVID, D. J., The determination of exchangeable sodium, potassium, calcium and magnesium in soils by atomic absorption spectrophotometry. *Analyst* 85 (1960) 495.
56. WILLIS, J. B., The determination of metals in blood serum by atomic absorption spectroscopy—II. Magnesium. *Spectrochim. Acta* 16 (1960) 273.
57. HALLS, D. J., and TOWNSHEND, A., Some interferences in the atomic absorption spectrophotometry of Mg. *Anal. Chim. Acta* 36 (1966) 278.
58. FIRMAN, R. J., Interference caused by iron and alkalis on the determination of magnesium. *Spectrochim. Acta* 21 (1965) 341.
59. HARRISON, W. W., and WADLIN, W. H., Magnesium spinel interferences in air–acetylene vs nitrous oxide–acetylene flames in atomic absorption spectrometry. *Anal. Chem.* 41 (1969) 374.

MERCURY

60. HINGLE, D. N., KIRKBRIGHT, G. F., and WEST, T. S., Some observations on the determination of mercury by atomic absorption spectroscopy in an air–acetylene flame. *Analyst* 92 (1967) 759.

MOLYBDENUM

61. KIRKBRIGHT, G. F., SMITH, A. M., and WEST, T. S., Rapid determination of molybdenum in alloy steels by atomic absorption spectroscopy in a nitrous oxide–acetylene flame. *Analyst* 91 (1966) 700.
62. MOSTYN, R. A., and CUNNINGHAM, A. F., Determination of molybdenum in ferrous alloys by atomic absorption spectrometry. *Anal. Chem.* 38 (1966) 121.
63. DAVID, D. J., The suppression of some interferences in the determination of molybdenum by atomic absorption spectroscopy in an air–acetylene flame. *Analyst* 93 (1968) 79.
64. DILLI, S., GAWNE, K. M., and OCAGO, G. W., Interferences by Acids in the Determination of Molybdenum by A.A.S. *Anal. Chim. Acta* 69 (1974) 287.

See also 91.

NIOBIUM

65. KIRKBRIGHT, G. F., SMITH, A. M., and WEST, T. S., An indirect amplification procedure for the determination of niobium by atomic absorption spectroscopy. *Analyst* 93 (1968) 292.
66. THOMAS, P. E., and PICKERING, W. F., Role of solution equilibria in atomic absorption spectroscopy. *Talanta* 18 (1971) 127.
67. ARENDT, D. H., The determination of niobium in Inconel 718 alloy by atomic absorption spectrophotometry. *Atomic Abs. Newsletter* 11 (1972) 63.

OSMIUM

68. (a) OSOLINSKI, T. W., and KNIGHT, N. H., Determination of osmium by atomic absorption spectrophotometry. *Appl. Spectroscopy* 22 (1968) 532.
69. (b) FERNANDEZ, F., Osmium wavelengths, sensitivities and operating parameters. *Atomic Abs. Newsletter* 8 (1969) 90.

PALLADIUM

70. LOCKYER, R., and HAMES, G. E., Quantitative determination of some noble metals by atomic absorption spectroscopy. *Analyst* 84 (1959) 385.
71. STRASHEIM, A., and WESSELS, G. J., The atomic absorption determination of some noble metals. *Appl. Spectroscopy* 17 (1963) 65.
72. ERINC, G., and MAGEE, R. J., The determination of palladium by atomic absorption spectroscopy. *Anal. Chim. Acta* 31 (1964) 197.
73. SCHNEPFE, M. M., and GRIMALDI, F. S., Determination of Palladium and Platinum by Atomic Absorption. *Talanta* 16 (1969) 591.
74. HARRINGTON, D. E., Colloidal Palladium determination by Atomic Absorption Spectroscopy. *Atomic Abs. Newsletter* 9 (1970) 106.

See also 37, 38, 40, and 41.

PHOSPHORUS

(See also Chapter 7 refs 140, 201 and 202)

75. KIRKBRIGHT, G. F., SMITH, A. M., and WEST, T. S., An indirect sequential determination of phosphorus and silicon by atomic absorption spectrophotometry. *Analyst* 92 (1967) 411.
76. KIRKBRIGHT, G. F., and MARSHALL, M., Direct determination of phosphorus by atomic absorption flame spectrometry. *Anal. Chem.* 45 (1973) 1610.

PLATINUM

77. LOCKYER, R., and HAMES, G. E., Quantitative determination of some noble metals by atomic absorption spectroscopy. *Analyst* 84 (1959) 385.
78. STRASHEIM, A., and WESSELS, G. J., The atomic absorption determination of some noble metals. *Appl. Spectroscopy* 17 (1963) 65.
79. PITTS, A. E., VAN LOON, J. C., and BEAMISH, F. E., The determination of Platinum by Atomic Absorption Spectroscopy. Part I. Air–Acetylene Flame. *Anal. Chim. Acta.* 50 (1970) 181.
80. PITTS, A. E., VAN LOON, J. C., and BEAMISH, F. E., The determination of Platinum by Atomic Absorption Spectroscopy. Part II. Nitrous Oxide–Acetylene Flame. *Anal. Chim. Acta.* 50 (1970) 195.
81. PITTS, A. E., and BEAMISH, F. E., Use of organic solvents in the determination of platinum by atomic absorption spectroscopy. *Anal. Chim. Acta.* 52 (1970) 405.

See also 37, 38, 40, 42 and 73.

THE RARE EARTHS

82. KIRKBRIGHT, G. F., SEMB, A., and WEST, T. S., Spectroscopy on separated flames—III. Use of the separated nitrous oxide–acetylene flame in thermal emission spectroscopy. *Talanta* 15 (1968) 441.
83. KRIEGE, O. H., and WELCHER, G. G., Determination of scandium by atomic absorption. *Talanta* 15 (1968) 781.
84. OOGHE, W. and VERBEEK, F., A.A.S. of the lanthanides in minerals and ores, *Anal. Chem. Acta* 73 (1974) 87.
85. THOMERSON, D. R., and PRICE, W. J., Observations of the A.A. behaviour of Some of the Rare Earth Elements. *Anal. Chim. Acta* 72 (1974) 188.

86. VAN LOON, J. C., GALBRAITH, J. H., and AARDEN, H. M., The determination of Yttrium, Europium, Terbium, Dysprosium, Holmium, Erbium, Thulium, Ytterbium and Lutetium in Minerals by Atomic-absorption Spectrophotometry. *Analyst* 96 (1971) 47.
87. SEN GUPTA, J. G., Determination of lanthanides and yttrium in rocks and minerals by A.A.S. and F.E.S. *Talanta* 23 (1976) 343.
88. DEAN, J. A., and RAINS, T. C. (Ed.) Flame emission and atomic absorption spectrometry. Vol. 3, *Marcel Dekker Inc.* New York 1975. (Chapter 5).

RHENIUM

89. ELLIOTT, E. V., STEVER, K. R., and HEADY, H. H., AA Determination of Rhenium in Ores and Concentrates, *Atomic Abs. Newsletter* 13 (1974) 113.
90. SCHRENK, W. G., LEHMAN, D. A., and NEUFELD, L., Atomic absorption characteristics of rhenium. *Appl. Spectroscopy* 20 (1966) 389.
91. KIM, C. H., ALEXANDER, P. W. and SMYTHE, L. E., Preconcentration of molybdenum, tungsten and rhenium and determination by A.A.S. *Talanta* 22 (1975) 739.
92. SMITH, R., and LAWSON, A. E., The atomic emission spectroscopy of rhenium in the nitrous oxide acetylene flame. *Analyst* 96 (1971) 631.

RHODIUM

93. LOCKYER, R., and HAMES, G. E., Quantitative determination of some noble metals by atomic absorption spectroscopy. *Analyst* 84 (1959) 385.
94. STRASHEIM, A., and WESSELS, G. J., The atomic absorption determination of some noble metals. *Appl. Spectroscopy* 17 (1963) 65.
95. HENEAGE, P., A brief study of rhodium absorption. *Atomic Abs. Newsletter* 5 (1966) 64.
96. DEILY, J. R., The determination of rhodium in organic solutions by atomic absorption spectrometry. *Atomic Abs. Newsletter* 6 (1967) 66.
97. KALLMANN, S., and HOBART, E. W., Vital parameters in the determination of rhodium by atomic absorption. *Anal. Chim. Acta.* 51 (1970) 120.
98. ATWELL, M. G., and HERBERT, J. Y., Rhodium Determination by Atomic Absorption Spectrometry using the Nitrous Oxide–Acetylene Flame. *Appl. Spectroscopy* 23 (1969) 480.

RUTHENIUM

99. HARRINGTON, D. E., and BRAMSTEDT, W. R., Determination of ruthenium and iridium in anode coatings by atomic absorption spectroscopy. *Talanta* 22 (1975) 411.
100. MOUNTFORD, B., and CRIBBS, S. C., Determination of ruthenium by atomic absorption spectrophotometry. *Anal. Chim. Acta.* 53 (1971) 101.

See also 47.

SELENIUM

101. DAGNALL, R. M., THOMPSON, K. C., and WEST, T. S., Studies in atomic fluorescence spectroscopy—IV. The atomic-fluorescence spectroscopic determination of selenium and tellurium. *Talanta* 14 (1967) 557.
102. NAKAHARA, T., MUNEMORI, M., and MUSHA, S., Determination of Selenium in Sulphur by Atomic Absorption Spectrophotometry. *Anal. Chim. Acta* 50 (1970) 51.

See also 6 and 8.

SILICON

103. KIRKBRIGHT, G. F., SMITH, A. M., and WEST, T. S., An indirect sequential determination of phosphorus and silicon by atomic-absorption spectrophotometry. *Analyst* 92 (1967) 411.
104. PRICE, W. J., and ROOS, J. T. H., The determination of silicon by atomic-absorption spectrophotometry with particular reference to steel, cast iron, aluminium alloys and cement. *Analyst* 93 (1968) 709.
105. BURDO, R. A., and WISE, W. M., Determination of silicon in glasses and minerals by atomic absorption spectrometry. *Anal. Chem.* 47 (1975) 2360.
106. MUSIL, J., and NEHASILOVA, M., Interferences in the A.A. determination of silicon. *Talanta* 23 (1976) 729.

107. DEVINE, J. C., and SUHR, N. H., Determination of silicon in water samples. *Atomic Abs. Newsletter* 16 (1977) 39.
See also 2.

SILVER
108. BELCHER, R., DAGNALL, R. M., and WEST, T. S., Examination of the atomic absorption spectroscopy of silver. *Talanta* 11 (1964) 1257.
109. TINDALL, F. M., Silver and gold assay by atomic absorption spectrophotometry. *Atomic Abs. Newsletter* 4 (1965) 339.
110. TINDALL, F. M., Notes on silver and gold assay by atomic absorption. *Atomic Abs. Newsletter* 5 (1966) 140.
See also 38.

STRONTIUM
111. CARTER, D., REGAN, J. G. T. and WARREN, J., A.A. determination of strontium in silicate rocks: A study of major element interferences in the nitrous oxide-acetylene flame. *Analyst* 100 (1975) 721.

TANTALUM
112. SCHLEWITZ, J. H., and SHIELDS, M. G., The determination of low level tantalum in zirconium and zirconium alloys. *Atomic Abs. Newsletter* 10 (1971) 43.

TELLURIUM (see also Chapter 7, Ref 203.)
113. WU, J. Y. L., DROLL, H. A., and LOTT, P. F., Determination of tellurium by atomic absorption spectrophotometry. *Atomic Abs. Newsletter* 7 (1968) 90.

TIN
114. AGAZZI, E. J., Determination of tin in hydrogen peroxide solutions by atomic absorption spectrometry. *Anal. Chem.* 37 (1965) 364.
115. CAPACHO-DELGADO, L., and MANNING, D. C., Determination of tin by atomic absorption spectroscopy. *Spectrochim. Acta* 22 (1966) 1505.
116. WELSCH, E. P., and CHAO, T. T., Determination of trace amounts of tin in geological materials by atomic absorption spectrometry. *Anal. Chim. Acta* 82 (1976) 337.
117. JULIANO, P. O., and HARRISON, W. W., Atomic absorption interferences of tin. *Anal. Chem.* 42 (1970) 84.
118. NAKAHARA, T., MUNEMORI, M., and MUSHA, S., A.A.S. Determination of Tin in Premixed Inert Gas (Entrained Air) Hydrogen Flames. *Anal. Chim. Acta* 62 (1972) 267.
See also 5.

TITANIUM
119. KIRKBRIGHT, G., F., SMITH, A. M., WEST, T. S., and WOOD, R., An indirect amplification procedure for the determination of titanium by atomic absorption spectroscopy. *Analyst* 94 (1969) 754.
120. SABA, C. S., and EISENSTRAUT, K. J., Determination of titanium in aircraft lubricating oils. *Anal. Chem.* 49 (1977) 454.
See also 66, and 127.

TUNGSTEN
See 66, 91 and 126.

URANIUM
121. MARTIN, M. J., Determination of uranium by atomic absorption spectrophotometry. *Analyst* 96 (1971) 843.
122. ALDER, J. F., and DAS, B. C., Indirect determination of uranium by A.A.S. using an air-acetylene flame. *Analyst* 102 (1977) 564.

VANADIUM
123. AMOS, M. D., and WILLIS, J. B., Use of high-temperature premixed flames in atomic absorption spectroscopy. *Spectrochim. Acta.* 22 (1966) 1325.

124. SACHDEV, S. L., ROBINSON, J. W., and WEST, P. W., Determination of vanadium by atomic absorption spectrophotometry. *Anal. Chim. Acta* 37 (1967) 12.
125. CHAKRABARTI, C. L., and MCNEIL, D. P., Evaluation of the flame parameters in the determination of vanadium by A.A.S. with a nitrous oxide-acetylene flame. *Can. J. Spectrosc*. 20 (1975) 90.

ZIRCONIUM
126. BOND, A. M., Use of Ammonium Fluoride in Determination of Zirconium and Other Elements by Atomic Absorption Spectrometry in the Nitrous Oxide–Acetylene Flame. *Anal. Chem.* 42 (1970) 932.
127. PANDAY, V. K., Interferences in the determination of titanium, zirconium and hafnium by atomic absorption spectrophotometry. *Anal. Chim. Acta* 57 (1971) 31.

INDIRECT METHODS OF A.A.S.
128. KIRKBRIGHT, G. F., and JOHNSON, H. N., Application of Indirect Methods in Analysis by Atomic Absorption Spectrometry. *Talanta* 20 (1973) 433.
129. Annual Reports on Analytical Atomic Spectroscopy 1976, vol. 6 and 1977, vol. 7. Section 1.1.3 *Chemical Society*, London.
130. SLAVIN, S., and LAWRENCE, D. M., Atomic Absorption Newsletter Bibliographies. Published in January–February and July–August issues. Indirect methods have been indexed since 1974.

HYDRIDE GENERATION TECHNIQUES
131. MANNING, D. C., A high sensitivity arsenic–selenium sampling system for atomic absorption spectroscopy. *Atomic Abs. Newsletter* 10 (1971) 123.
132. FERNANDEZ, F. J., and MANNING, D. C., The determination of arsenic at sub-microgram levels by atomic absorption spectrophotometry. *Atomic Abs. Newsletter* 10 (1971) 86.
133. POLLOCK, E. N., and WEST, S. J., The determination of antimony at sub-microgram levels by atomic absorption spectrophotometry. *Atomic Abs. Newsletter* 11 (1972) 104.
134. POLLOCK, E. N., and WEST, S. J., The generation and determination of covalent hydrides by atomic absorption. *Atomic Abs. Newsletter* 12 (1973) 6.
135. SULLIVAN, E. A., Sodium borohydride handling/uses/properties/analytical procedures. Published by the Ventron Corporation, 44 Congress Street, Beverly, Mass. 01915, U.S.A.
136. FERNANDEZ, F. J., Atomic absorption determination of gaseous hydrides utilising sodium borohydride reduction. *Atomic Abs. Newsletter* 12 (1973) 93.
137. THOMPSON, K. C., and THOMERSON, D. R., Atomic absorption studies on the determination of Sb, As, Bi, Ge, Pb, Se, Te and Sn by utilising the generation of covalent hydrides. *Analyst* 99 (1974) 595.
138. SMITH, A. E., Interferences in the determination of elements that form volatile hydrides with $NaBH_4$ using AAS and the A–H_2 flame. *Analyst* 100 (1975) 300.
139. SIEMER, D. D., KOTEEL, P., and JARIWALA, V., Optimisation of arsine generation in atomic absorption arsenic determinations. *Anal. Chem.* 48 (1976) 836.
140. WAUCHOPE, R. D., Atomic absorption determination of arsenic: application of a rapid arsine generation technique to soil, water and plant samples. *Atomic Abs. Newsletter* 15 (1976) 64.
141. VIJAN, P. N., and WOOD, G. R., An automated sub-microgram determination of selenium in vegetation by quartz tube furnace atomic absorption spectrophotometry. *Talanta* 23 (1976) 89.
142. VIJAN, P. N., RAYNER, A. C., STURGIS, D., and WOOD, G. R., A semi-automated method for the determination of arsenic in soil and vegetation by gas phase sampling and atomic absorption spectrometry. *Anal. Chim. Acta* 82 (1976) 329.
143. VIJAN, P. N., and WOOD, G. R., Semi-automated determination of lead by hydride generation and atomic absorption method. *Analyst* 101 (1976) 966.
144. FLEMING, H. D., and IDE, R. G., Determination of volatile hydride forming metals in steel by atomic absorption spectrometry. *Anal. Chim. Acta* 83 (1976) 67.
145. DRINKWATER, J. E., Atomic absorption determination of bismuth in complex nickel base alloys by generation of its covalent hydride. *Analyst* 101 (1976) 672.
146. PIERCE, F. D., and BROWN, H. R., Inorganic interference study of automated arsenic and selenium determination by atomic absorption spectrometry. *Anal. Chem.* 48 (1976) 693.

147. PIERCE, F. D., LAMOREAUX, T. C., BROWN, H. R., and FRASER, R. S., An automated technique for the sub-microgram determination of selenium and arsenic in surface waters by atomic absorption spectroscopy. *Applied Spectroscopy* 30 (1976) 38.
148. FIORINO, J. A., JONES, J. W., and CAPAR, S. G., Sequential determination of As, Se, Sb and Te in foods via rapid hydride evolution and atomic absorption spectrometry. *Anal. Chem.* 48 (1976) 120.
149. FISHMAN, M., and SPENCER, R., Automated A.A.S. determination of total arsenic in water and streambed materials. *Anal. Chem.* 49 (1977) 1599.
150. PIERCE, F. D. and BROWN, H. R., Comparison of inorganic interferences in A.A.S. determinations of arsenic and selenium. *Anal Chem.* 49 (1977) 1417.
151. ROONEY, R. C., Determination of bismuth in blood and urine. *Analyst* 101 (1976) 749.
152. CLINTON, O. E., Determination of selenium in blood and plant material by hydride generation and A.A.S. *Analyst* 102 (1977) 187.
153. THOMPSON, K. C., The atomic fluorescence determination of Sb, As, Se and Te using the hydride generation technique. *Analyst* 100 (1975) 307.

5 *Applications*

The units of concentration used in this Chapter are those most commonly accepted in the various fields of application, e.g., ppm *for measurement of concentrations in solids and* μg/ml *for measurement of concentrations in liquids for most industrial and research applications,* mg/*100* ml *and* μg/*100* ml *for certain clinical applications, etc.*

Introduction

Atomic Absorption Spectroscopy is now one of the most widespread and firmly established instrumental techniques of chemical analysis.

The outstanding advantage offered by the technique is that, because for all practical purposes, it is immune to spectral interferences caused by overlap of resonance absorption lines sample preparation can generally be confined to mere dissolution or extraction. Handling errors are therefore minimized, and analyses can be performed more rapidly than by procedures requiring more elaborate preparative steps.

Standard commercial instruments have reached a very high level of performance, and the types of analyses for which the technique is finding application are now far wider than was originally envisaged.

It was initially anticipated that the usefulness of atomic absorption spectroscopy would be confined to estimations of metals present as minor constituents. The extension of atomic absorption to the determination of trace constituents at one extreme, and of major components at the other is due not only to the improved performance of commercial equipment, but also to the sometimes awe-inspiring determination of users to apply the technique to an increasingly wide number of estimations. When devising a new method the reader should consider possible interelement effects (see page 31) and always check for the presence of non-specific background absorption. If significant background absorption is detected it should be corrected either manually or automatically by using a hydrogen or a deuterium hollow-cathode lamp (see page 239).

Complementary Methods of Analysis

In certain cases the atomic absorption detection limit using flames is not adequate. In these cases one of the complementary techniques can often be used—flameless (electrothermal) atomization, hydride generation techniques or atomic fluorescence techniques. The advent of flameless electrothermal atomization (see p. 216) has improved the detection limits for many types of analysis. Where a time-consuming pre-concentration step was required when performing certain flame analysis, samples can often be directly applied to flameless devices. A short bibliography on this technique is given on p. 233. It should be stressed that if an analysis can be readily performed using a flame technique this is usually to be preferred to the flameless technique as less operator skill is required, better precision can be obtained and easier automation can be achieved.

The hydride generation technique (see p. 79) for As, Bi, Sb, Se, Sn and Te is at least 100 times more sensitive than the flame and does not suffer from non-specific background absorption. The main requirement is that the final sample solution should contain 1–5 M acid (HCl, HNO_3, H_2SO_4, HCOOH, etc.). The main limitation is caused by the presence of large amounts of easily reducible elements such as silver and copper which inhibit hydride formation. Good results have been obtained from various types of samples including wet ashed plant and animal tissue samples; alloys containing large amounts of Cr, Fe, Ni, W; sea and river waters; effluents and foodstuffs. The technique can be readily automated.

The precision (R.S.D.) is about 3–5% and the rate of analysis is somewhat slower than that of flame techniques.

Atomic fluorescence techniques are useful for trace levels of elements which have their main resonance lines in the u.v. below 260 nm (e.g., As, Cd, Hg, Sb, Se, Te, Zn, etc.) and are suitable for matrices that cause severe background (molecular) absorption in atomic absorption (5000–20 000 $\mu g/ml^{-1}$ solutions of Cu, Pb, Zn, etc.) but are not very suitable for matrices that result in the formation of refractory oxide species in the flame and consequently cause appreciable scattering of source radiation (e.g., Al, Ca, Zr, etc.).

Trace Metal Estimations

Atomic absorption has proved to be a valuable tool in agricultural, biological and food analysis. In these fields it offers the most rapid, convenient and reliable procedure for the estimations of toxic and nutritional metals in food stuffs and natural products. In the areas of water treatment, effluent control, chemical production and soil science it has attained a position of major importance for the estimations of trace metals generally.

Solid Food Stuffs and Organic Materials

The main problems associated with trace metal estimation in organic materials are (i) the quantitative extraction of the metallic species from the matrix, as rapidly as possible, (ii) the avoidance of mechanical losses, (iii) the avoidance of contamination from reagents.

Digestion and Extraction Procedures

Dry ashing of biological materials is still practiced, but many authorities condemn this as being too vulnerable to mechanical losses in the smokes evolved and also to loss of volatile elements. In fact losses of cadmium, lead and selenium from certain types of biological tissue have been observed when the tissue was dried at 60°C.[19] Appreciable quantities of lead tetraalkyls (10–90% of the total lead present) have been found in a variety of fish tissues[20] and any form of drying is likely to lead to loss of lead.

Wet ashing, with mixtures of perchloric, nitric, and sulphuric acids is also widely practiced and in the hands of an experienced operator gives reliable results.

A procedure that has been developed,[7] and that possesses the advantage of speed of attack and complete destruction of the carbonaceous material, is the use of 50 per cent hydrogen peroxide. This has been employed by the author (R.J.R.) in combination with concentrated sulphuric acid, and also in combination with concentrated nitric acid, with very satisfactory results for the digestions of blood, liver, meat products, wood, leaves, fruit, tallow, soap, and animal feeds.

It should be stressed that the procedures involving the use of 50% m/m hydrogen peroxide can be dangerous and the interested reader should consult the reports entitled 'The Use of 50 per cent hydrogen peroxide for the destruction of organic matter' (First and Second Reports) produced by the Analytical Methods Committee of the Analytical Division of the Chemical Society. (First report *Analyst* 92 (1967) 403, Second report *Analyst* 101 (1976) 62.

It is not, of course, valid to describe one standard procedure that will be suitable for the extraction of every element from every substance. The quantities of hydrogen peroxide and acid required will vary with the nature of the material to be attacked, and with the sample size, which itself will depend on the trace metal concentration in the sample. The decision whether to use nitric or sulphuric acid in conjunction with the hydrogen peroxide will depend on the nature of the trace metal; e.g., sulphuric could be unsuitable for the extraction of lead since lead sulphate is insoluble. Vegetable oils are frequently best attacked with sulphuric acid/hydrogen peroxide.

We have recently found 5 M hydrochloric acid with a small addition of 100 volume hydrogen peroxide to be a very effective reagent for the digestion of meat products.

Adequate extraction can frequently be attained by digesting a conveniently sized sample (1, 5 or 10 g) of the material with nitric acid. The cooled digest is filtered and the filtrate and washings transferred to a graduated flask and adjusted to the mark. Separated oil or fat is excluded or can be removed by adding a small quantity of chloroform, shaking and subsequently nebulising the upper aqueous layer.

Liquid Samples

Beverages such as beers, mineral waters, etc., must be degassed before aspiration to an atomic absorption spectrophotometer. This is frequently accomplished by warming and shaking, but an even more effective procedure is to pour the beverage twice through a filter paper. Comparison is made against standards that match the test samples in alcohol content. Experience has shown that such analyses must be performed upon freshly degassed samples. Erroneously low results are obtained if the degassed beverages are allowed to stand for prolonged periods (24 hours) before estimation.

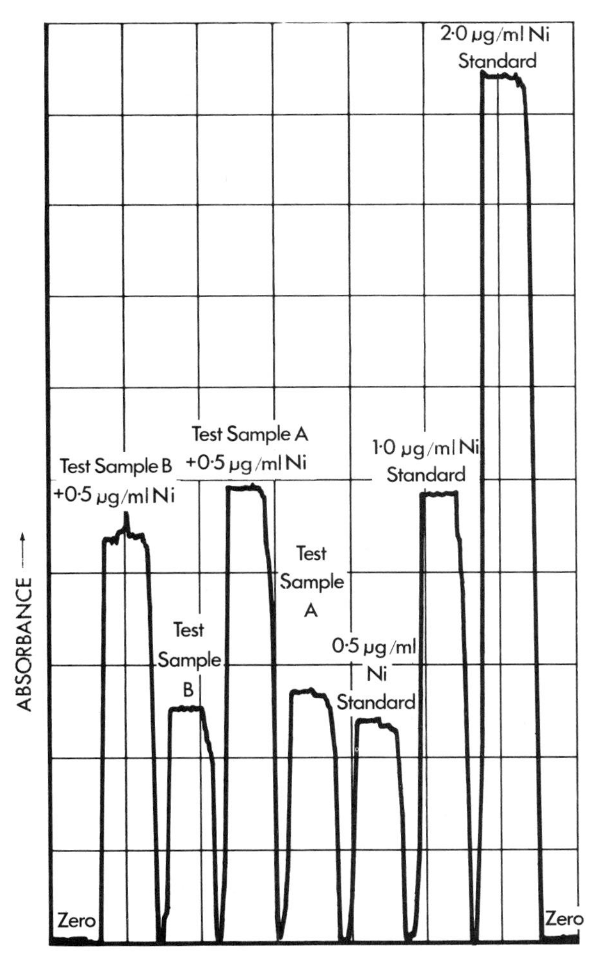

Fig. 5.1 Determination of nickel in digests from vegetable oils with a simple direct-reading atomic absorption spectrophotometer

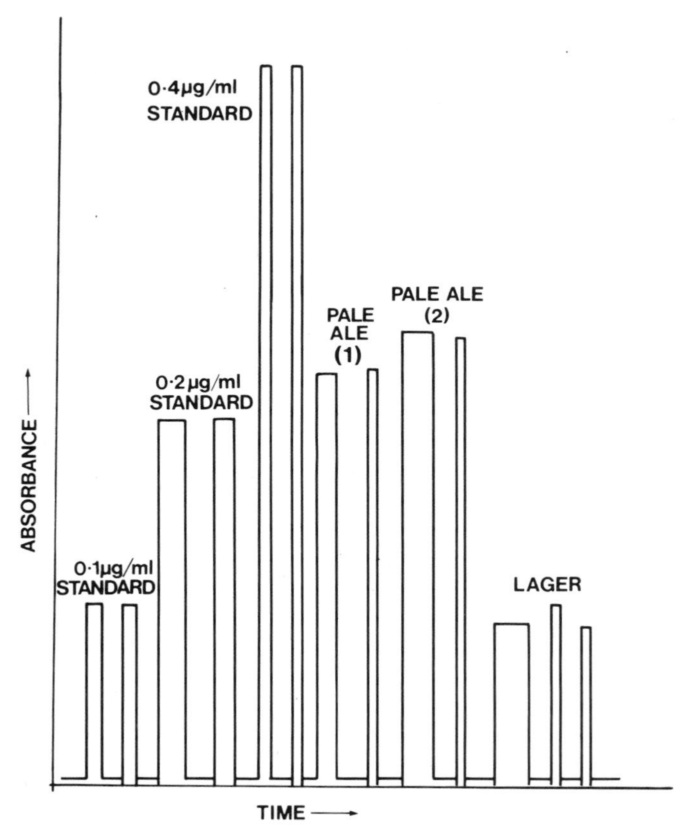

Fig. 5.2 Direct determination of copper in beers using scale expansion and signal integration

Levels of determination. The limits of detection in aqueous solution for the elements that can be determined by atomic absorption are stated in Table 4.1, Chapter 4.

For most determinations of this nature it will be necessary to use scale expansion, and greater precision is attainable if integrated readings are taken.

Concentration Procedures. If the extract from the digested biological or organic material contains the trace metals at a concentration still too low for estimation, it will be necessary to resort to extractive concentration. A convenient and generally applicable procedure is described in Chapter 3 (method 6).

Lead, copper, zinc and nickel are perhaps the most common toxic elements of interest to the food analyst, while the estimation of iron is often necessary, particularly in the brewing and distilling industries for reasons of taste and colour control. The procedures outlined above are applicable to trace metals generally.

Figures 5.1–5.4 illustrate a number of common trace estimations that are conveniently performed by atomic absorption. The two important nutritional elements calcium and magnesium are very easily determined by the procedure, but for their determination it is advisable to add lanthanum chloride to the test and standard solutions to overcome the effects of probable phosphate suppression.

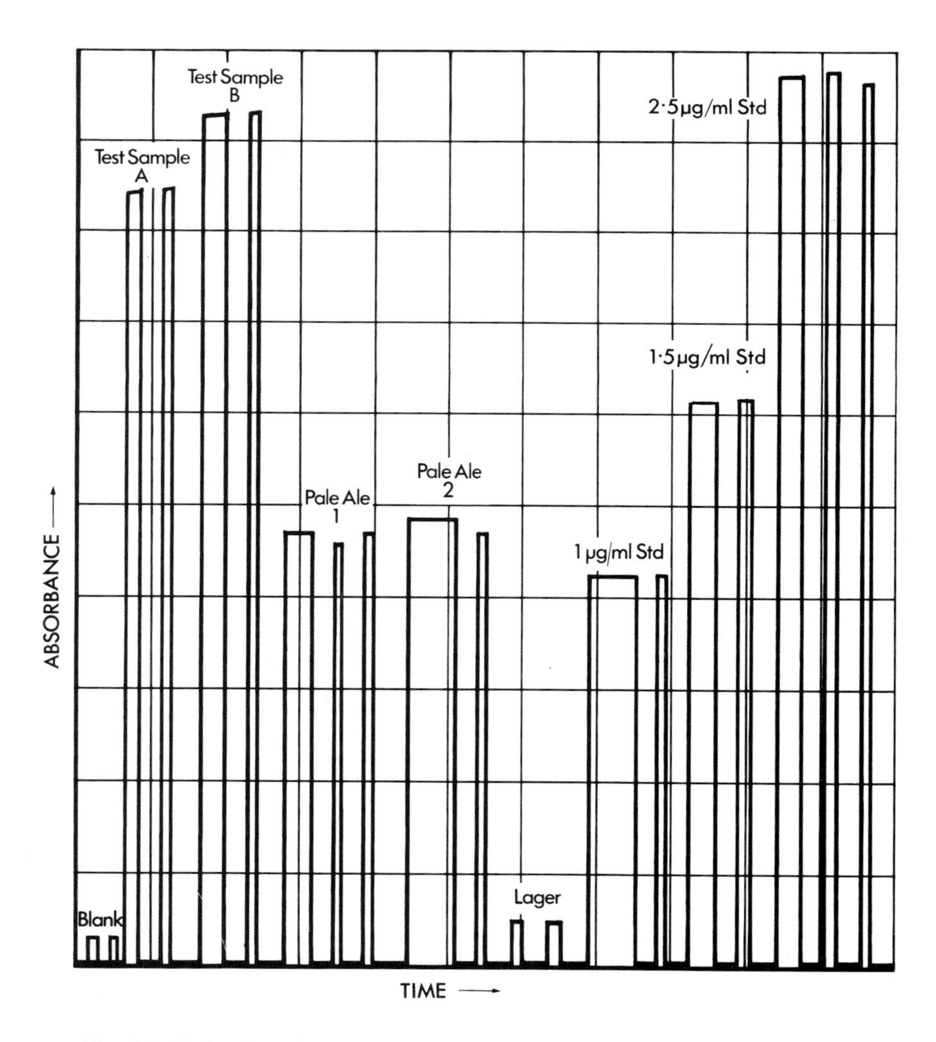

Fig. 5.3 Estimation of zinc in beers at ×1 scale expansion using signal integration

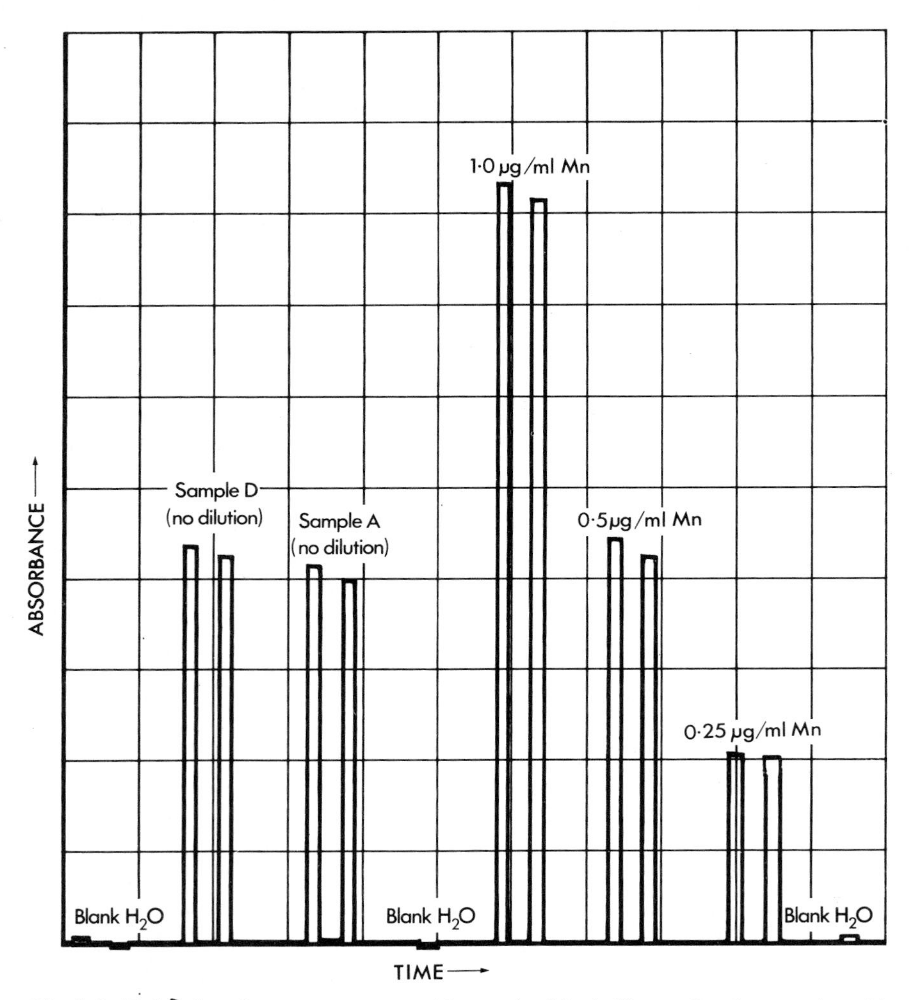

Fig. 5.4 Estimation of manganese extracted from animal feed. About ×5 scale expansion with signal integration

The Estimation of Mercury by Flameless Atomic Absorption Spectroscopy

Mercury levels in fish and natural waters have recently received considerable publicity in the popular press. The most convenient and accurate procedure to perform this estimation is by flameless A.A.S.

In principle a solution containing mercury at low level is treated with stannous chloride solution, which reduces the mercury present to the metallic state. This metallic mercury is then carried as a vapour to the flameless absorption cell, by passing a stream of air through the solution.

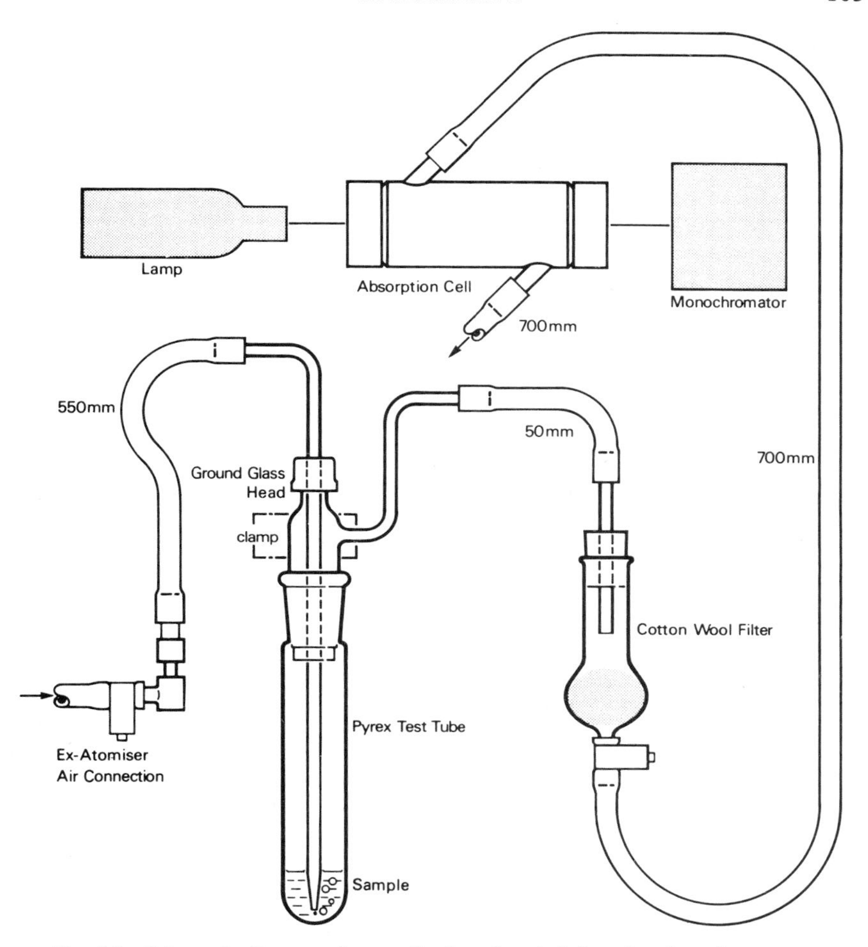

Fig. 5.5 Schematic diagram of set-up for flameless A.A.S. estimation of mercury

The apparatus required is simple, and consists of a $18 \times 2{\cdot}5$ cm Pyrex test tube fitted with a ground glass head through which pass the inlet and outlet tubes. The inlet is drawn to a fine jet and reaches almost to the bottom of the test tube (Fig. 5.5).

A cotton-wool filter can be connected to the outlet so as to ensure that spray is not carried to the absorption cell.

The absorption cell is usually constructed from Pyrex tubing (typically 150 mm length by 8 mm internal diameter) with silica end windows cemented to the ends of the tube. It is advisable to heat the tube by placing a 60–100-watt lamp bulb directly above it in order to prevent condensation of water vapour on the silica windows. If condensation occurs an unstable baseline is observed.

In some units the air line can conveniently be removed from the atomizer and connected to the inlet of the Pyrex test tube. The absorption cell can be clamped to the burner grid, and is connected to the exit tube from the Pyrex test tube.

Readings are most conveniently taken on a chart recorder and a standard hollow-cathode mercury lamp is used as the light source. For most estimations a scale expansion factor of 5 to 20 times is necessary.

A known volume of mercury containing solution is placed in the Pyrex test tube and 0·5 ml of 20 per cent m/v stannous chloride solution added. The head is firmly replaced into the test tube, and the air flow started.

After a short pause the mercury vapour is flushed from the test tube into the absorption cell. The absorption, as indicated by the main meter reading and the recorder response, rises quickly to a maximum and slowly falls as the

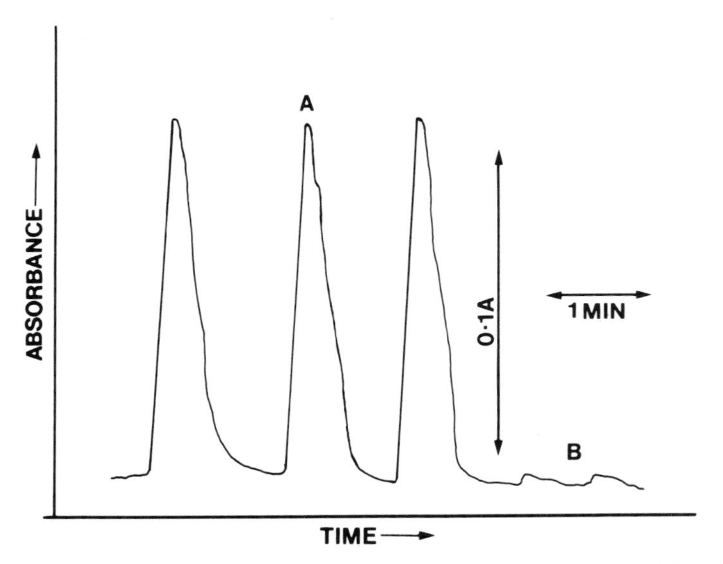

Fig. 5.6 Flameless A.A.S. estimation of mercury. Typical responses. Chart speed 3 cm/min. A 25 ml 0·01 μg/ml mercury; B 25 ml Blank

mercury is flushed from the cell. A series of typical responses is shown in Fig. 5.6. The damping control switch should be set to give a time constant, as measured at the signal output of the instrument, of 1–2 seconds. This will usually result in the optimum signal: noise ratio being obtained.

The impression can be gained from some literature that since this technique measures absolute quantities of mercury, as distinct from concentrations, it should be possible to determine the element when present at increasingly low concentrations, simply by increasing the sample size. This is only partially true. The response for a constant quantity of mercury decreases as the sample size, in which it is dispersed, increases. The time taken for the response to reach peak height also increases as the sample size is increased.

The optimum-sized samples for estimation is 5 or 10 ml. The maximum sample size is about 30 ml. Samples less than 5 ml in volume give rise to higher responses but slight differences in volume have a more pronounced effect upon the absorption, and can cause inaccuracies.

The rate of air flow also influences the sensitivity for the estimation. Higher flow rates reduce the absorption, but the time taken to reach the peak responses is less than at low air flows. The optimum air flow for good responses and acceptable analysis time is about 2 litres per min.

The main disadvantages of the absorption technique are: an enclosed cell is required (this can lead to memory effects); other substances (e.g., acetone) absorb radiation at 253·7 nm and can be recorded as mercury unless background correction is applied; the sensitivity and linearity of the calibration curve of an absorption technique are somewhat limited.

The cold vapour fluorescence technique effectively overcomes the above limitations[41,42] and is depicted in Figs 5.7 and 5.8. Mercury vapour is generated in the glass mercury cell and carried through to the L-shaped glass delivery tube, on emerging from which it is excited by the radiation from a simple low-pressure 4-watt mercury lamp positioned to the side of and just above the delivery tube. This tube is sheathed by a flow of argon to prevent ingress of air.

2 ml of acidic 2% m/v stannous chloride solution are placed in the glass mercury cell (Fig. 5.7) and argon is bubbled through it until a stable baseline is obtained. Then a 0·1–5 ml sample containing 0·0001–0·2 μg mercury is added through the side arm of the cell. The generated mercury vapour energes from the end of the L-shaped glass delivery tube and the mercury vapour fluoresces in the unenclosed region above the argon-sheathed delivery tube. The recorded highly electronically damped peak height (time constant 5 s) is proportional to the amount of mercury introduced into the sample cell over the concentration range 0·0001–0·2 μg mercury. A typical trace is shown in Fig. 5.9. The detection limit for this technique is governed by the light throughput of the monochromator and the stability of the output at high E.H.T. voltages. With a Shandon Southern (Baird Atomic) A3400 it is in the range 0·00002–0·00005 μg. It is essential to use argon rather than nitrogen (approximately 8 times less sensitive) or air (approximately 32 times less sensitive). This is due to the occurrence of fluorescence quenching reactions (see p. 264) such as:

$$Hg^* + N_2 \rightarrow Hg + N_2^*$$
$$Hg^* + O_2 \rightarrow Hg + O_2^*$$

The asterisk indicates an excited atom or molecule.

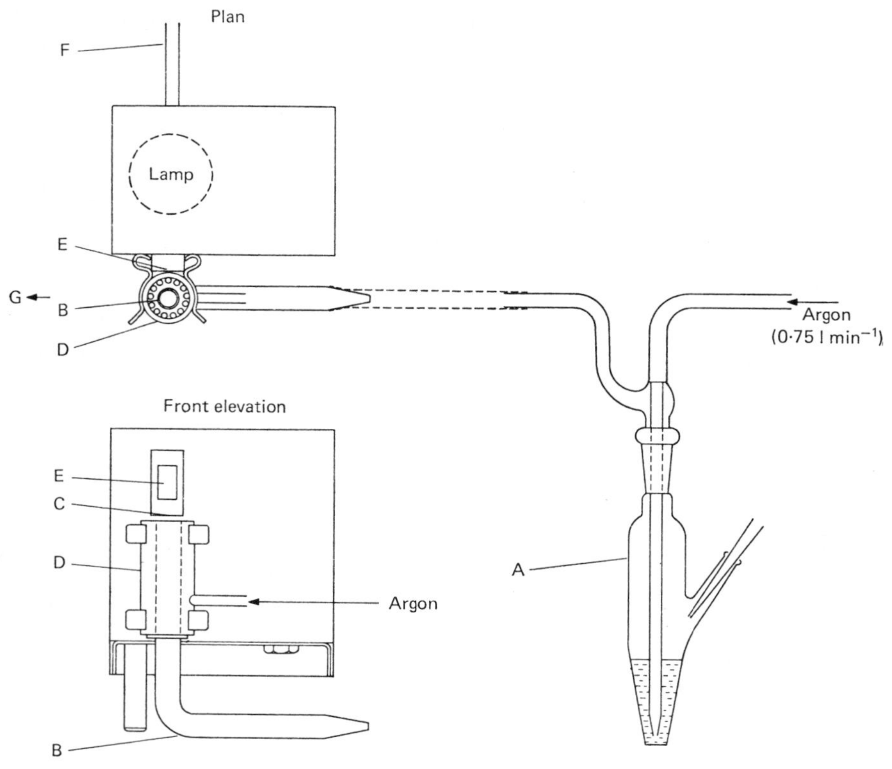

Fig. 5.7 Mercury generation system and excitation source for cold vapour fluorescence technique for mercury. A, cell; B, 7 mm i.d. Pyrex tube; C, region where fluorescence occurs; D, sheath unit; E, lamp aperture; F, argon gas inlet to lamp (to stabilize lamp output and prevent ozone formation); G, optical axis

The presence of acetone (up to 50%) does *not* affect the baseline, but it does result in some depression in the mercury signal. This is probably caused by quenching of excited mercury atoms by acetone molecules.

The breakdown of organic mercury species (especially methyl mercury compounds) prior to analysis by the cold vapour absorption or fluorescence techniques, requires digestion of the sample solution with permanganate for 24 hours.[22,29,31] This is followed by removal of the excess permanganate with hydroxylamine hydrochloride prior to the addition of the stannous chloride required to liberate the mercury. A much more rapid breakdown can be achieved by bubbling an oxygen–ozone gas stream (containing about 1·5 milli-moles/min of ozone) through the solution for 15 minutes.[43] Another simple method is to pipette 50 ml of the sample into a 100-ml volumetric flask, add 2 ml of a solution containing 1% m/v potassium bromide and 0·27% m/v potassium bromate followed by 5 ml hydrochloric acid

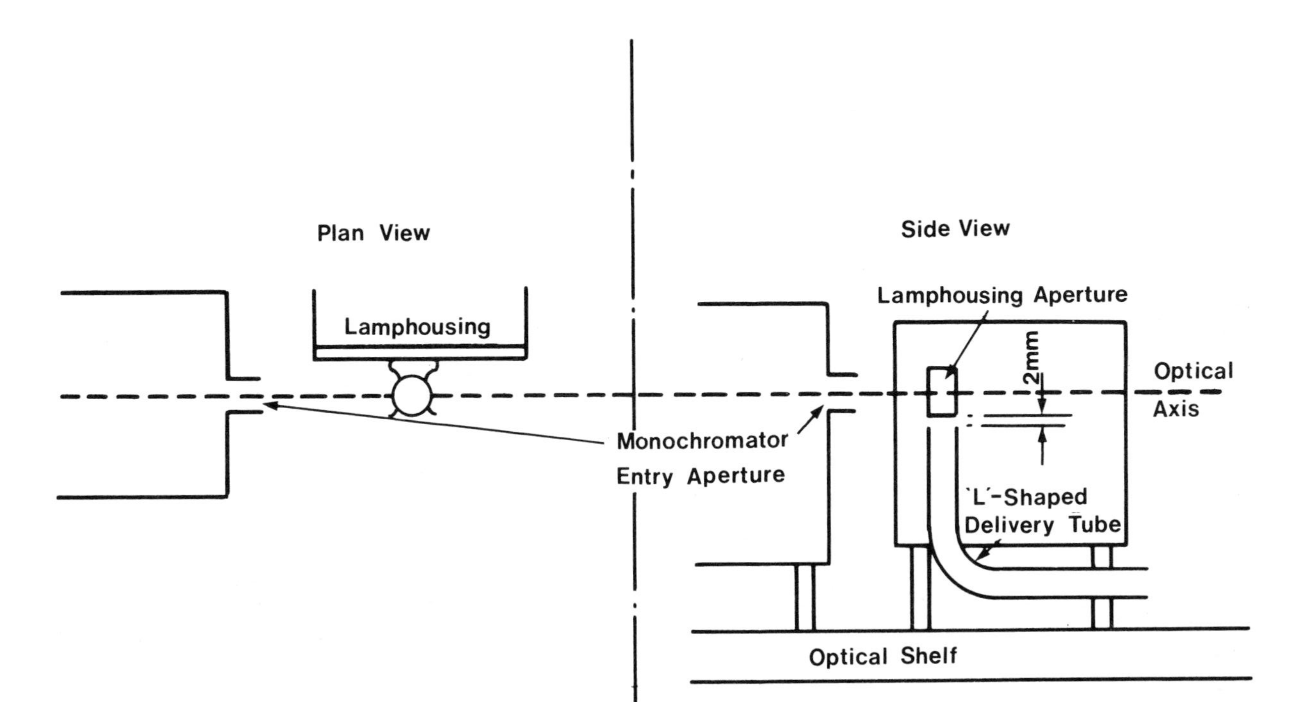

Fig. 5.8 Diagrammatic arrangement of cold vapour fluorescence technique for mercury. The argon sheath unit surrounding the delivery tube has been omitted for the sake of clarity

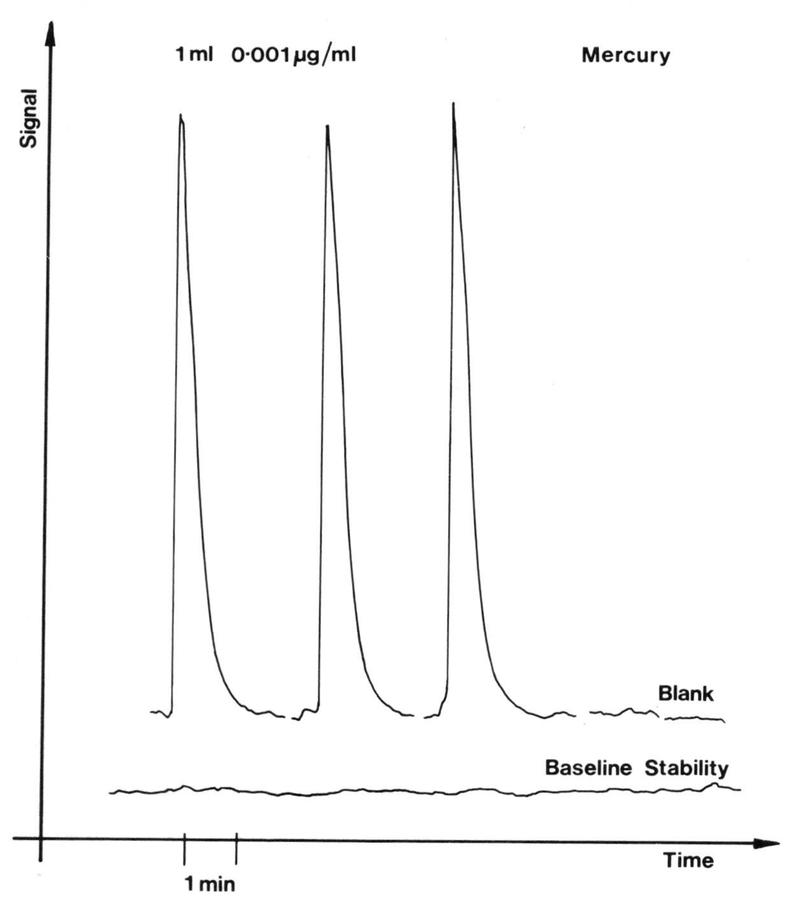

Fig. 5.9 Typical trace obtained by the cold vapour mercury fluorescence technique

(36% m/v). Allow to stand for 15 minutes then add sufficient hydroxylamine hydrochloride to remove the excess bromine and dilute to volume.[44] This latter method resulted in 90–100% recoveries from river water and sewage effluents spiked with methyl mercuric chloride.

A very important consideration with low level mercury determination is loss of mercury from the sample solution on to the container or to the atmosphere. It has been found that at levels of around 1 ng/ml severe losses can occur within one hour of sampling. These losses can be minimised by pretreating the polythene sample containers with typical river waters in order to block the active sites. After collection of the sample it should be frequently agitated to maintain a high dissolved oxygen level and analysed within 10 hours.[48] For longer term storage the addition of an oxidising agent such as bromine[44] (see above), 1% v/v nitric acid and 16 ng/ml gold (III)[49] or 5% v/v nitric acid and 0·01% m/v potassium dichromate should be used,[50] the sample being collected in a borosilicate glass bottle. With this latter procedure there was negligible mercury loss even at the 0·1 ng/ml level after five months storage.

The Estimation of Toxic Metals in Paint Films

A trace estimation that has assumed considerable importance is the Toys (Safety) Regulations 1974 – soluble toxic metals in paint films. (Statutory instrument number 1367, 1974). This regulation specifies that the soluble toxic metal content of paint films on toys must not exceed 500 ppm of barium, 250 ppm each of antimony and chromium and 100 ppm each of arsenic, cadmium and mercury. The statutory procedure for extracting the metals consists essentially of shaking a finely powdered sample of the paint film with 50 times its weight of 0·07M (0·25% m/m) hydrochloric acid at ambient temperature (20–22°C). The solution is allowed to stand for 1 hour. The toxic metal level in the hydrochloric acid is then determined after filtration. The maximum allowable concentrations of antimony, barium, cadmium and chromium in the filtrate are 5, 10, 2 and 5 μg/ml respectively and the traces in Figs. 5.10, 5.11, 5.12 and 5.13 indicate the capability of standard commercial equipment to perform low level analysis for these elements. Arsenic and mercury can easily be determined using the hydride generation technique and the cold vapour technique respectively (see pages 79 and 102).

Trace Metals in Water and Effluents

The mineral contents of natural waters, feed waters, and drinking supplies vary widely in both the nature and concentration of the metals present in them.

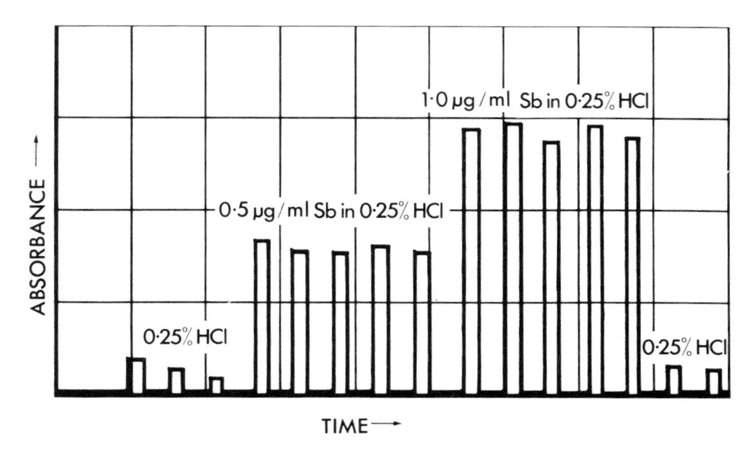

Fig. 5.10 Estimation of antimony in extract from paint film. Integrated mode. ×10 scale expansion 217·6 nm.

This metal content can affect the potability, and suitability of a water as a source for public supply.

The figures in Table 5.1 indicate the criteria for drinking waters. All concentrations are expressed in μg/ml.

Table 5.1 Criteria for Drinking Waters (μg/ml)

	Permissible I	Desirable I	USPHS II	WHO III
As	0·05[4,5]	0	0·01–0·05[b]	0·05[c]
Ag	0·05[1,4]	0	0·05[b]	
Ba	1·0[1]	0	1·0[b]	1·0[c]
B	1·0[2]	0	1·0	
Cd	0·01[1,4]	0	0·01[b]	0·01[c]
Cr(VI)	0·05[e]	0	0·05[b]	0·05[c]
Cu	1·0[1,2,4]	VA	1·0	0·05
Ca + Mg	60[1]	<60		
Ca				75, 200[d]
Fe[a]	0·3[1,2,4]	VA	0·3	0·1
Mg				50
Mn[a]	0·05[1,2,4]	0	0·05	0·05
Pb	0·05[3,4,5]	0	0·05[b]	0·1[c]
Se	0·01[4,5]	0	0·01[b]	0·01
Zn	5[1,2,4]	VA	5	5
TDS	500	200		500

[a] Filterable.
[b] Grounds for rejection.
[c] Recommended maximum allowable concentration.
[d] Permissible and excessive concentrations, also defined as maximum acceptable and maximum allowable.
[e] This level determinable by flame A.A.S. or flameless techniques, but extraction should be employed to isolate Cr(VI).
TDS Total dissolved solids.
VA Virtually absent.
[1] Directly determinable by flame A.A.S.
[2] Determinable by flame A.A.S. after 10 × preconcentration step, e.g., evaporation.
[3] Close to flame A.A.S. detection limit. Use of flameless techniques or 10 × preconcentration step is recommended.
[4] Directly determinable by flameless electrothermal techniques.
[5] Directly determinable by hydride generation techniques.
I Report of Committee on Water Quality Criteria, F.W.P.C.A. United States Department of Interior 1968.
II United States Public Health Service Standards.
III International Standards for Drinking Water, World Health Organisation 1970.

WATERS AND EFFLUENTS

Effluent waters and sewage can contain an even larger variety of metals covering a much wider range of concentrations. The measurement and control of these pollutants constitutes an important part of the activities of water authorities, manufacturing companies, and health authorities. Atomic absorption spectroscopy offers the most versatile, reliable, simple, and rapid procedure available to the chemist for many of these important analyses.

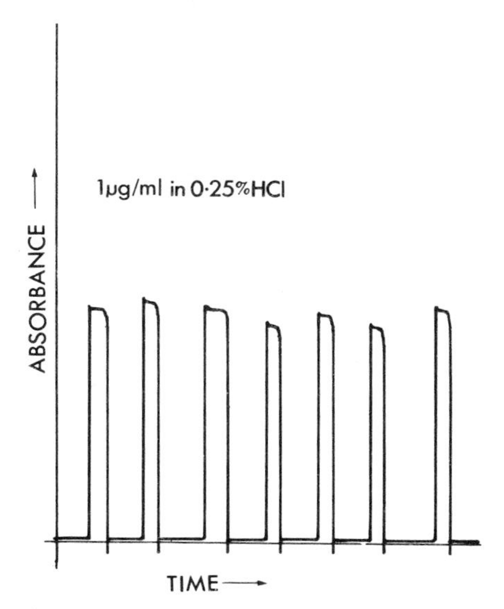

Fig. 5.11 Estimation of barium in extract from paint film. Integrated mode. ×10 scale expansion 455·4 nm ionic line N_2O–Acetylene

Interferences. Even with heavily polluted effluents it is unlikely that any of the metals will be present in so high a concentration as to exert a significant interference effect upon the absorption of others present—a very significant advantage to the analyst. It is nevertheless convenient to use combined standards, each containing all the metals that are to be analysed, in ranges that cover the anticipated concentrations in the effluent, because this procedure will reduce the number of flasks required. Any small inter-element effects should be minimized by the adoption of this procedure.

Reference to the relevant section of Chapter 4 will give a good indication of the few serious interferences likely to be encountered. For convenience they are listed below:

(i) Depression caused by the presence of phosphate upon the absorption of the alkaline earths.
(ii) Depression caused by the presence of aluminium upon the absorption of the alkaline earths.
(iii) Bicarbonate interference upon calcium absorption.
(iv) Depression caused by the presence of iron upon the absorption of chromium.

All four of these interferences are overcome by the addition of lanthanum chloride to the test and standard solutions.

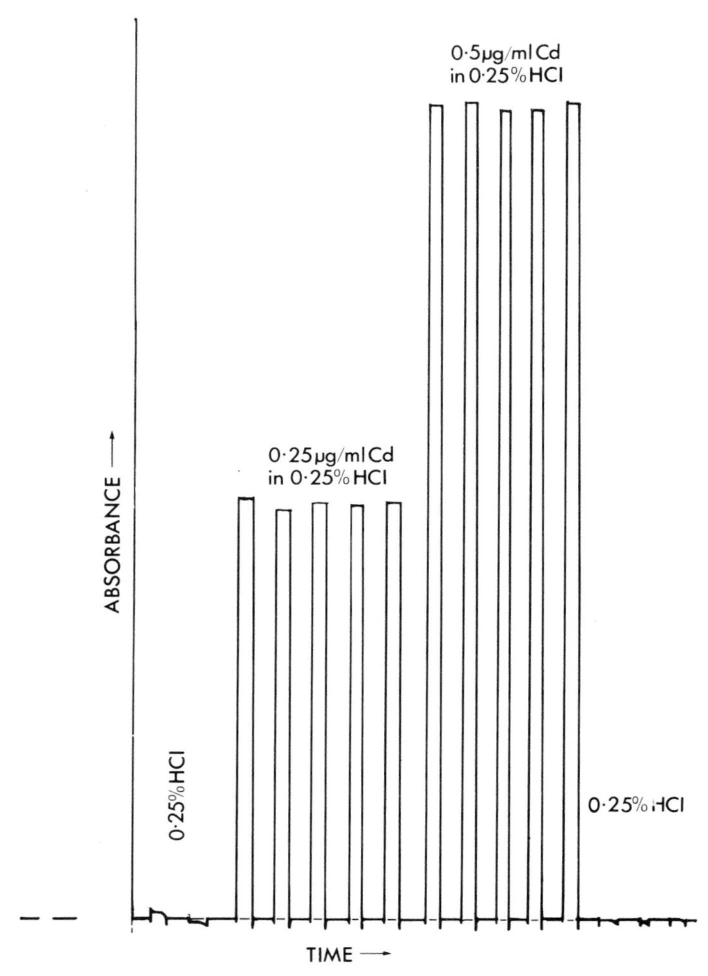

Fig. 5.12 Estimation of cadmium in extract from paint film. Integrated mode. ×10 scale expansion

Types of Determination. Analyses for metallic constituents of waters and effluents may conveniently be divided into three categories.

(i) Those in which element to be determined is present in sufficient quantity for a reading to be made directly usually after dilution of the sample (Ca, Mg, Na, K). A rapid method for the determination of calcium and magnesium is given on p. 204.

(ii) Those in which the element is present in solution in only sufficient

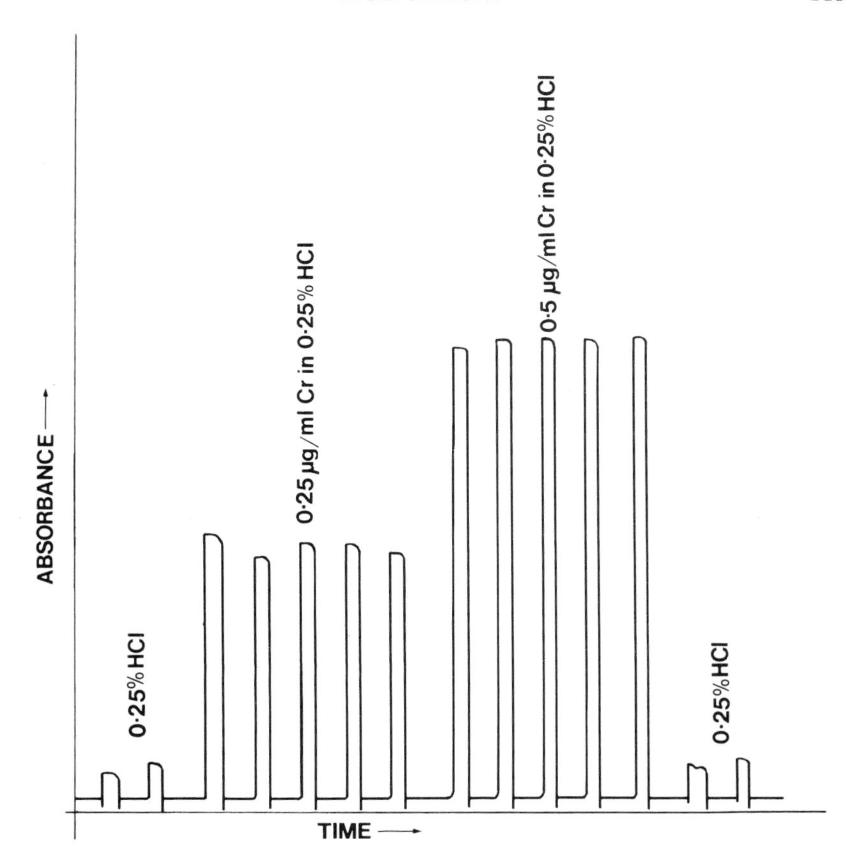

Fig. 5.13 Estimation of chromium in extract from paint film. Integrated mode. ×10 scale expansion

quantity for a reading to be made directly upon the sample by scale expanding the signal (Fe and Zn).

(iii) Those in which the metal is usually present in so low a concentration that a preconcentration step is usually necessary (Al, Cu, Ni, Pb). Alternatively flameless electrothermal atomization techniques can be employed.

The concentration ranges over which metals can be directly determined and the limits of detection in aqueous solution are given in Chapter 4 under the characteristics of each element.

Concentration Procedures. The most valuable and commonly used procedure for extractive concentration of heavy metals is that employing chelation with ammonium pyrrolidine dithiocarbamate (A.P.D.C.) followed by extraction into methyl isobutyl ketone (M.I.B.K.). This is described in Chapter 3.

Concentration by evaporation is a useful technique for non-saline water analysis. 250 ml of the sample are taken and 2·5 ml of hydrochloric acid is added, the sample is then carefully evaporated down to about 5–10 ml. It is essential not to allow the beaker to boil dry or adsorption losses can occur. If appreciable organic matter is present, 2–3 ml of '100-volume' (30% m/m) hydrogen peroxide should be added to the simmering solution, then the solution is allowed to cool and made up to a volume of 25 ml. Suspended matter can be removed by centrifuging. It is essential to use background

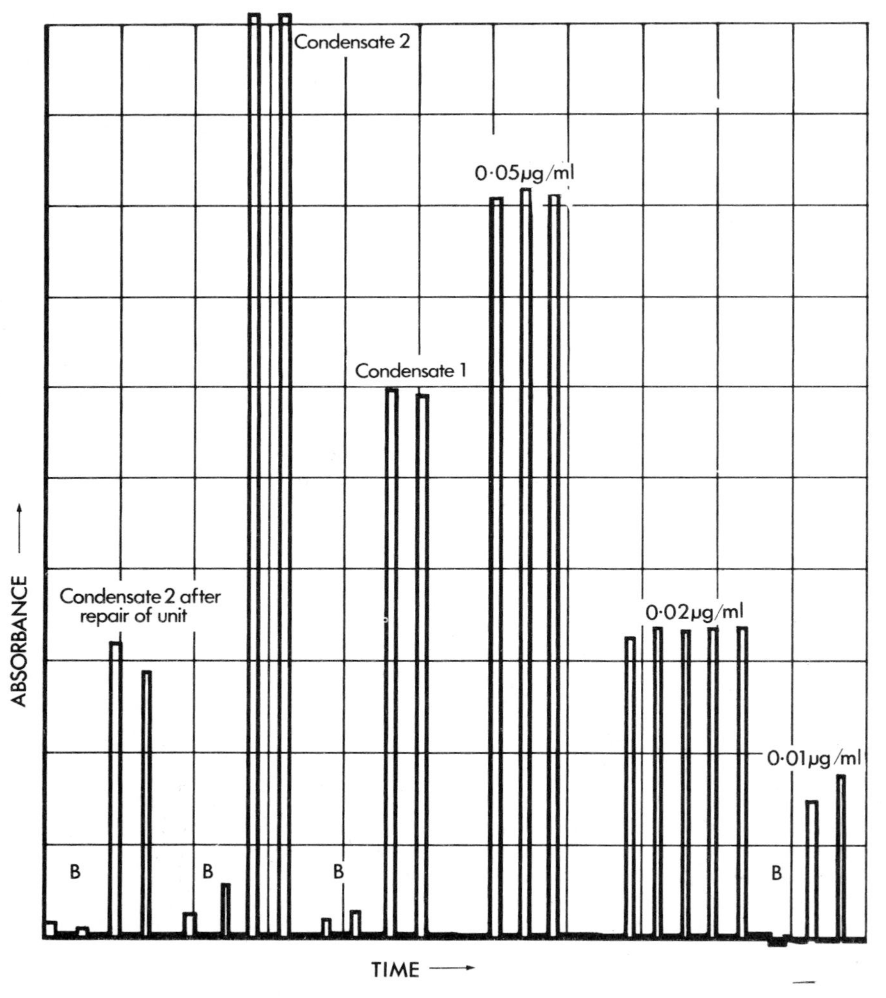

Fig. 5.14 Estimation of magnesium in power station condensates. Integrated mode. ×20 scale expansion
('B' corresponds to the blank)

correction to correct for the presence of the concentrated matrix elements (mainly calcium and magnesium). This simple procedure will result in an approximately tenfold increase in detection limit. Calibration should be made against standards prepared in 10% v/v hydrochloric acid.

POWER STATION CONDENSATES

An example of an estimation which is readily performed by atomic absorption spectroscopy, but which would be difficult to perform by any other technique is that of magnesium in power station steam condensates.

A leak in the steam condensers of an electricity generating station can have a serious effect upon the turbines.

It is essential, therefore, to trace leakages at a very early stage. This is frequently accomplished by checking condensates for an increase in sodium content. Very special care has to be exercised in order to prevent the inclusion of sodium from external sources (hands, atmosphere, etc.) during the analysis.

The estimation of magnesium, though, is much less open to errors from these sources, and has been used as a more reliable alternative for monitoring condensates.

The magnesium is always present at a very low level in a satisfactory unit. Samples of the condensates are aspirated directly to the atomic absorption spectrophotometer. The trace, shown in Fig. 5.14, indicates the capability of standard equipment to perform this type of check analysis. Alternatively the original sample can be added directly to a flameless electrothermal device (see p. 216) which is about 100 times more sensitive than flame atomic absorption.

Water to be used as a feed to modern high-pressure boilers must contain less than 5 μg per litre of copper or iron.

Analyses of Fine and Heavy Chemicals, Fertilizers, etc.

Atomic absorption is widely used in the chemical industry. Because prior chemical separations are unnecessary it is particularly valuable for estimations of, for instance, trace impurities of one alkaline earth in salts of another, e.g., traces of calcium in barium salts.

Estimations of traces of cobalt in nickel salts and vice-versa are examples of analyses which are more accurately and rapidly performed by atomic absorption than by any other procedure.

The capability of standard atomic absorption equipment to perform analyses of this type is illustrated in Figs 5.15, 5.16, and 5.17.

Soil Analysis

This was one of the first areas in which atomic absorption spectroscopy found application. The agricultural chemist is required to estimate a surprisingly large number of metals, of which calcium, magnesium, potassium, sodium, lithium, manganese, cobalt, nickel, copper, zinc, iron, lead, stron-

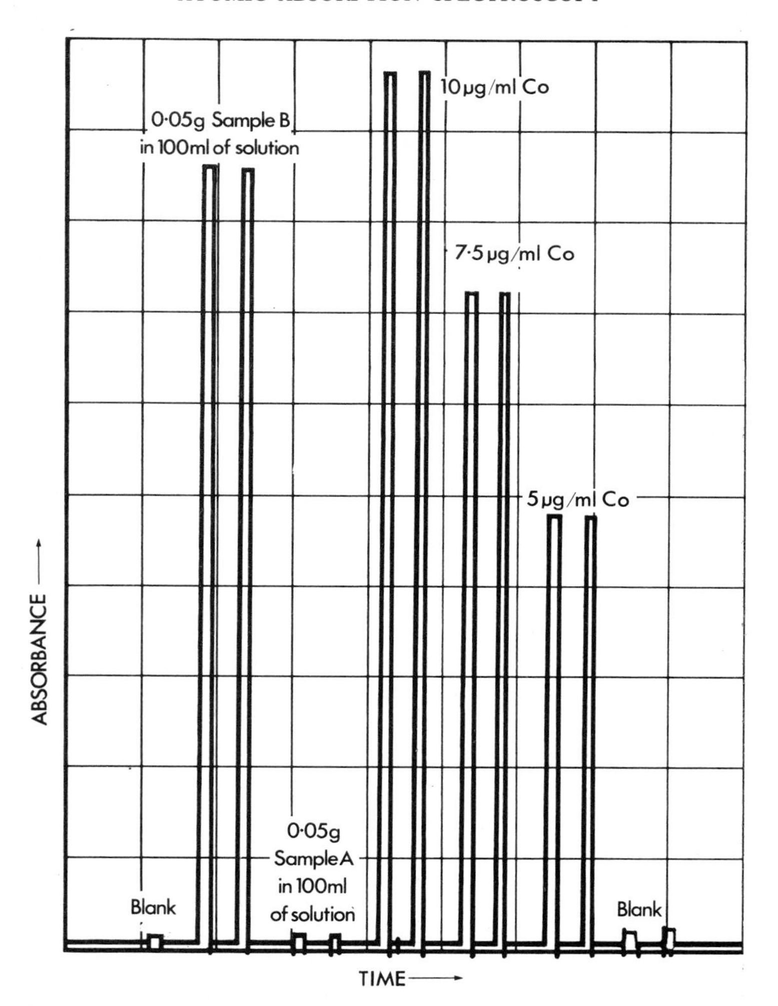

Fig. 5.15 Cobalt in catalyst samples. Integrated mode. Scale expansion ×1·3 (approx.)

tium, cadmium, chromium, aluminium, barium, mercury, and molybdenum can be determined reliably and quickly in trace quantities by atomic absorption. The freedom of the technique from spectral interferences allows metals to be determined in extract solutions and run-off waters without prior chemical separations; indeed at the low levels present in these solutions even chemical interferences are usually absent.

Soil analysis can be divided into two main sections (a) total elemental analysis and (b) cation exchange analysis (the leachable metal contents of the soil).

Total Elemental Analysis

The method is described in more detail under silicate analysis (p. 150) and consists essentially of digesting the dried or ignited soil with hydrofluoric–perchloric–nitric acid mixture, so that silicon is volatilized as the tetrafluoride. The residue is dissolved in hydrochloric acid, lanthanum chloride added to overcome suppression of the calcium response by aluminium, and adjusted accurately to a suitable volume.

Complex standard solutions are prepared to cover the expected concentration range of the constituents, and the determinations are made by aspirating standards and test solutions to the spectrophotometer and comparing the response. The determinations of all the elements likely to be required are straightforward and the reader is referred to Chapter 4 for further detailed information that may be required.

For this application, calcium and magnesium are most reliably determined with the nitrous oxide–acetylene flame system.

If manganese is to be determined in solutions that contain perchloric acid it is essential to reduce the metal to the divalent state by the action of sodium

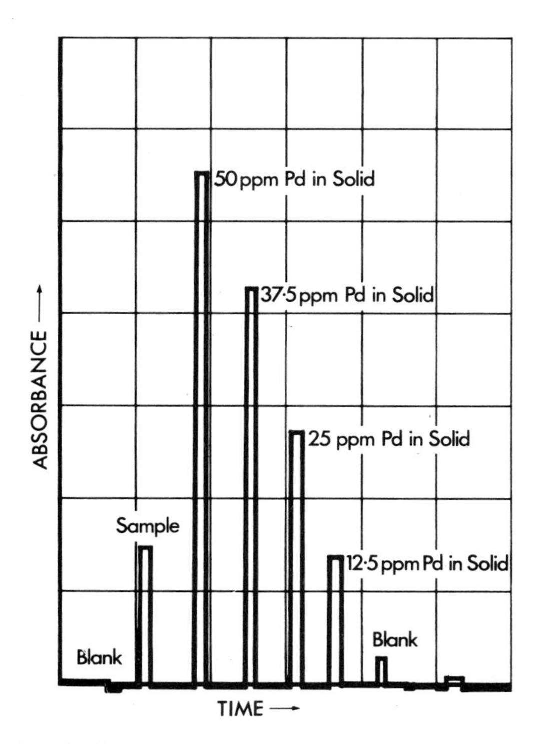

Fig. 5.16 Estimation of palladium in penicillin. Integrated mode. 'Pd free' penicillin used as a blank (1 g in 25 ml H_2O). ×10 scale expansion

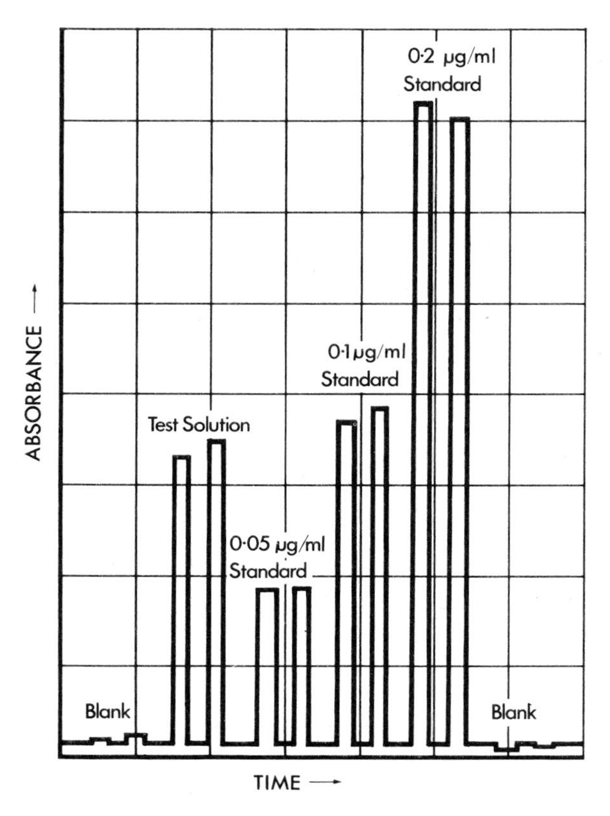

Fig. 5.17 Estimation of cadmium in zinc oxide. Integrated mode. ×10 scale expansion

nitrite or hydroxylamine hydrochloride. See the section on manganese in Chapter 4.

CATION EXCHANGE ANALYSIS

The cation exchange determinations that can be advantageously performed by atomic absorption spectroscopy are:

1. Measurement of the exchangeable metallic cations.
2. Measurement of cation exchange capacity of the soil.

These measurements then allow the chemist to determine the percentage exchange saturation of the soil and the percentage of saturation with alkali cations.

Exchangeable Metallic Cations are eluted from the soil either by continuous or repeated washing of a sample. Continuous washing requires the use of a percolation device such as a funnel or porous porcelain crucible. Repeated washings may be accomplished by repeatedly dispersing the sample in fresh portions of solvent and removing the eluate by centrifuging. Full practical details of the procedures employed and precautions to be taken are to be found in *Soil Chemical Analysis* by M. L. JACKSON (University of Wisconsin, 1965).

The exchangeable metallic cations most frequently determined are calcium, magnesium, potassium, manganese, and sodium. They may all be estimated in an extract obtained by eluting the soil with 1 M ammonium acetate solution. This solvent possesses the advantages that it effectively wets a soil, efficiently replaces the cations, and is suitable for flame photometric determinations.

The less frequently determined exchange metals (aluminium, iron, copper, zinc, and cobalt) may also be estimated in the ammonium acetate extract, but are more reliably determined when extracted by alternative procedures, e.g., copper is best extracted with a solution of E.D.T.A. disodium salt.

In the procedure employing ammonium acetate, the soil sample, preferably in the field-moist condition, is passed through a 2 mm sieve and 50–100 g weighed into a 250 ml flask. 100 ml of 1 M ammonium acetate solution are added, the flask stoppered, shaken, and the contents allowed to stand overnight. The contents are transferred to a Buchner funnel fitted with a Whatman No. 42 filter paper. After the filtrate has been separated from the solid, washing is carried out, over a period of about 1 hour, with a further 300 ml of ammonium acetate solution.

The extract is transferred to a 500-ml graduated flask, treated with 50 ml of lanthanum chloride solution (65 000 μg/ml La) and made up to the mark with 1 M ammonium acetate solution (the lanthanum chloride addition is made to overcome possible phosphate interference). This test solution is used for the determinations of all the commonly sought metals. Complex standards, prepared so as to resemble approximately the composition of the test solution and to cover the estimated concentration ranges of the metals to be determined, are used.

The results are normally related to the soil in 'the 100°C oven-dried condition' by calculation.

Cation Exchange Capacity. The eluted soil from the Buchner funnel is conveniently used for this determination, which involves the measurement of the total quantity of negative charges per unit weight of the soil. This capacity is measured by reloading the ammonium acetate leached soil with a 0·05 M solution of calcium chloride, buffered to a neutral or slightly alkaline *p*H, and then washing out the excess salt with an electrolyte-free solvent, such as 80 per cent v/v acetone. The calcium is now replaced by further washing with 1 M ammonium acetate and is determined in the eluate as described to give the total exchange capacity of the soil.

An extremely large number of methods for estimating the cation exchange capacity of soils is provided by different combinations of pretreatment cation salt, solvent, and washing procedure. Atomic absorption spectroscopy provides a reliable and rapid method for the final estimation in every case.

Applications in Clinical Chemistry

INTRODUCTION

There is in the field of clinical chemistry a large and growing demand for accurate, reproducible and rapid trace metal analyses. Clinicians generally recognise not only the desirability of performing such analyses on more samples from more patients, but would also like to increase the number of metallic elements routinely determined in biological fluids.

Many of the people called upon to perform these determinations are not chemists by training, and are interested only in the clinical significance of the final results. For these reasons the superiority of an instrumental procedure over classical wet analytical methods is becoming increasingly accepted.

Atomic absorption spectroscopy possesses many features that must be attractive to the clinical chemist. It is accurate, versatile, rapid, simple to operate and cheap to install. Because of its freedom from spectral interferences prior chemical separation of constituents is not required, preparation usually being limited to placing the sample in solution, or, perhaps, extracting the required species by simple elution. Lastly the very high degree of reproducibility that is easily attainable by the procedure is of enormous value in the clinical laboratory.

Table 5.2 gives an indication of element levels in blood, serum, urine and tissue and also suggests the most suitable method of analysis.

The Determinations of Calcium and Magnesium in Clinical Samples

IN BLOOD SERUM

The estimation of calcium in blood serum is one of the most important and frequently performed determinations in the clinical laboratory. Serum magnesium determinations are much less frequently required; but because it is possible, by atomic absorption spectroscopy, to determine both elements in a serum sample at a single dilution, it is convenient to describe their estimations together.

The calcium content of serum normally lies between 8·5 and 11 mg/100 ml.

An increase in serum calcium level (hypercalcemia) can be diagnostic of many pathological conditions such as hyperparathyroidism and neoplastic processes of bone. A reduction in the serum calcium concentration (hypocalcemia) is frequently seen in hypoparathyroidism, kidney diseases and obstructive jaundice.

The most likely source of error in this determination is the possible under-estimation of calcium due to phosphate interference. This is overcome by the addition of lanthanum chloride during sample preparation, which

Table 5.2 Typical element levels in blood, serum, urine and wet tissue

Element	Blood $\mu g\ ml^{-1}$	Serum $\mu g\ ml^{-1}$	Urine $\mu g\ ml^{-1}$	Organs (general) $\mu g\ g^{-1}$	Suggested methods of analysis*
Al	0·1	0·1	0·05	0·5 (lung 50)	3, 4, 5
As	0·1	0·1	0·02	0·2 (hair 1)	7
Ba		0·02		0·1	4
Bi	0·005				7
Ca		100	200		1
Cd		0·01	0·001	1 (kidney 10)	4, 5, 6
Co	0·005		0·005	0·01	3, 4, 5
Cr		0·02	0·002	2	4
Cu	1	1	0·2	1 (kidney 5)	1, 4
Fe	500	1	0·5	100 (spleen 200)	1, 4
Hg	0·005	0·002	0·01	0·1 (kidney 2)	8
K	2000	200	2000		1, 2
Li		0·02	0·005		1, 2
Mg	50	20	50	100 (bone 1000)	1
Mn	0·02	0·01	0·01	0·2 (liver 2)	4
Mo	0·01	0·005		0·02 (liver 0·2)	5
Na		3200	2000		1, 2
Ni	0·05	0·01	0·05	0·5	3, 4
Pb	0·2	0·05	0·05	1 (bone 20)	3, 4, 5, 6
Sb	0·0005				7
Se		0·01		0·2	7
Si		5			4
Sn		0·05	0·02	0·2	7
Sr		0·05	0·2	1 (bone 50)	2
V		0·02	0·005	0·01 (lung 0·5)	
Zn	10	1	0·5	20 (cerebral spinal fluid 0·2)	1

Most of the concentration values in this Table have been extracted, with kind permission, from an exhaustive Table compiled by Dr. J. B. Dawson† of the Department of Medical Physics, The General Infirmary, Leeds, U.K. The values have been calculated from a large number of literature sources.

† Dawson, J. B., *The Scientific Foundations of Clinical Biochemistry*, Vol. 1, Analytical Techniques, Edited by Marps, V., Williams, D. L., and Nunn, F. S. Heinemann, 1975.

Notes

Numerical values are rounded off to 1, 2 or 5.

To convert the concentrations from $\mu g\ ml^{-1}$ (or $\mu g\ g^{-1}$) to $\mu g\ 100\ ml^{-1}$ (or $\mu g\ 100\ g^{-1}$) the figures consists simply of diluting the serum. Because of the simplicity of sample preparation required, the estimation of serum calcium by atomic absorption spectroscopy is more reliable and faster than by any of the other techniques currently used.

An exhaustive investigation carried out by R.J.R. on the calcium estimation revealed the following facts:

1. No evidence of interference from protein or from any of the other cations present in serum could be detected.
2. Very severe depression of calcium absorption was observed to occur in solutions that were more than 1 M with respect to hydrochloric acid. This observation led us to suspect that differences hitherto reported between samples of serum that had been deproteinized, and those that had been simply diluted, and which had been attributed to 'protein enhancement', could have been due to *depressed* results being obtained from the deproteinized test solutions due to their higher acidity.

   It is essential that the acid concentration in the test solutions should be less than 0·5 M.
3. By resorting to zero suppression followed by scale expansion, it is possible to work over the range 0·1–0·25 mg/100 ml and to obtain a standard deviation of about 0·002 mg/100 ml, leading to a precision of about 0·1 mg/100 ml in the original serum.

The normal level for serum magnesium is between 1·9 and 2·5 mg/100 ml.

An elevated serum magnesium level is observed in renal insufficiency, hypertension, arteriosclerosis and comatose diabetes. Reduced serum magnesium occurs with severe diarrhoea, malabsorption, kwashiokor, alcoholic cirrhosis, acute pancreatitis, chronic nephritis and congestive heart failure.

The estimation of magnesium by atomic absorption spectroscopy is more convenient and reliable than by any other presently used method.

---

should be multiplied by 100. To convert concentrations from $\mu g\ ml^{-1}$ to meq. l^{-1} the figures should be divided by the atomic weight of the element.
The concentration values (especially for non-essential elements) can vary considerably between individuals.
The urine concentrations refer to a typical 24 h urine sample.

* Suggested methods of analysis;
1. Suitable dilution followed by flame atomic absorption.
2. Suitable dilution followed by flame emission.
3. Pre-concentration using solvent extraction followed by flame atomic absorption.
4. Direct flameless (electrothermal) atomization.
5. Separation from matrix using solvent extraction followed by flameless (electrothermal atomization.
6. Delves cup or punched disc technique.
7. Wet ashing followed by hydride generation technique.
8. Cold vapour generation of mercury.

Further general notes on determinations of calcium and magnesium in blood serum:

1. Standard solutions are best kept in polythene bottles. Standards below about 5 mg/100 ml should be freshly made up each week.
2. During a long run (more than about 20 sera determinations) it sometimes happens that the burner slit becomes partially blocked. This fault can cause serious drift of readings and is quickly recognized from the appearance of the flame, which becomes hollow at the centre. The burner is conveniently cleaned by turning off the acetylene supply, and removing the burner. The slit can then be cleaned with a razor blade, and the interior washed with 1 M hydrochloric acid followed by deionized water. This cleaning should be carried out at the end of each run, as a routine. A wide slot ('high solids') air–acetylene burner can prove useful for this type of determination.
3. The difference in viscosity between the prepared sample and the standard solutions had no effect upon the determinations.
4. The lanthanum chloride reagent must be of a grade that is low in calcium and magnesium.
5. Lanthanum chloride is most frequently used as the suppressant for phosphate interference in serum calcium determinations. E.D.T.A. diammonium salt is preferred by some workers, and is equally acceptable. The concentration of the salt in the test solution should be 0·75% m/v.
6. Although individual laboratories will continue to use their own diluents, the following procedure has been recommended by the National Bureau of Standards (U.S.A.). This involves a 1 + 49 dilution of serum or plasma with a diluent solution containing 0·01 M lanthanum(III) chloride and 0·05 M hydrochloric acid.

The calcium content of blood serum is about 10 mg/100 ml and the magnesium content about 2 mg/100 ml. The optimum ranges for the estimations of calcium and magnesium at ×1 scale expansion are 0·1 to 0·8 mg/100 ml and 0·01 to 0·1 mg/100 ml.

The estimation of calcium and magnesium can thus readily be performed upon a serum sample diluted between 20 and 50 times. At lower dilution factors the magnesium level in the test solutions will lie at the top end of the optimum range for estimation at ×1 scale expansion with a standard 10 cm burner, and it will be necessary to resort to some method of reducing the sensitivity such as rotation of the burner. Figure 5.18 shows a typical trace estimation of calcium in blood serum.

Calcium and Magnesium in Urine

During a 24-hour period less than 150 mg of calcium and about 100 mg of magnesium are normally removed in the urine. The ranges, though, especially for magnesium are far wider than for serum.

A decrease in urinary calcium is noted in tetany and myxoedema and an increase is observed with some bone diseases. Estimations of calcium and magnesium are generally considered to be of limited diagnostic value.

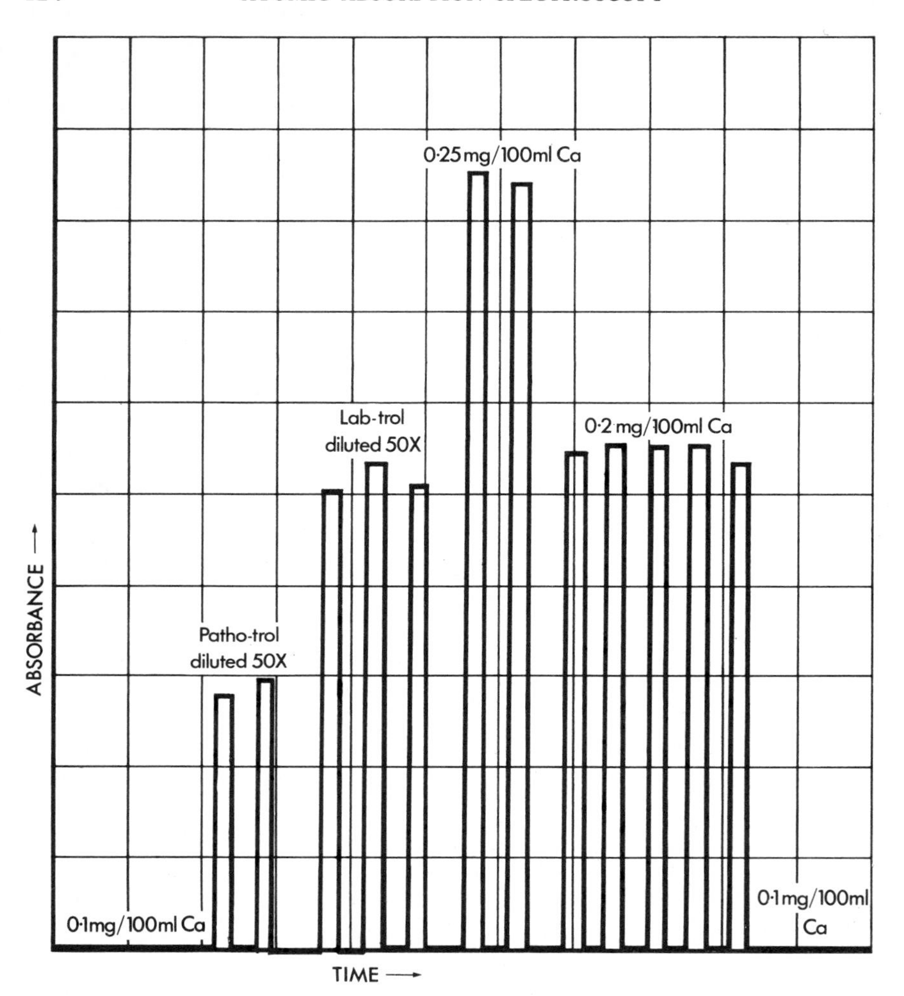

Fig. 5.18 Estimation of calcium in blood serum. Using zero-suppression followed by scale expansion. (Integrated mode)

Urine samples should be acidified with hydrochloric acid so as to ensure that they remain clear.

An aliquot of the urine is treated with lanthanum chloride solution and diluted with deionized water so that the calcium and/or magnesium contents lie in the optimum range(s) for estimation.

Calcium and Magnesium in Body Tissues

These elements are present in nearly all human tissues. The general procedure for their estimation is to accurately weigh about 1 g, desiccate the sample overnight, and then obtain a dry weight.

Sample preparation consists of wet ashing with a suitable reagent, e.g., nitric–perchloric acid mixture, sulphuric acid and hydrogen peroxide, etc., followed by treatment of the digest with lanthanum chloride solution and dilution so that the elements lie in the optimum ranges for estimation.

The Determination of Iron in Clinical Samples

The Estimation of Iron in Blood Serum

Most of the iron present in blood occurs in the haemoglobin. Estimation of the lesser quantity, though, that is present in serum is of more value to the clinician for diagnostic purposes. Serum iron deficiency can be indicative of an increased rate of haemoglobin synthesis (following haemorrhage) and post infection transport of iron to tissues. Elevated serum iron concentrations are associated with acute hepatitis, malignant tumours and iron toxicity.

The normal level of iron in blood serum is from 50 to 170 μg/100 ml and is higher in men than in women. The red cells contain between 700 and 1000 times more iron than in serum. It is difficult to obtain serum that is totally devoid of red cells (either damaged or intact) so that the contribution of iron from this source can be significant. Serious errors can arise from the presence of haemoglobin and grossly haemolysed serum, characterized by a pink tinge, should not be used for iron estimations.

In addition to the possible inclusion of iron in the test samples by haemolysis and from external sources the estimation of serum iron is subject to two controlling factors, the low content in the serum, and the fact that only about 2 ml of serum are likely to be available in practice.

Until recently the procedures employed for this estimation by atomic absorption resorted to attack of the serum with trichloroacetic acid, separation of the solution from the solid debris, complexing the iron, and concentrative extraction into an organic solvent. Such procedures offered no advantages in simplicity, speed of operation and accuracy over the established colorimetric methods.

The dramatic improvements in lamp technology that have taken place, together with the use of scale expansion and signal integration, now make it possible to perform this estimation upon a sample of serum simply diluted 1 + 1 or 1 + 3 with deionized water.

Such a procedure is obviously much more rapid and freer from handling errors than one involving a series of preparative steps.

Fig. 5.19 illustrates the capability of standard equipment possessing the facility of signal integration to perform this analysis.

From the series of responses for the 20 μg/100 ml Fe standard a standard deviation of about 1 μg/100 ml is calculated for the estimation.

This would lead to a standard deviation of 2 μg/100 ml in the undiluted

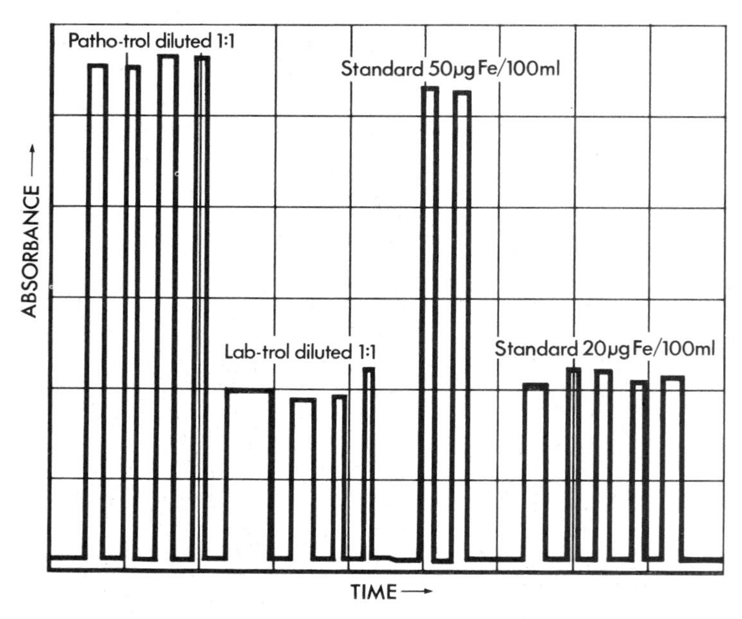

Fig. 5.19 Estimation of iron in blood serum. Integrated mode. ×10 scale expansion

serum if it were diluted 1 + 1 for the estimation, or 4 μg/100 ml if a 1 + 3 dilution were used.

The use of flameless techniques can be used for this estimation (see p. 233).

OLSON and HAMLIN[83] draw attention to the fact mentioned above, that serious errors can occur if the serum is significantly haemolysed. They exploit the fact that the serum iron is present in the ferric state whereas the haemoglobin derived iron will be mainly in the ferrous form, to effect a separation. Their recommended procedure is to digest a 1 or 2 ml sample of serum for 15 minutes at 90°C with an equal volume of trichloroacetic acid. The solid debris containing the precipitated ferrous iron is removed by centrifuge. The ferric iron remains in solution and is determined by aspirating the supernatant liquor directly to the atomic absorption spectrophotometer.

ESTIMATION OF TOTAL IRON BINDING CAPACITY

This can be determined by the procedure of OLSON and HAMLIN[83] in which 2 ml of serum are treated with an equal volume of a solution of ferric chloride (containing 500 μg of iron per 100 ml). The excess reagent is removed by the addition of magnesium carbonate (200 mg) followed by separation of the solid matter by centrifuge.

The iron content of the supernatant liquid can then be estimated directly, from which the Total Iron Binding Capacity can be calculated; or if the original serum was suspected to have suffered significant haemolysis an aliquot can be treated with trichloroacetic acid and the ferric iron content estimated as described above.

The Estimation of Lead in Urine and Whole Blood

URINE

The estimation of lead in urine is of most value to the clinician when applied to patients undergoing therapy with chelating agents. During this treatment the quantity of lead excreted (even from a healthy subject) is well above the normal level (500 μg per 24 hours). It is therefore possible to perform the estimation by aspirating urine directly to the atomic absorption spectrophotometer. The clinical analyst can thus rapidly provide the physician with information that allows him to monitor the progress of the treatment with confidence.

About 2 ml of hydrochloric acid (36% m/m) should be added to each 100 ml of urine to prevent the formation of precipitates.

The large concentrations of inorganic ions present in the urine produce considerable non-specific background absorption which can be corrected manually or automatically by the use of a deuterium hollow-cathode lamp (see p. 239).

WHOLE BLOOD

Lead is mainly transported through the body by the red blood cells so that clinical estimation of the metal is best performed on whole blood. This determination constitutes the most important single test for lead intoxication.

Clinical interest in the determination of lead in blood may be broadly divided into two areas.

There is firstly what may be termed the 'resident hospital' interest. Here a fairly limited number of patients resident in a hospital are known to be suffering from lead poisoning and are receiving the appropriate treatment. Tests are periodically carried out on blood or urine samples to monitor the progress of the treatment. Adequately large samples of either fluid are normally available and the precision of the analyst is of major concern.

For this field of work a procedure based upon a digestive preparation is acceptable.

The second area is that concerned with the proposed periodic screening of lead-workers, on site. Here the object is to diagnose early symptoms of poisoning, and a procedure capable of dealing rapidly with large numbers of subjects is desirable. A reasonable, but not exacting level of accuracy is required. For screening tests of this nature a rapid procedure capable of utilizing ear- or finger-prick samples is desirable.

PROCEDURES UTILIZING A 2 TO 5 ml SAMPLE OF WHOLE BLOOD INVOLVING ACID DIGESTION

The original and still practised method involves three processes:

1. The precipitation of blood-proteins by digestion with trichloroacetic acid, followed by removal of the solid debris from the liquid phase by centrifuge.
2. Adjustment of *p*H, and then the formation of a chelated lead complex by the addition of ammonium pyrrolodine dithiocarbamate (A.P.D.C.).
3. Extraction of this organo-complex into iso-butyl-methyl ketone and separation from the aqueous phase.

This procedure was initially developed by ELEANOR BERMAN(72) and has been modified by other workers. Its major advantage is that, because the lead is extracted from the other constituents in the digest (in particular sodium), complications attributed to background absorption interferences are avoided. Enhanced absorption is also derived from the greater volatility of the organic medium, but this is of less value now than it was when the method was first developed. Its main disadvantages are vulnerability to handling losses, contamination and the length of time required, both due to the multiple operations involved. Also, if the patient is undergoing treatment with a chelating agent (such as E.D.T.A.), there is a risk that not all the lead will be extracted.

A second general procedure(76) attempts to eliminate the chelation and extraction stages, and consists essentially of precipitation of blood proteins with trichloroacetic acid and their removal by centrifuging. The supernatant liquid is then adjusted to a suitable volume and aspirated directly to the atomic absorption spectrophotometer.

The method involves fewer operations and is less time consuming than BERMAN's procedure, but the absorption readings obtained are affected by non-specific background absorption interference caused by the heavy loadings of sodium in the test solutions. This effect has to be corrected (see p. 239). It should be noted also that low-level determinations performed under such conditions are subject to much higher noise levels than would be obtained from aqueous solutions containing only a lead salt, so that the precision of the estimation is diminished.

The procedure described by EINARSSON and LINDSTEDT(85) has proved very satisfactory and is outlined below:

Digestion Reagent. Prepare a 5 per cent (m/v) aqueous solution of trichloroacetic acid. Mix 75 ml of this solution with 25 ml of perchloric acid (60% m/m).

Sample Preparation. Pipette 5 ml of heparinized blood into a 15-ml centrifuge tube. EINARSSON and LINDSTEDT consider that weighing is more exact than pipetting. They then use a density factor of 1·06 g/cm^3 to convert the weight to volume.) The test samples can be reduced to about 2 ml, if only a restricted quantity of blood is available. Add 3 ml of digestion reagent and 1 ml of deionized water and stir the contents of the tube with a glass rod. Allow the treated sample to stand for 1 hour.

Remove the solid debris by centrifuging for 8 min. Decant the supernatant liquid into a second centrifuge tube and spin for a further 4 min to remove all traces of suspended solid matter.

The clear solution now obtained is ready for aspiration to the atomic absorption spectrophotometer.

Preparation of Standards. A pooled blood of low lead content is used to prepare standards. Blood and digestion reagents are mixed as described above, but instead of water, 1 ml of aqueous lead solutions containing 1, 2 and 4 μg of lead are added, so as to increase the effective level of lead in the 5 ml blood sample by 20, 40 and 80 μg/100 ml.

Measurement. Measurement may be made at either 217·0 or 283·3 nm. Maximum scale expansion will be required.

The reading observed for the standard solutions is plotted against the effective increase in lead concentration of the pooled blood, and over the range used an approximately linear relationship will be obtained. A reagent blank should be run.

To correct for non-specific background absorption, readings are taken for a pooled serum sample containing no added lead at 220·3 or 287·3 nm (depending on whether the estimation is being performed at 217·0 or 283·3 nm). Alternatively, a deuterium hollow-cathode lamp can be used (see p. 239).

The sum of the readings for the reagent blank and the background absorption is noted on the Y-axis and a line parallel to the original calibration drawn through this point. This second line is used to obtain the lead contents of the test samples.

Notes

The author (R.J.R.) investigated the use of nitric acid to digest the blood but found that with the reagent an excessively bulky precipitate was produced, which was very difficult to separate effectively from the liquor. The trichloroacetic/perchloric acid mixture was found to give rise to a denser precipitate which could be more effectively separated from the liquor, and so gave rise to a larger solution volume suitable for the analysis.

Check analyses ascertained that a negligible quantity of lead was retained in the solid debris.

The error for a single determination, calculated from duplicate analyses of 127 blood samples was stated to be ±1·2 μg/100 ml.

The procedure allows about 20 blood samples together with the required

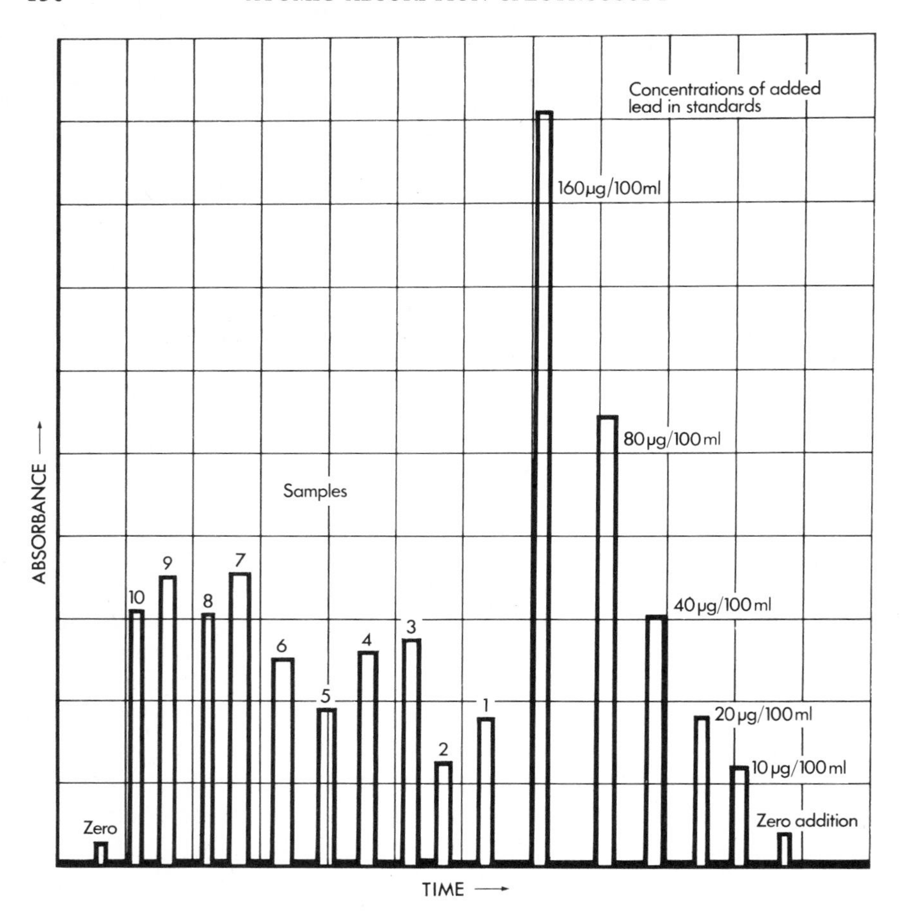

Fig. 5.20 Estimation of lead in blood. Procedure of EINARSSON and LINDSTEDT, *Scan. J. Clin. Lab. Invest.* 23 (1969) 367. Integrated mode. ×20 scale expansion. Wavelength 217 nm

standards to be estimated in duplicate by one person in a normal working day. Figure 5.20 shows typical results using this technique.

RAPID MICROSAMPLING PROCEDURES

a) Delves Cup

Several procedures designed to perform the analysis of lead in whole blood on ear-prick samples were developed but in their original forms were not readily available to a convincingly reliable, easily operated low-priced commercial unit.

In both the 'tantalum-boat' technique of KAHN, PETERSON and SCHALLIS[80] and WHITE's[86] platinum-loop/nickel absorption tube method a recorder trace was taken of the transient absorption signal produced when a dried sample was introduced into the flame. The measurement was complicated by the fact that two signals were obtained when a biological material was introduced into the flame system. The first was attributed to light scatter by the combustion products of the sample, the second being produced by the atomic absorption of the lead atoms.

WHITE[86] was unable to resolve these two signals when a sample of whole blood was fed directly into the flame, and resorted to wet oxidation on a spotting tile in an attempt to overcome the difficulty. An aliquot of the digest was then transferred to the platinum loop and plunged into the flame. The additional operations involved lengthened the procedure and reduced its accuracy.

KAHN *et al.*[92] partially oxidized the sample in the tantalum boat by the addition of nitric acid, dried the residue and then introduced the boat into the flame. They were able effectively to resolve the two signals but unfortunately a very wide variation in sensitivity was found between different boats and more seriously by varying the sampling position in the same boat. This latter effect was considered to be due to the difference in the cross-section of the radiation from the hollow-cathode lamp as it traversed the flame above the boat.

To summarize, WHITE's procedure loses accuracy by the multiple handling and volumetric operations entailed in carrying out wet oxidation on a spotting tile, and then transferring an aliquot to a platinum loop. The tantalum boat procedure does much to eliminate error from this source.

By utilizing the nickel-absorption tube of WHITE's method (which is effectively a flame cell) the most serious source of error in the tantalum boat procedure (variation in sampling position) is diminished.

The procedure developed by DELVES[87] is designed to incorporate the advantageous features of both the above systems and to eliminate their disadvantages. The tantalum boat is replaced by a micro crucible with the objective of eliminating intra-boat variation in sensitivity. Oxidation is effected within the crucible and a nickel absorption tube is mounted above the flame of a standard air–acetylene burner. The flame and the atomized sample enter the tube through a hole half-way along the length of the tube.

Comparison is made against standards prepared by adding 10 or 20 μl aliquots of standard lead solutions, containing 0, 20, 40, 80, 120, 160 and 200 μg/100 ml Pb to crucibles containing 10 or 20 μl of normal blood. The lead content of this blood is of a low known value and is ascertained by extrapolating the graph back to the negative X-axis.

Considerable variation in sensitivity can occur between individual crucibles, and to reduce error from this source 'matched' crucibles that show a sensitivity variation of less than $\pm$10 per cent between each other should be selected.

The microsampling positioner and a burner assembly are illustrated in Figs 5.21 and 5.22.

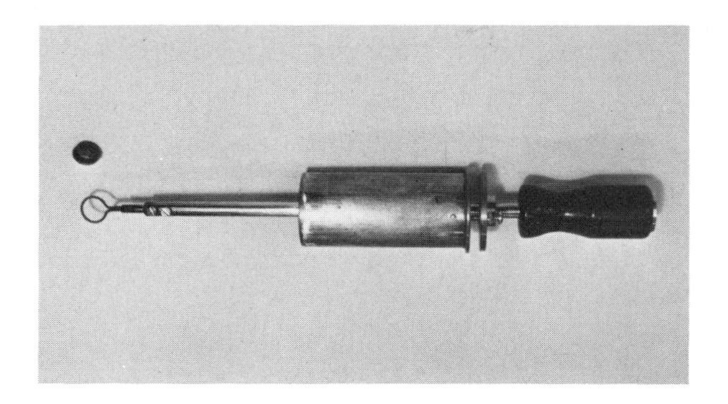

Fig. 5.21 A microsampling positioner

The sensitivity and reproducibility attainable are indicated on the typical recorder trace, Fig. 5.23 and the corresponding calibration curve is shown in Fig. 5.24.

After a short practice period (two or three calibration runs) an experienced clinical chemist could expect to obtain meaningful results of satisfactory reliability for screening tests and therapy-progress checks.

It is essential to recognize the fact that the procedure is *NOT absolutely* reproducible, due to inconsistencies in pipetting samples as small as 20 μl and the difficulty of positioning the samples at precisely the same place in the flame

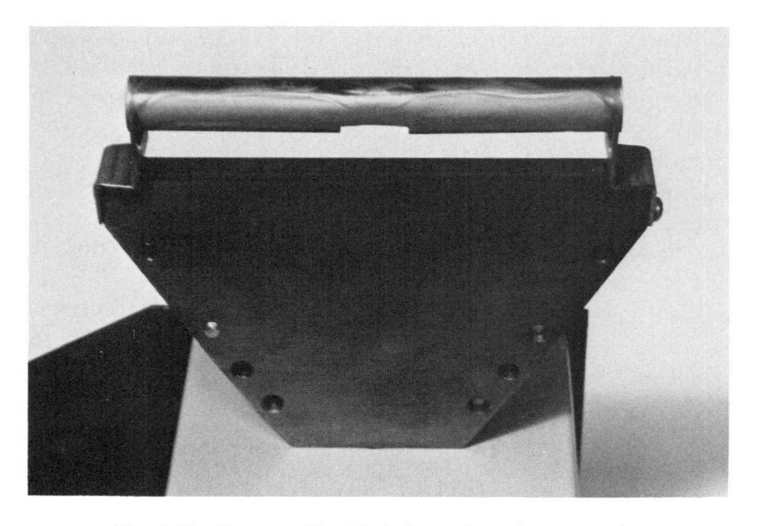

Fig. 5.22 Burner with nickel absorption tube mounted

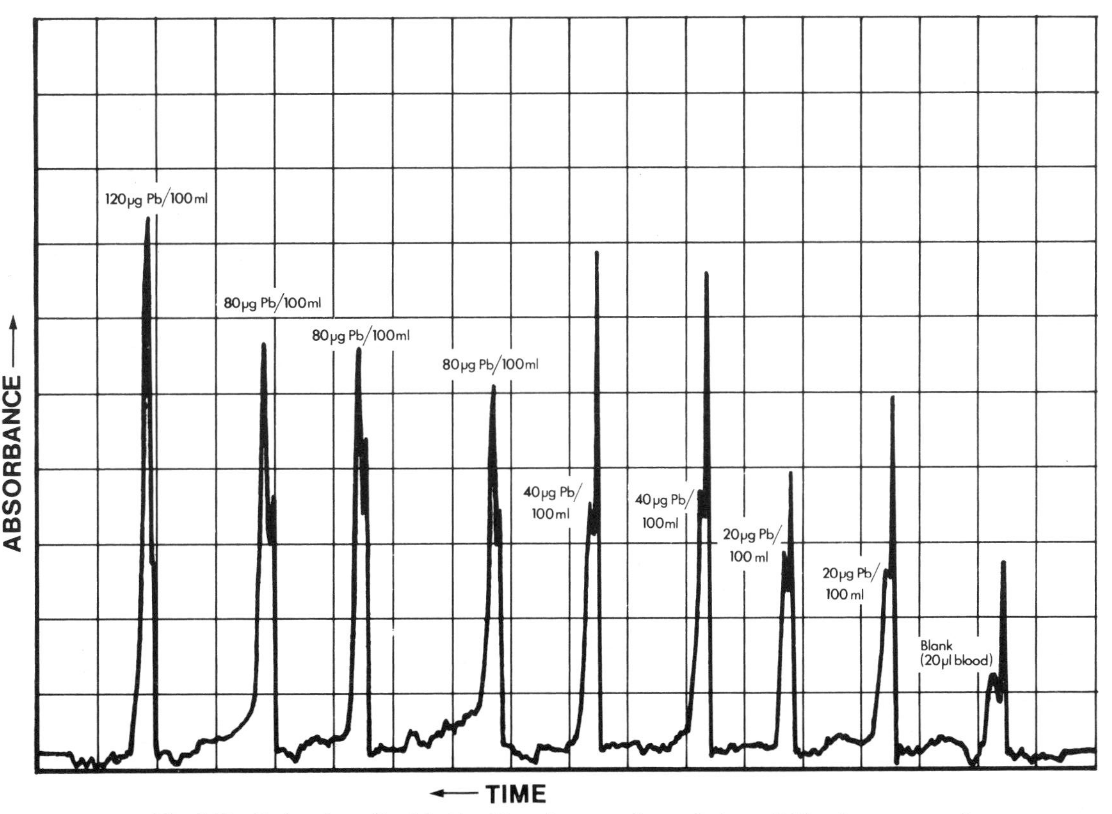

Fig. 5.23 Estimation of lead in blood by micro-sampling technique. Calibration responses for 20 μl of blood +20 μl of lead solutions. Direct readings at $\times 2$ scale expansion, 283·3 nm

each time. Furthermore, the sensitivity for the estimation obtainable from a particular nickel tube and set of crucibles will vary from day to day, and will in general drop-off after prolonged use. The great advantages are the speed and simplicity of the procedure, and the small sample sizes required.

A variation of this method using silica crucibles and a preliminary[88] oxidation with aqua regia has been reported.

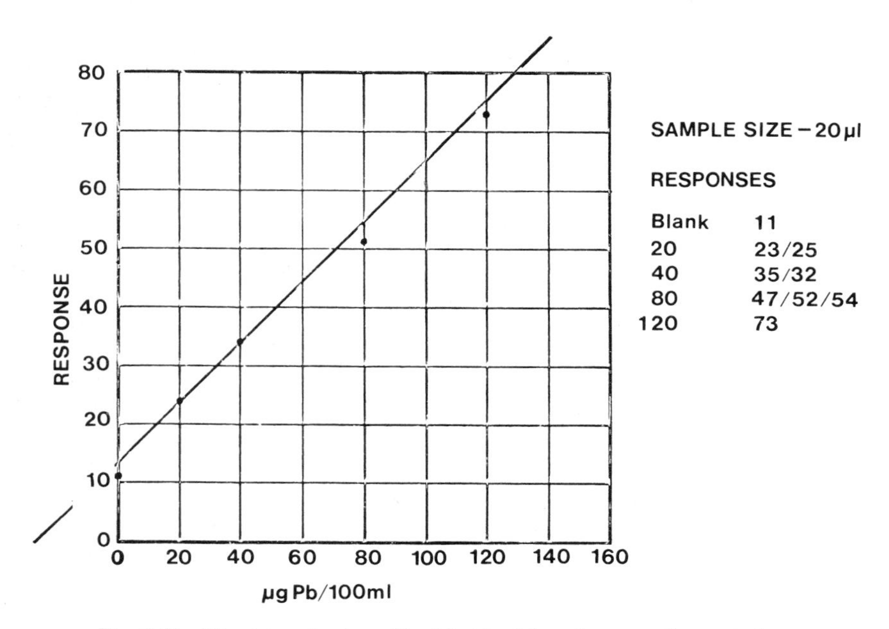

Fig. 5.24 The determination of lead in blood by micro-sampling technique

The standard Delves Cup technique will not work satisfactorily for cadmium because it is impossible to resolve the matrix smoke peak from the atomic cadmium signal. However DELVES[89] has extended the method to cadmium by the addition of 20 μl of 3% m/v diammonium hydrogen orthophosphate and 10 μl of blood into the nickel cup, followed by drying on a hotplate at 150°C for three minutes. This results in the formation of involatile cadmium phosphate and allows resolution of the cadmium atomic absorption signal from the smoke peak when the cup is inserted into the flame.

b) The Punched Disc Technique

Accurate pipetting of 10–20 μl volumes of whole blood can be difficult owing to the high viscosity of blood. A method that does not require any accurate pipetting has been described by CERNIK and SAYERS.[98,99] 30–60 μl of whole blood is simply spotted on to a filter paper (Whatman No. 4) and allowed to spread and dry at room temperature. A spot of 13–18 mm diameter is obtained with uniform spread of lead except at the very edges of the spot. A 9 mm disc of filter paper is then punched out from the central region of the spot and simply placed in the nickel micro crucible (Delves Cup) described above.

The main advantages of the punched disc technique are summarised below:

1. It is possible to take free-flowing capillary blood samples from scrupulously cleaned ear lobes into a heparinised capillary tube and then expel 30–60 μl of the blood directly on to a filter paper (contained in a clean Petri dish) at the time of sampling. This avoids the necessity of taking venous samples with subsequent accurate pipetting.
2. Standard blood samples, accurately analysed by more time-consuming techniques (e.g. polarography or anodic stripping voltammetry), can be used to prepare standard blood spots on filter paper. These have been found to be stable for periods of over two years. Blanks are negligible.
3. By punching a 4 mm diameter disc both lead and cadmium can be measured using the flameless electrothermal atomization technique.[99–101] The disc is simply placed in the graphite cup of the Mini Massmann furnace (see Chapter 7 and Fig. 7.8). It is then dry ashed at 300°C and atomized at 1400°C for lead. For cadmium the corresponding figures are 285°C and 650°C. It is essential to use automatic background correction.

c) Pulse nebulization technique

One of the authors (K.C.T.) has recently reported a simple method for monitoring excessive lead levels in whole blood using a direct pulse nebulization technique[91] (see p. 201). The standard nebulizer was replaced by a nebulizer containing a wide-bore capillary (0·70 mm bore). A 1 ml sample of blood was placed in a small test tube and diluted with 1 ml of 0·02% m/v Triton-X-100 (a non-ionic surfactant) and placed in a water-cooled ultrasonic bath for 10 min in order to effect haemolysis. 200 μl (0·2 ml) of the ultrasonically treated sample was then nebulized into an air–hydrogen flame maintained on a 'high solids' wide-slot air–acetylene burner. The 217·0 nm lead line was used with background correction being made at the lead 220·4 nm non-resonance line. The damped output (2 s time constant) was monitored on a pen recorder. The limit of detection was about 10 μg lead per 100 ml of whole blood.

Flameless electrothermal atomization (see p. 233) can also be used for the analysis of lead and cadmium in blood and urine.

The Estimation of Copper in Clinical Samples

Atomic absorption spectroscopy provides a reliable and convenient procedure for the estimation of copper in clinical samples. In the body the highest concentrations of copper occur in the liver, heart, kidneys, hair and brain. The element forms metalloproteins and is believed to participate in the utilization of iron.

The copper level in serum is reduced in several disorders including Wilson's disease, while evaporated serum copper is associated with acute leukemia, Hodgkin's disease and rheumatoid arthritis.

Increased urinary copper is characteristic of Wilson's disease and is a valuable indicator for diagnosing this disorder.

The Estimation of Copper in Serum

The normal values for copper in serum are 106 μg/100 ml (range 81–137) for men and 120 μg/100 ml (range 87–153) for women.

The estimation is readily performed at the 324·7 nm line upon a sample of serum diluted five times with deionized water. Scale expansion is necessary, and comparison can be made against straight aqueous standards containing 10, 20 and 50 μg/100 ml Cu.

The Estimation of Copper in Urine

The normal quantity of copper excreted in urine is 48 ± 16 μg in 24 h. Measurements directly in urine from subjects, therefore, require considerable scale expansion and are unreliable. For realistic estimation upon such samples it is necessary to resort to chelation with A.P.D.C. and concentrative extraction into M.I.B.K. (p. 23). The use of flameless techniques can be used for this estimation (see p. 233). With patients suffering from Wilson's disease the copper level in the urine is sufficiently elevated to permit direct estimation. It is essential to correct for background absorption.

The copper level in urine from patients undergoing therapy with chelating agents is also elevated.

Copper in soft Tissue

The estimation of copper in a liver sample taken by needle biopsy can be a valuable diagnostic test for Wilson's disease.

Between 0·5 and 1 g of tissue is accurately weighed and desiccated overnight to obtain a dry weight. The sample is ashed at 120°C with 10 ml of perchloric-nitric acid mixture; or can be digested with hydrochloric acid and hydrogen peroxide. The clear digest is transferred to a 25 ml graduated flask and diluted to the mark with deionized water. Comparison is made against aqueous standards covering the range 10 to 100 μg/100 ml Cu.

Estimation of Other Metals in Clinical Samples

The estimation of Zinc in Serum

Zinc occurs in a number of metalloenzymes and the normal range of the element in serum is 80 to 160 μg/100 ml. An elevated serum zinc level is associated with hyperthyroidism, hypertension and eosinophilia. Certain acute infections and malignant tumours can depress the serum zinc concentration.

Estimation is performed upon a sample of serum simply diluted ten times with deionized water and comparison is made against straight aqueous standards covering the range 5 to 20 μg/100 ml Zn.

Zinc in Urine

The normal quantity of zinc excreted in urine is 110 to 500 μg in 24 h. Grossly elevated quantities are excreted by patients with cancer of the prostate gland.

Due to the great variability in the quantity of this element excreted it is essential to utilize a 24 h sample for this analysis. The sample is acidified with hydrochloric acid. An aliquot is then diluted by a factor of five times and comparison made against straight aqueous zinc standards that cover the range 1 to 20 μg/100 ml Zn.

The estimation of Gold in Serum and Urine

The treatment of rheumatoid arthritis by gold injections is practised by a considerable number of hospitals. Effective monitoring of the gold level in serum and that excreted in urine is essential.

Atomic absorption provides a reliable and convenient procedure for this estimation, which would be difficult to perform by any other means.

For serum, comparison is made against straight aqueous standards, the serum samples being diluted 2, 5 or 10 times with 2 M hydrochloric acid. For estimations in urine, the standards can conveniently be prepared by the addition of gold to a known blank urine, so as to cover the range 25 to 200 μg Au/100 ml. The urine samples and standards must be acidified by addition of 2 ml hydrochloric acid (36% m/m) to each 100 ml of urine.

Correction for non-specific background absorption should be made.

Fig. 5.25 shows a typical estimation.

The estimation of Lithium in Serum and Urine

After Cade's work, in 1949, the use of lithium therapy to treat the manic phase of manic depressive psychosis became widespread in all countries except the U.S.A., where it took longer to gain acceptance. The therapeutic level for lithium in serum (3·5–12 mg/l) approaches the toxic level (14 mg/l) so that careful monitoring is essential.

The measurement of lithium in urine is a valuable guide to the maintenance of the dose.

Flame emission procedures are frequently used for this estimation but atomic absorption provides an equally convenient means for monitoring with the added advantage that it is less open to possible spectral interference effects.

Evaluation trials established that the presence of protein and other cations exerted negligible effect upon the lithium absorption. The estimation can,

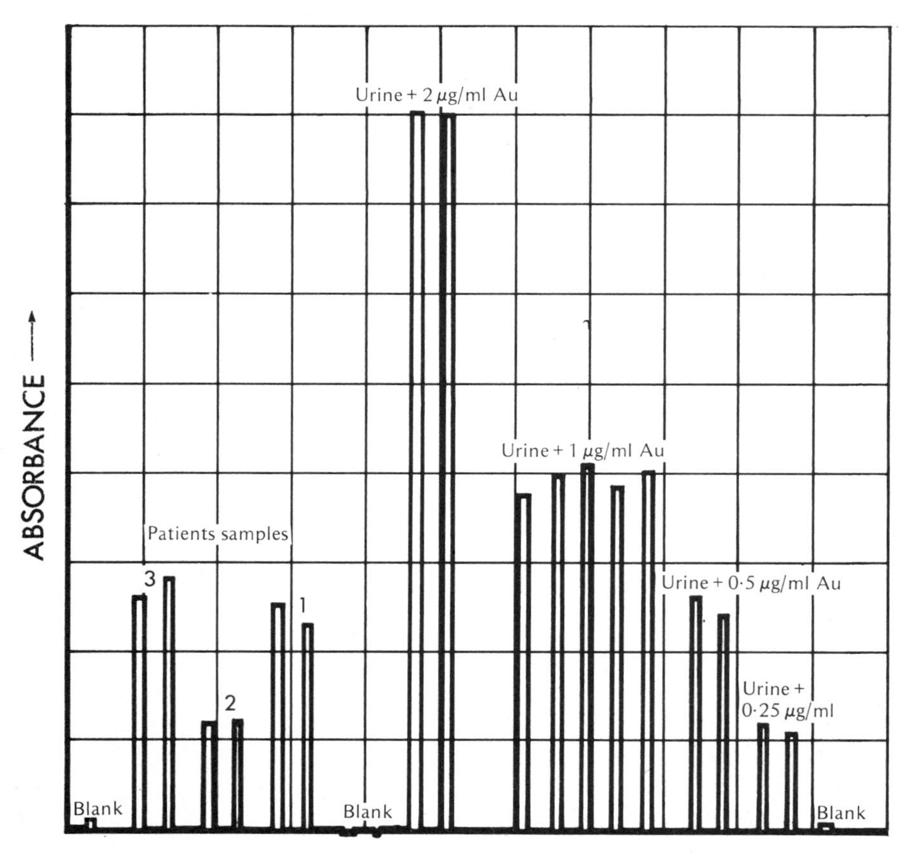

Fig. 5.25 Estimation of gold in urine from patients receiving treatment for rheumatoid arthritis. Integrated readings at about ×10 scale expansion

therefore, be performed directly upon sera diluted ten times with deionized water, comparison being made against straight aqueous lithium standards.

The estimations of Sodium and Potassium

Although these two important alkali metals can be reliably estimated in clinical samples by atomic absorption spectroscopy the technique is seldom used for their measurement. This is because their estimation by flame emission photometry is well established. In the clinical laboratory estimations of both elements are required in sufficiently large numbers to warrant a separate unit for the work. Over the fairly limited ranges at which the elements occur in clinical samples, automatic flame photometers capable of estimating both sodium and potassium simultaneously provide the most convenient and least costly means for their analysis.

Metallurgical Analysis, including Plating Solutions

Introduction—General Advantages

All the common metals, and some important non-metallic alloying constituents can be reliably estimated by atomic absorption spectroscopy. The technique is virtually immune to spectral interferences caused by overlap of resonance line profiles, so that sample preparation can generally be confined to mere dissolution, or extraction of the species to be determined. Handling errors are therefore minimized and analyses can be performed more rapidly than by procedures requiring more elaborate preparative steps. Metallurgical analysis was one of the first branches of analytical chemistry to exploit these outstanding practical advantages; and striking examples of estimations which would be difficult and unreliable to perform by any other procedure are provided by the estimations of cadmium in zinc, cobalt in nickel, and nickel in cobalt.

It is easier to specify the metals that cannot be estimated by atomic absorption spectroscopy than to enumerate fully the analyses that can be performed by the technique.

Arsenic is difficult to estimate with confidence at low levels using conventional flames, but reasonably reliable results can be obtained over the range 5 to 150 ppm in aqueous solution using the nitrous oxide–acetylene flame. However, the hydride generation technique is 500 to 1000 times more sensitive. Cerium cannot be satisfactorily determined directly by atomic absorption spectroscopy.[103] The sensitivities attainable for hafnium, zirconium, niobium, tantalum and uranium are insufficient for their direct estimations at trace levels in routine analysis by atomic absorption spectroscopy. This is primarily due to the stability of the metal oxide species. The energy available from the nitrous oxide–acetylene flame is insufficient to cause dissociation.

Some common metallurgical analyses that can be performed using atomic absorption are listed below. The list indicates not only the large number of metals that can be determined, but also the wide range of compositions over which determinations can be carried out.

1. Co, Al, Ni, Cu, Pb, Mn, B, Mo, Cr, V, Si and W in steels.
2. Al, Co, Fe in high-nickel alloys.
3. Cu, Sb, Sn in white metals.
4. Mg and Zn in cadmium metal.
5. Zn, Pb, Ni, Fe, Mg in high-copper alloys.
6. Cu, Pb, Ni, Fe, Cd, Zn, Mg, Mn, Sn in high-aluminium alloys.
7. Al, Cu, Zn in brass.
8. Cu, Zn, Mn in magnesium alloys.
9. Fe, Ca, Al, Mg in iron-ore sinter.
10. Au, Ag, Cu, Ni, Co, Zn, Rh, Pd, Pt and Fe in plating solutions and recovery liquors.

11. Traces of Pb, Cu, Ag, Ni in high-purity Bi, In and Ga, In/Ga and In/Al alloys.
12. Traces of Zn, Cu, Cd and Fe in tin/lead solders.
13. Traces of Sb in lead.
14. Si in Al alloys.
15. Cu, Sb, Sn and Pb in Babbit metal (SAE 13).
16. Cu, Pb, Sn, Zn in high-copper-bearing metal (SAE 798).
17. Au, Ni and Co in bright gold plate.
18. Al in aluminium bronze.

All of these analyses can be performed more rapidly and conveniently by atomic absorption spectroscopy than by classical wet procedures, and in many cases atomic absorption gives more reliable results.

Chemical Interference Effects

Although interference of one metal upon another, or of excess acid upon a metal's absorption may occur, these effects can usually be controlled or eliminated by the analyst. A rough guide to the interferences to be expected between metals are indicated in Chapter 4.

Interference-free Determinations

When it is required to develop an atomic absorption procedure for large numbers of similar analyses, e.g., the routine control of an alloy (particularly when only a few of the elements present in it are to be estimated) it is always worthwhile examining the possibility of using simple standard solutions. This is especially likely to be possible when the nitrous oxide–acetylene flame is being used, and it is worth noting that in the air–acetylene flame Ni, Zn, Fe, Cu and Pb are remarkably free of interferences from elements with which they are likely to be found in association.

The analyst is indeed sometimes prohibited from using complex standards; an example of this situation occurs in the trace determinations of copper, cadmium, iron and zinc in tin/lead solders. These estimations are complicated by the fact that even the purest sources of lead and tin (readily available to the chemist) contain substantial quantities of cadmium and zinc, so that the preparation of 'matching standards' that contain the two major constituents of the alloy is impracticable. A series of recovery trials, using the method of additions, though, established that accurate results could be obtained merely by making comparison against standards that contained copper, cadmium, iron and zinc in concentrations that covered the ranges of these elements in the test solutions.

The procedure employed to establish this fact is outlined below.

The trace metals are usually present in the solders at the levels shown, which, when 2 g are dissolved and the resulting solution adjusted to 100 ml, give rise to the concentrations indicated in the test solution.

Element	% in solder	μg/ml in test solution
Cu	0·01–0·03	2–6
Cd	0·001	0·2
Fe	0·001–0·002	0·2–0·4
Zn	0·001–0·002	0·2–0·4

Test solutions of the alloy(s) are prepared by dissolving 2 g in either hydrochloric or nitric acid, followed by filtration from the precipitated lead chloride or metastannic acid. The precipitate is washed with a hot 0·1 M solution of the relevant acid and the washings added to the filtrate.

A 'straight' test solution is prepared as above, and diluted to 100 ml in a graduated flask. This is solution 'A'.

A second test solution, 'B', having further additions of 200 μg Cu, 100 μg Cd and 100 μg Zn made to it, is also prepared. The concentrations of copper, cadmium, and zinc in solution 'B' are thus in excess of those in solution 'A' by 2 μg/ml Cu, 1 μg/ml Cd and 1 μg/ml Zn.

Standard solutions are prepared as follows:

Standard solution I (μg/ml)		Standard solution II (μg/ml)		Standard solution III (μg/ml)	
Cu	1·0	Cu	4·0	Cu	8·0
Cd	0·1	Cd	0·2	Cd	0·5
Zn	0·2	Zn	0·4	Zn	0·8

The standard and test solutions are now aspirated to the atomic absorption spectrophotometer and the results obtained for the concentrations of the three elements, as derived from calibration curves constructed from the responses for the simple standard solutions, compared with those obtained from solutions 'A' and 'B' by the method of additions (p. 28). If these two sets of results are identical, as they proved to be in this case, it may be concluded that the use of simple standards is valid for the determination.

General Procedure

The most general, reliable and convincing procedure used by the atomic absorption spectroscopist for the analysis of alloys is to prepare a range of standards that approximately match in composition the anticipated test solution that will be obtained from the alloy to be analysed.

The standard solutions may be 'synthetically' prepared from pure single-element master solutions, but in many cases, for example, the estimations of silicon in aluminium alloys, alloying elements in steel, etc., it is more convenient and reliable to prepare them from British Chemical Standard Alloys, by the same procedure used for preparing the test sample.

An example of an analysis for which this general procedure offers the most suitable approach is provided by the estimation of the minor constituents of a Duralumin alloy.

The sample is prepared by dissolving 0·5 g of the alloy in hydrochloric acid, and adjusting the volume accurately to 1 litre.

The essential data for this estimation is summarized in the following table:

Element	Concentration in alloy		Optimum range for estimation (μg/ml)	Concentration in solution 0·5 g alloy to 1 litre (μg/ml)
	%	ppm		
Cu	4·42;	44 200	0·5 –10	22·1
Mg	0·74;	7 400	0·05–1	3·7
Mn	0·73;	7 300	0·25–5	3·65
Fe	0·40;	4 000	1–20	2·0
Zn	0·11;	1 100	0·1–2	0·55
Al	92·55;	925 500		462·8
Other Constituents	1·05;	10 500		

Standard solutions are prepared that match the constitution of this test solution, and cover the anticipated concentration ranges of the elements in it; as shown in the second table.

Element	Standard solution I (μg/ml)	Standard solution II (μg/ml)	Standard solution III (μg/ml)
Cu	20·0	22·5	25·0
Mg	2·0	3·0	4·0
Mn	2·0	3·0	4·0
Fe	1·0	2·0	3·0
Zn	0·2	0·5	1·0
Al	400·0	450·0	500·0

(The minor constituents whose analyses are not required are omitted from these standards. Aluminium, although its analysis is not required, is included as the major possible source of interference.)

The estimations of all the elements can now be made by aspirating the standard and test solutions to the atomic absorption spectrophotometer, constructing concentration/response curves and reading off the concentrations of the test solutions.

The estimations of iron, manganese and zinc are most conveniently made with the air–acetylene flame. Scale expansion can be used to improve the accuracy for the iron determination. Burner rotation can be used to reduce the sensitivity of the instrument for the copper determination, or alternatively measurement can be made at the less absorbing 327·5 nm line, so as to avoid the necessity for further dilutions. Magnesium is best determined with the nitrous oxide–acetylene flame, and again burner rotation can conveniently be used.

This general method for performing the analysis of alloys by atomic absorption possesses the following advantages:

1. The number of standard solutions is much fewer than if a range were prepared for each element separately and
2. any slight interelement effects are compensated for.

It is particularly useful for non-routine analyses.

The determination of Silicon in Aluminium and Aluminium alloys

Aluminium alloys can contain silicon over a range from about 1 per cent to 15 per cent. The silicon content influences the properties of such alloys and it is, therefore, necessary to control its level accurately.

The estimation of silicon in aluminium alloys is, thus, of considerable interest to the non-ferrous metallurgist. Classical procedures for this analysis are tedious, but atomic absorption provides a convenient, rapid and reliable means of determination.

Procedure. Weigh a sample of the alloy (0·25 g) in the form of turnings or filings into a stainless steel beaker. Add 10 ml of 10 M sodium hydroxide solution and gently heat the mixture on a hotplate until a paste is formed. The beaker should be covered with a polythene disc to prevent spray losses.

Add 35 ml of deionized water, boil gently for about 1 min, cool and add 60 ml of 1 + 1 nitric acid + water. Warm to dissolve the aluminium hydroxide, cool and transfer the solution to a 250-ml graduated flask. Dilute to the mark with deionized water.

A more effective solution for the attack and dissolution of some high-silicon aluminium alloys can be prepared as follows:

Nitric Acid (70% m/m)—150 ml.
Hydrofluoric Acid (40% m/m)—20 ml.
Mercurous Nitrate—0·01 g.
Deionized Water to make final volume up to 200 ml.

The solution is stored in polythene. The dissolution should be carried out in

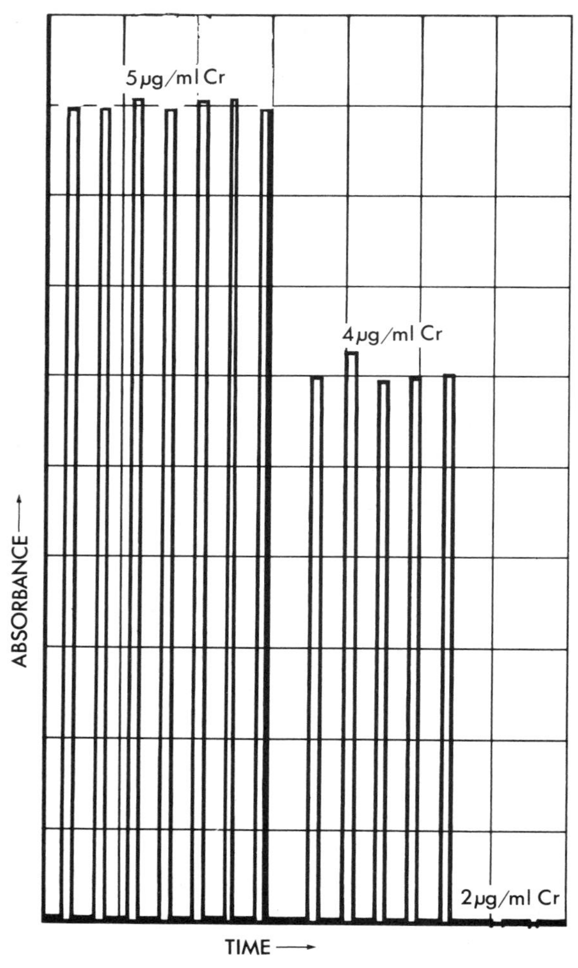

Fig. 5.26 Typical integrated responses for the estimation of chromium in stainless steel. Stainless steel contains about 18% Cr. Effective dilution for the estimation is 50 000 times. Responses for the lowest standard suppressed to read zero × 10 scale expansion then applied. Standard deviation calculated from the response at 4 μg/ml level is 0·04 μg/ml, leading to a precision of 0·2% at the 18% level

P.T.F.E. beakers. 20 ml of this solution will dissolve 0·5 g of an aluminium alloy. When dissolution is complete add 30 ml of a 5-per-cent solution of boric acid to neutralize the excess hydrofluoric acid, transfer to a graduated flask (250 ml) and dilute to the mark with deionized water.

Standard solutions are prepared from British Chemical Standard alloys.

The instrumental conditions for silicon determinations are given on p. 70.

For low-level silicon, it is possible to nebulize a 2·5% m/v aluminium solution using the pulse nebulization technique (see p. 201).

THE DETERMINATION OF CHROMIUM IN STEEL

The determination of chromium in steel is recognized as one of the most difficult metallurgical analyses to perform reliably by atomic absorption spectroscopy. It provides an example of an estimation which has been made easier by the improvements in hollow cathode lamps (see p. 45).

The two most absorbing lines in the chromium spectrum are located at 357·9 and 359·4 nm.

In the air–acetylene flame, the presence of iron strongly depresses the absorption of chromium, and this depression is itself dependent upon the acidity of the solution and fuel richness of the flame.

With the nitrous oxide–acetylene flame, iron does not interfere with the absorption of chromium, although excess acid in the solutions does still exert an effect.

Procedures for the estimation of chromium in steel that employ the air–acetylene flame, and attempt to overcome the interference of iron by the addition of ammonium or lanthanum chloride, have been described. These procedures can give reliable results, but are not absolutely infallible in all circumstances.

Using the nitrous oxide–acetylene flame comparison against standards prepared from pure solutions of iron and chromium salts or even straight chromium solutions can under most, but not absolutely all, conditions lead to acceptable estimations.

The most generally satisfactory procedure that we have as yet examined for this important determination is to use the nitrous oxide–acetylene flame and to make comparison against standard solutions prepared from B.C.S. steels of similar composition to the material under test.

The chromium content of most common steels varies from 1 to 19 per cent. The optimum range for chromium estimation, with the nitrous oxide–acetylene flame in aqueous solution, is between 2 and 20 μg/ml at 357·9 nm, and the sample size is selected so that the test solutions will contain chromium at this concentration.

Figure 5.26 shows the instrumental precision attainable for a high-chromium steel in the nitrous oxide-acetylene flame. The technique of zero suppression followed by scale expansion was employed.

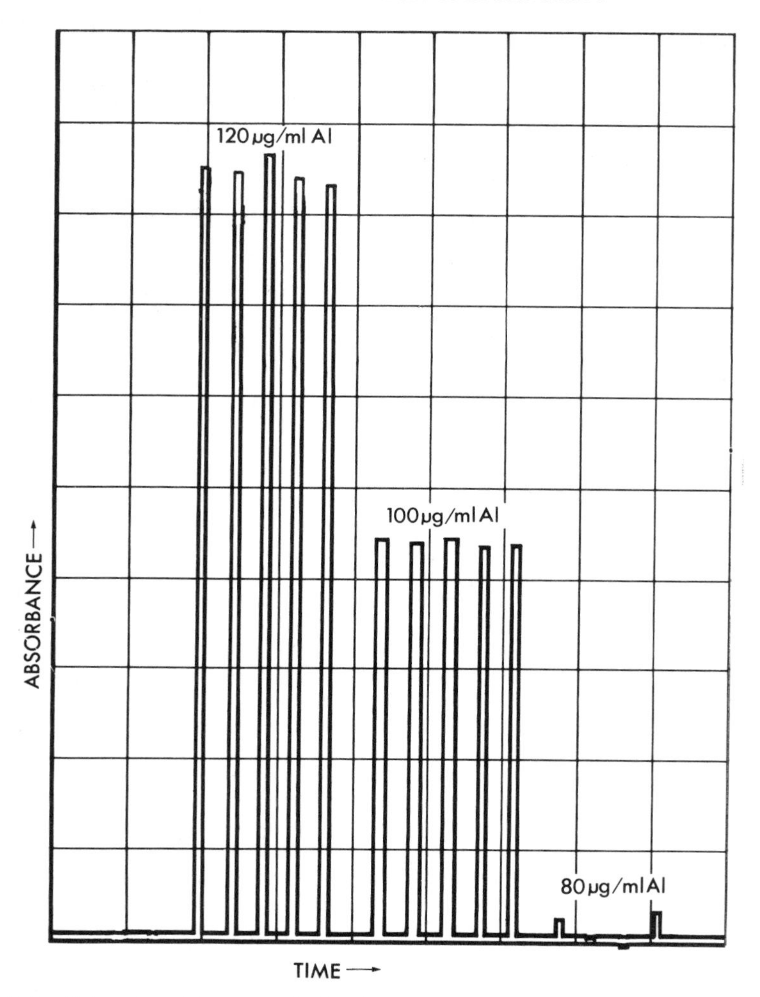

Fig. 5.27 Estimation of aluminium in aluminium bronze. Integrated mode. Aluminium bronze contains about 10% Al. Effective dilution required for estimation is 1000 times. Response for lowest standard suppressed to zero. Scale expansion applied. Standard deviation calculated from responses at 100 μg/ml level is 1·5 μg/ml. Precision at 10% level is therefore about 0·15%

Estimations of the Noble Metals in Alloys and Plating Solutions

The absorption for the noble metals might reasonably be expected to be free from chemical interference effects, but in fact very pronounced interferences do occur, especially with cooler flame systems. These interferences are detailed in the data for the individual elements in Chapter 4.

No proven explanation for the mechanism of these interferences has been advanced. It is necessary for the plating and refinery assayer to be aware of their existence and to develop *ad hoc* procedures that allow him to perform his specific analyses with confidence.

Estimations of gold in plating solutions are particularly prone to error when sample preparation is confined to simple dilution and acidification. Reliable results can frequently be obtained by acidifying an aliquot with hydrochloric acid evaporating to dryness and redissolving the residue in aqua regia.

Tables 5.3 and 5.4 below, summarize the interelement effects of some common metals upon some noble metals with which they are likely to be found in association. They are only a general guide.

Also see Chapter 4 references, nos 34–42.

Table 5.3 Some interferences upon noble metals in the air–acetylene flame

	Noble metal, major anion present, and effect of interferant					
	Gold		Palladium	Platinum	Rhodium	Iridium
Interferant	Cl^-	CN^-	Cl^-	Cl^-	Cl^-	Cl^-
Cu	n	E	n	D	E	D
Au	—	—	n^+	n	E	n
Ir			D	D	n	—
Fe	D	n^+	n^-	D	D	D
Ni	D	n	D	D	D	D
Pd	D	D	—	D	E	D
Pt	n^-	n^+	n^-	—	E	n
Rh			n^+	D	—	D
Zn	n^-	n^+	n	n^+	n	D

Code: n Negligible interference
n^+ Slight enhancement
n^- Slight depression
E Enhancement
D Depression

Table 5.4 Some interferences upon the noble metals in the N_2O–acetylene flame

	Noble metal, major anion present, and effect of interferant					
	Gold		Palladium	Platinum	Rhodium	Iridium
Interferant	Cl^-	CN^-	Cl^-	Cl^-	Cl^-	Cl^-
Cu	n	n	n	n	n	n
Au	—	—	E	n	n	n
Ir			E	D	D	—
Fe	D	n	n	D	n	D
Ni	n^-	n^-	n	D	D	D
Pd	n	D	—	D	n	D
Pt	n	D	n	—	n	
Rh			E	D	—	D
Zn	n	n	n	n	n	D

Code: n Negligible interference
D Depression
E Enhancement

ANALYSIS OF HIGH-SILVER ALLOYS. POTENTIAL HAZARDS DUE TO FORMATION OF SILVER ACETYLIDE[141]

When solutions containing silver at high concentrations, as for example, in the estimation of trace impurities in silver metal, are aspirated to an acetylene flame system silver acetylide can be formed in the mixing chamber of the instrument. The formation of this unstable compound occurs in the aerosol phase even when the solutions contain excess nitric acid.

For many applications (e.g., the estimations of copper and lead) the air–hydrogen system can be used to avoid this hazard. If it is essential to use acetylene as fuel the silver should be removed before aspirating the test solution to the atomic absorption spectrophotometer.

SOME EXAMPLES THAT ILLUSTRATE THE CAPABILITY OF STANDARD A.A.S. EQUIPMENT IN METALLURGICAL ANALYSES

Traces that indicate the instrumental precision attainable with standard atomic absorption equipment for a number of fairly common metallurgical analyses are reproducible below (Figs 5.26–5.29).

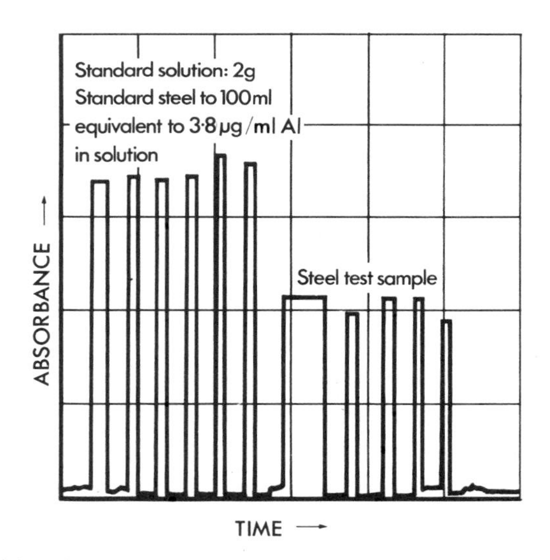

Fig. 5.28 Estimation of aluminium in steel. Typical integrated responses at × 10 scale expansion

Brief details of the preparative and instrumental conditions are stated in the captions.

Another useful technique, especially for the determination of trace elements (e.g., Al, As and Sn, etc.) in steel or other alloys, is the pulse nebulization technique (see p. 201).

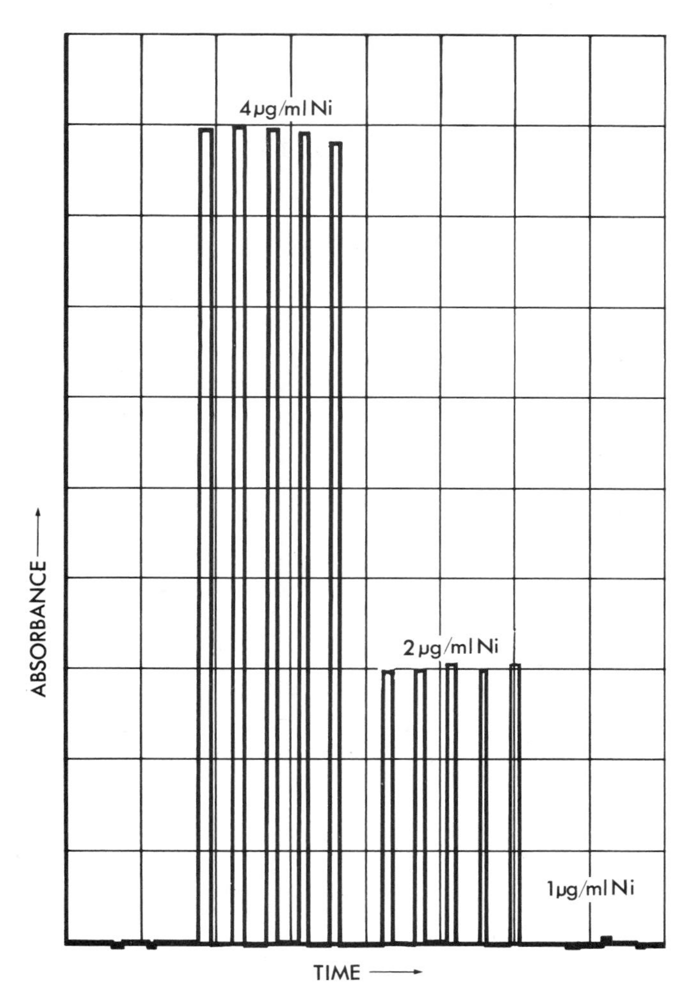

Fig. 5.29 Estimation of nickel in stainless steel (232·0 nm). Integrated mode. Stainless steel contains about 8% Ni. Effective dilution for estimation is 50 000 times. Response for lowest standard suppressed to read zero. Scale expansion then applied. Standard deviation calculated from responses for 2 μg/ml standard is 0·018 μg/ml. Precision at 8% level is 0·075%

A novel 'reagentless' dissolution method for metallic samples has been described by GHIGLIONE and co-workers.[156] A water immersed spark was used to produce a colloidal type suspension of the metal or alloy in water. The suspension was then directly nebulized into the nitrous oxide-acetylene flame. The suspensions were reasonably stable and for a given metal concentration gave similar signals to conventional aqueous standards.

Silicate Analysis, including Glass, Ceramics, Coal Ash, Minerals, etc.

Note: For methods involving the direct analysis of solid materials see page 235.

The analysis of ceramic materials provides an outstanding example of a field where the application of atomic absorption has been continually extended by the determination of users to exploit the advantages of speed and simplicity that it offers.

Initially, it was anticipated that the technique would provide a convenient, rapid and precise method for the estimations of calcium, magnesium and perhaps iron, present at the 0·1 to 2 per cent level in the silicate materials. This objective was effectively accomplished, as also were the estimations of sodium and potassium over the same concentration ranges. Having enjoyed so much success it was inevitable that some of the more demanding workers in the field would attempt to apply atomic absorption to the estimation of the major constituents, alumina and silica.

For these macro estimations the ceramic samples are most conveniently attacked by fusion with a mixture of sodium carbonate and boric acid. The cooled melt is dissolved in diluted hydrochloric acid, adjusted to analytical volume and aspirated to the atomic absorption spectrophotometer.

Comparison is made against either 'synthetic standards' or solutions prepared from British Chemical Standard materials. (Lithium metaborate is considered by some workers to be superior to sodium carbonate–boric acid mixture as the attacking flux.)[169,170]

The precision attainable for the estimations of silica and alumina in ceramic materials by atomic absorption spectroscopy is of course lower than that of classical procedures. (0·4 per cent for Al_2O_3; and 0·7 per cent for SiO_2.) It is, nevertheless, surprisingly impressive and adequate for many purposes. Figures 5.30 and 5.31 indicate the performance levels attainable.

A scheme of analysis based upon the principles outlined above for the estimations of SiO_2, Al_2O_3, Fe_2O_3, CaO and MgO is described below.

THE DETERMINATION OF SiO_2, Al_2O_3, Fe_2O_3, CaO AND MgO IN SILICATES AND HIGH ALUMINA REFRACTORIES

Reagents

1. *Lanthanum Stock solution (65 000 μg/ml La).* Dissolve 76·22 g of lanthanum oxide (calcium-free grade) La_2O_3 in 150 ml of hydrochloric acid (36% m/m). Make up to 1000 ml with deionized water.
2. All other reagents, Na_2CO_3 and HCl must be of AnalaR grade or its nearest equivalent in purity.
3. Deionized water should be used as the diluent.

Standards. For this analysis standard solutions are best prepared from British

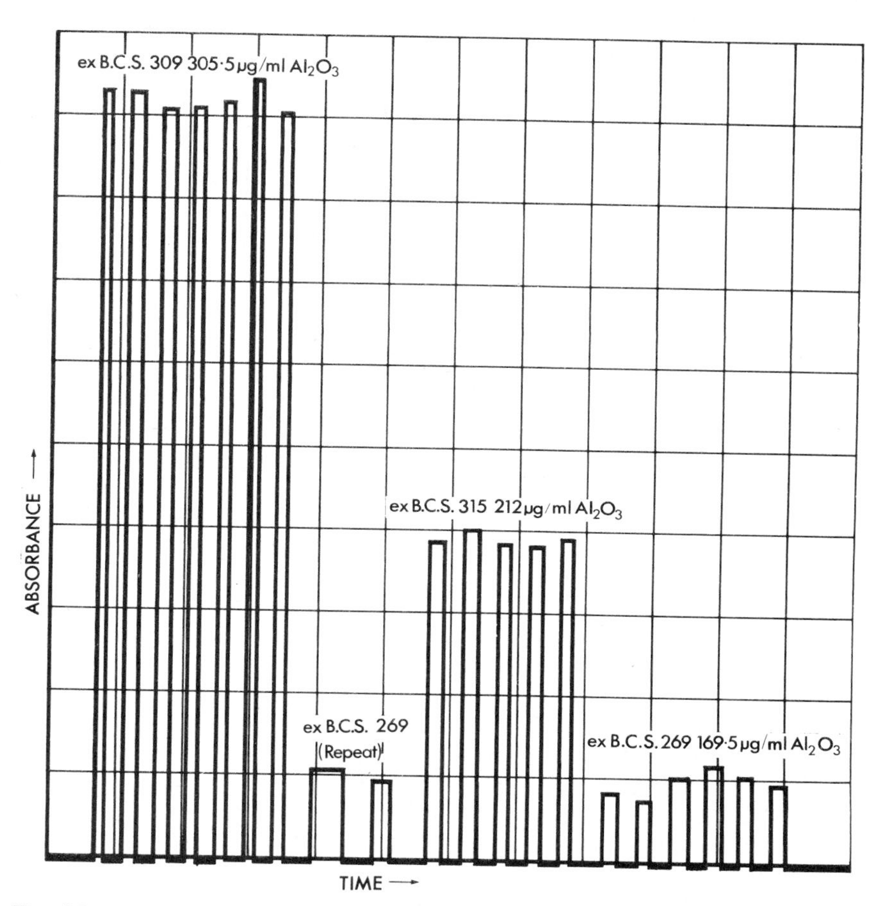

Fig. 5.30 Estimation of Al_2O_3 in refractories. Integrated mode. Effective dilution for the estimation is 2000 times. Response for lowest standard suppressed to read about 10 chart divisions. Scale expansion applied. Standard deviation calculated from the response for B.C.S.269 is 2 μg/ml Al_2O_3. Precision at level in refractory is about 0·4%

Chemical Standard materials, by the same procedure used for the test samples. Suitable B.C.S. materials are:

B.C.S. No. 269 Firebrick
B.C.S. No. 309 Sillimanite
B.C.S. No. 315 Firebrick

Sodium Carbonate, Boric Acid Matching Solutions. Dissolve a mixture of 100 g of sodium carbonate (Na_2CO_3) plus 20 g of boric acid (H_3BO_3) in the minimum

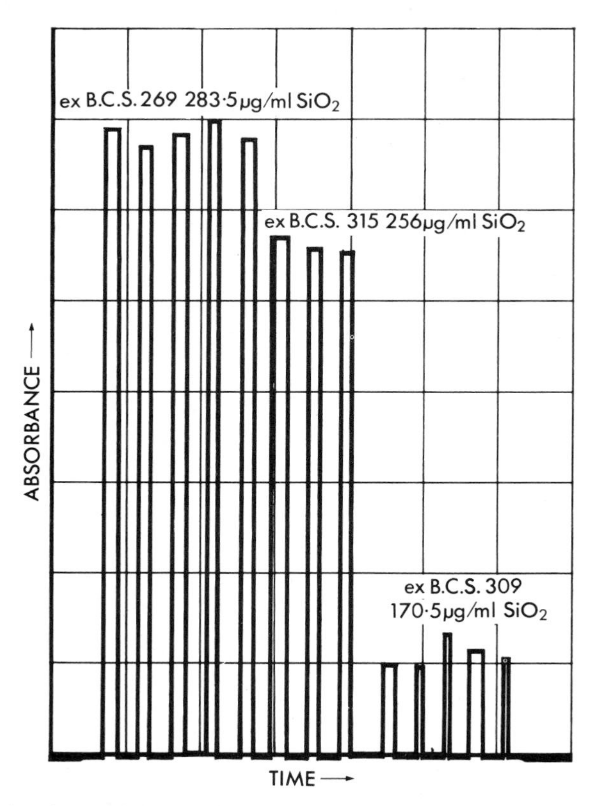

Fig. 5.31 Estimation of SiO_2 in refractories. Integrated mode. Effective dilution for the estimation is 2000 times. Response for lowest standard suppressed to read about 10 chart divisions. Scale expansion then applied. Standard deviation for B.C.S.309 is 3·5 μg/ml. This is equivalent to a precision of 0·7% at the SiO_2 level in the refractory.

quantity of 1 + 1 concentrated hydrochloric acid + water. Transfer to a 1 litre graduated flask and dilute to the mark with deionized water.

Blank solution (for estimations of CaO, MgO, Fe_2O_3). Transfer 5 ml of the above matching solution to a 100 ml graduated flask. Add 10 ml of lanthanum chloride stock solution and dilute to the mark with deionized water.

Preparation of Silicate materials and Standards. Weigh 0·1 g of the powdered silicate into a platinum crucible. Mix with 0·5 g of Na_2CO_3 plus 0·1 g of H_3BO_3. Fuse at 1000 to 1200°C for about 10 min. Cool the crucible and add 10 ml of water plus 4 ml of hydrochloric acid (36% m/m). Warm gently to assist dissolution.

Transfer the solution to a 200-ml graduated flask; add 20 ml of lanthanum chloride solution (65 000 μg/ml La) and make up to the mark with deionized water.

The conditions for measurement for the various elements (wavelength, optimum range, flame, etc.) are detailed in Chapter 4. The procedure of zero suppression followed by scale expansion is used for the estimations of alumina and silica. To avoid the interference of aluminium upon calcium and magnesium the nitrous oxide–acetylene flame is employed for their estimations.

The above procedure permits the estimation of SiO_2. It is also necessary to resort to the fusion technique for the estimation of Al_2O_3 in high-alumina refractories, due to the sparingly soluble nature of aluminium fluoride which is produced if the hydrofluoric acid attack described below is used.

The determination of Al_2O_3 (below 20%), Fe_2O_3, CaO, MgO, Na_2O and K_2O in Silicates

The method consists essentially of digesting the mineral with hydrofluoric–perchloric–nitric acid mixture, so that silicon is volatilized as the tetrafluoride. The residue is dissolved in hydrochloric acid and lanthanum chloride added to overcome suppression of the calcium response by aluminium. This solution is adjusted to volume and used for determinations.

Procedure. The procedure is best illustrated by considering the analysis of a typical earthenware body, which is made up as follows:

SiO_2	71·25%		
Al_2O_3	18·00%	Al	9·52%
K_2O	1·50%	K	1·25%
CaO	0·80%	Ca	0·57%
Na_2O	0·75%	Na	0·56%
Fe_2O_3	0·50%	Fe	0·35%
MgO	0·30%	Mg	0·18%
TiO_2	0·25%		
Loss on ignition	6·65%		

If 0·5 g of such a body is attacked and the contents eventually placed in 500 ml of solution, that solution will contain

Al	95·2 μg/ml
K	12·5 μg/ml
Ca	5·7 μg/ml
Na	5·6 μg/ml
Fe	3·5 μg/ml
Mg	1·8 μg/ml

Reagents 1 + 1 nitric acid 70% m/m + water
1 + 4 perchloric acid 60% m/m + water
Hydrofluoric acid 40% m/m (A.R.)
Hydrochloric acid 36% m/m (A.R.)
Lanthanum chloride solution containing 65 000 μg/ml lanthanum.

Sample Preparation. Weigh 0·5 g of the silicate into a platinum crucible, and treat with 10 ml of hydrofluoric acid, 5 ml of 1 + 1 nitric acid + water, and 5 ml of 1 + 4 perchloric acid + water.

Evaporate the digest almost to dryness on a sand bath, treat the residue with a further 5 ml of 1 + 4 perchloric acid + water, and again evaporate to near dryness.

Add about ten drops of hydrochloric acid (36% m/m) to the crucible followed by 15 ml of water, and digest on a steam bath for 20 min.

Cool the solution, and transfer to a 500-ml graduated flask. Add 50 ml of lanthanum chloride solution, and make up to the mark with deionized water.

Standard Solutions. Prepare complex standard solutions. For the earthenware body described above, convenient standards would have the following compositions:

Standard solution I (μg/ml)		Standard solution II (μg/ml)		Standard solution III (μg/ml)	
La	6500	La	6500	La	6500
Al	50	Al	100	Al	150
K	5	K	10	K	15
Ca	2·5	Ca	5	Ca	7·5
Na	2·5	Na	5	Na	7·5
Fe	2	Fe	5	Fe	10
Mg	0·5	Mg	1·0	Mg	2·0

Standard Solutions could even more conveniently be prepared from suitable British Chemical Standard (B.C.S.) Materials by the same procedure used for attack of the test samples.

Determination. The determinations of all the elements are straightforward, and the analyst is referred to Chapter 4 for specific details of their characteristics. For this application it is necessary to use the nitrous oxide–acetylene flame for the estimation of calcium and magnesium, in order to avoid the chemical interference of aluminium.

THE DETERMINATION OF Al_2O_3 (UP TO 50%), Fe_2O_3, CaO, MgO, Na_2O, TiO_2, SiO_2, MnO IN SILICATES BY USE OF A PRESSURE DIGESTION VESSEL

PRICE and WHITESIDE[167] have devised a rapid dissolution method that will rapidly dissolve silicate samples containing up to 50% m/m of ignited alumina. For a full description of the method and the wide range of application of this method, the interested reader is urged to consult the original paper. The basic method is as follows:

Reagents. Aqua regia
[1 + 3 nitric acid (70% m/m) – hydrochloric acid (36% m/m)]
Hydrofluoric acid (40% m/m)
Boric acid 4% m/v solution
Caesium chloride 2% m/v solution
or Potassium chloride 2% m/v solution.

Sample preparation. 0·2 g of the sample is ground to pass a 200-mesh sieve and transferred to a polytetrafluoroethylene (P.T.F.E.) lined stainless steel pressure vessel. 5 ml of water, 2 ml aqua regia and 1 ml hydrofluoric acid are added. The vessel is then sealed and heated to 160°C for 30 minutes and allowed to cool. It is then opened and 10 ml of 4% m/v boric acid solution is added. It is then sealed and reheated at 160°C for 20 minutes. It is again cooled and the contents transferred to 100 ml calibrated flask and 5 ml 2% m/v of a caesium or potassium chloride solution added (to suppress ionisation), prior to dilution to volume with deionised water. In the presence of fluoroboric acid there is no requirement to add lanthanum as a releasing agent when the nitrous oxide-acetylene flame is used. The method has also been successfully adapted for use with microsamples (~10 mg).

Standard solutions. Prepare complex standard solutions containing an equivalent amount of hydrofluoric acid, boric acid and potassium or caesium chloride as the sample solution. B.C.S. materials can be used (see page 154).

Determination. The nitrous oxide-acetylene flame should be used for all elements and the analyst is referred to Chapter 4 for specific details of their characteristics.

Analysis of Cement*

To a first approximation, the composition of a typical cement is as follows:

CaO	64%	Ca	46·45%
SiO_2	21%		
Al_2O_3	6%	Al	3·2%
Fe_2O_2	3%	Fe	2·1%
MgO	1%	Mg	0·60%
SO_3	1%		

* The pressure digestion vessel method described on page 155 is eminently suitable for this analysis.

K_2O	1%	K	0·83%
Na_2O	0·25%	Na	0·19%
TiO_2	0·25%		
P_2O_5	0·25%		
MnO	0·10%	Mn	0·08%
Loss on ignition	1·15%		

Of these constituents aluminium, iron, calcium, magnesium, sodium, potassium, and manganese can be determined in solution by atomic absorption spectroscopy, without separation from other constituents or from one another. Silica can also be determined, but in order to do so it is necessary to get the element into solution. It has been suggested that the lithium borate fusion, outlined in the section on silicate analysis, should prove suitable for this step.

In an alternative procedure[183] that has been described for the determination of silicon in cements, a 0·4 g sample is weighed into a polythene beaker, wetted with 10 ml of water and attacked with 6 ml of (36% m/m) hydrochloric acid and 0·5 ml of (40% m/m) hydrofluoric acid. Twenty ml of a solution containing 2·5% m/v vanadium trichloride is 0·2M hydrochloric acid and 10 ml of a 5000 μg/ml Na solution are added to the digest, which is then made up to 100 ml in a calibrated polythene flask. Measurement is made against standards comparable with the test solution. The other constituents of the cement can also be determined.

Principle of a Method in which SiO_2 is not determined

For a cement approximating to the above composition, 1 g of cement is digested with hydrochloric acid and the solution filtered from the insoluble silica. The solution is adjusted to a volume of 200 ml with deionized water, and will then contain:

Ca	2322·5 μg/ml
Al	160 μg/ml
Fe	105 μg/ml
K	41·5 μg/ml
Mg	30 μg/ml
Na	9·5 μg/ml
Mn	4 μg/ml

Complex standard solutions are prepared to resemble the cement extract, and to cover the anticipated range of concentrations for the elements.

Aluminium and manganese can then be determined by spraying the solution prepared from the cement against these standards.

By diluting the complex standards and test solution(s) ten times, and adding lanthanum to suppress possible interference from aluminium on the magnesium response, the estimations of iron, magnesium, sodium, and potassium may be effected.

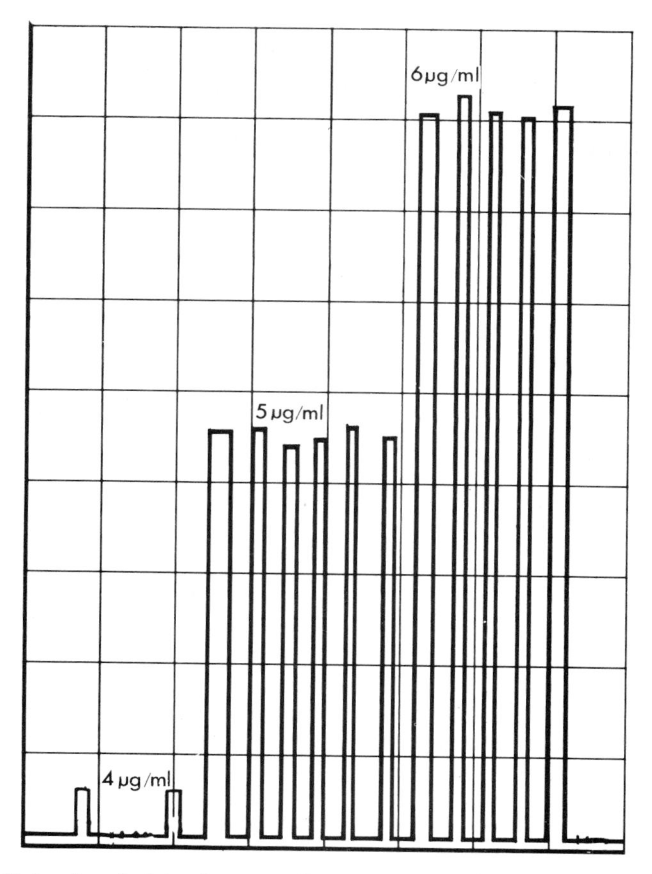

Fig. 5.32 Estimation of calcium in cement. Integrated mode. Effective dilution for estimation is 100 000 times. Response for 4 μg/ml solution suppressed to near zero. Standard deviation from response at 5 μg/ml level is about 0·02 μg/ml, leading to a precision of 0·2% at the Ca level of the cement

The determination of calcium, to the required degree of precision, is the most difficult feature of this analysis. Experimental work established that the optimum concentration range for the determination of calcium was, for this application, between about 4 and 6 μg/ml in solution. In order to achieve an accuracy of $\pm$1% on the CaO content (i.e., $\pm$0·7% on the calcium content of the cement) it is necessary to be able to discern differences better than 0·07 μg/ml calcium in the test solutions. This requirement is achieved by employing the procedure described in Chapter 3 in the section on the determination of metals present as macro constituents of a sample.

Procedure

Sample Preparation (taken from C.E.R.I.L.H. technical publication No. 172, p. 32)

Reagents Hydrochloric acid (A.R.) 36% m/m
Gelatine solution 2·5% m/v

Weigh 1 g of cement into a 250-ml tall-form beaker. Add 15 ml of hydrochloric acid and, when the initial reaction has subsided, boil gently for about 6 min. Cool to 70°C and add 5 ml of gelatine solution also at a temperature of 70°C. Filter the flocculated silica from the clear liquor, wash the precipitate with hot water, cool, transfer the filtrate and washings to a 200-ml graduated flask and dilute to the mark. This solution is the master extract.

Standard Solutions. Prepare complex standard solutions. For the cement described above convenient standards would have the following compositions:

Standard solution I (μg/ml)		Standard solution II (μg/ml)		Standard solution III (μg/ml)	
Ca	2000	Ca	2500	Ca	3000
Al	100	Al	150	Al	200
Fe	50	Fe	100	Fe	200
K	10	K	20	K	50
Mg	10	Mg	20	Mg	50
Na	5	Na	10	Na	20
Mn	1	Mn	2	Mn	5

The determination of the elements other than calcium is straightforward. Aluminium and manganese would be determined on the sample solution, prepared as described above, by comparison with the standard. Magnesium, iron, sodium, and potassium would be determined upon aliquots prepared by transferring 20 ml of the master extract and above standards, to 200 ml graduated flasks, adding to each flask 20 ml of stock lanthanum chloride solution (65 000 μg/ml lanthanum) and diluting to the mark with deionized water.

For this application the magnesium determination is best carried out with the nitrous oxide–acetylene flame system.

Preparation of Test and Standard Solutions for Calcium Determination. Re-dilute the diluted extracts prepared for the determinations of iron, etc., fifty times to yield solutions containing from 4–6 ppm calcium. Pipette 5 ml of each diluted extract and diluted standard to a separate 200 ml graduated flask. Add 20 ml of lanthanum chloride stock solution (65 000 μg/ml lanthanum) and dilute to the mark with deionized water.

The atomic absorption determination is then carried out with the nitrous oxide–acetylene flame, utilizing scale expansion in conjunction with suppression of the zero, so as to set the 4 μg/ml standard to zero on the expansion unit and the 6 μg/ml standard to full scale, as described in Chapter 3 (method 7).

It is advantageous if the expanded signal can be fed to a chart recorder. Under normal routine conditions an accuracy of about 0·2 per cent upon the Ca content of the cement is attainable. Fig. 5.32 shows a typical trace.

Petroleum Analysis

Introduction

A knowledge of the concentrations of metallic constituents of oils is essential to the petroleum technologist. Such constituents can be divided into three main categories.

1. Additives, e.g., Ca, Ba, Mg and Zn.
2. Injurious impurities, e.g., Na, Al and V in crude and fuel-oils.
3. Wear metals, e.g., Fe, Ni, Cu, Cr, etc.

One of the major advantages that atomic absorption spectroscopy offers is that solutions need not be confined to aqueous media, so that sample preparation can be limited to dispersion of a weighed aliquot of oil in a suitable solvent, such as a mixture of 1 part propan-2-ol and 9 parts white spirit.

Time-consuming ashing, digestion and extraction procedures, necessitating the running of reagent blanks, and prone to handling losses, are thus avoided.

General Procedure

Diluents

Oils and greases must be diluted with a suitable solvent before they can be aspirated to an atomic absorption spectrophotometer. References will be found in the older literature that recommend the use of hydrocarbons such as iso-octane (2,2,4-trimethyl pentane), heptane and xylene for this dilution. Such solvents in our experience are frequently unsuitable since they give rise to such a rich smoky flame that it is found necessary (in order to establish non-luminous conditions) to restrict the acetylene flow to the point where the flame lifts off the burner.

The use of an oxygenated diluent such as cyclohexanone, 1 + 9 propan-2-ol + white spirit or 1 + 9 MIBK + iso-octane overcomes this difficulty. With these diluents, solutions may be prepared that allow determinations to be carried out safely with both the air–acetylene and nitrous oxide–acetylene flames.

A generally applicable procedure for the analysis of crude, fuel and lubricating oils is given in 'I.P Standards for Petroleum and its Products'.[201]

The diluent recommended is 1 + 9 propan-2-ol + white spirit (B.S. 245: 1956).

It is essential that white spirit of this or an equivalent specification is used since this solvent can vary in composition. Some grades contain high levels of aliphatics and are unsuitable for atomic absorption work.

The standard solutions are prepared from oil-soluble compounds, or from previously analysed oils. Ideally the compound selected as a standard should be of a similar type to the metal species expected to be present in the sample.[204,205] Some suitable organo-metallic-compounds (for the more commonly required elements) are listed below:

Barium naphthenate (solution)
Calcium naphthenate (solution)
Calcium 2-ethylhexanate
4 cyclohexylbutyric acid salts of Ag, Al, Cd, Cu, Fe, Mg, Na, Ni, Pb, Zn
Nickel naphthenate
Octaphenylcylco-tetrasiloxane
Sodium naphthenate
Sodium tetraphenylborate
Potassium naphthenate
Potassium erucate
Vanadium naphthenate
Bis-(1-phenyl 1,3-butanediono)-oxovanadium(IV).
Vanadyl tetraphenylporphyrin

These, together with suitable compounds of other metals, should be obtainable from the following suppliers:

1. Angstrom, Inc., P.O. Box 248, Belleville, Mich. 48111, U.S.A.
2. Baird-Atomic, Inc., 125 Middlesex Turnpike, Bedford, Mass. 01730, U.S.A.
3. J. T. Baker Chemical Co., 222 Red School Lane, Phillipsburg, N.J. 08865, U.S.A.
4. B.D.H. Chemicals Ltd., Poole, Dorset BH12 4NN, England.
5. Messrs. Burt and Harvey Ltd., Brettenham House, Lancaster Place, Strand, London W.C.2, England.
6. Carlo Erba, Divisione Chimica Industriale, Via C Imbonati 24, 20159 Milano, Italy.
7. Conostan Div., Continental Oil Co., P.O. Drawer 1267, Ponca City, Okla. 74601, U.S.A.
8. Durham Raw Materials Ltd., 1–4 Great Tower Street, London EC3R 5AB, England.

9. Eastman Organic Chemicals, Eastman Kodak Co., 343 State Street, Rochester, N.Y. 14650, U.S.A.
10. Hopkin and Williams Ltd., P.O. Box 1, Romford, Essex RM1 1HA, England.
11. E. Merck, D 61 Darmstadt, West Germany.
12. National Spectrographic Laboratories, Inc., 19500 South Miles Road, Cleveland, Ohio 44128, U.S.A.
13. Office of Standard Reference Materials, National Bureau of Standards, Washington, D.C. 20234, U.S.A.
14. Research Organic/Inorganic Chemical Corp., 11686 Sheldon Street, Sun Valley, Calif. 91352, U.S.A.
15. Ventron Corp, Alfa Products, 44 Congress Street, Beverly, Mass. 01915, U.S.A.
16. M. B. H. Analytical Ltd., Station House, Potters Bar, Herts., EN6 1AL, England.

The metal content of the various compounds is specified by the suppliers.

Intermediate standard solutions containing not less than 500 μg/ml of the metals are prepared by weighing aliquots of the compounds into clean dry beakers, dissolving in the relevant solvent, transferring to graduated flasks and adjusting to the marks with the solvent. Solutions of most metals at this concentration can be stored for at least two months.

Specific Notes for Preparation of Crude, Fuel and Lubricating Oils

Propan-2-ol/white spirit mixture possesses the ability to extract sodium from glass, so that sodium standards and oil samples to be analysed for this element, should never be retained for prolonged periods in glass containers.

Difficulty may be experienced with sodium determinations upon some samples of crude oils due to the element being present in a particulate form (mainly suspended water droplets). In extreme cases it may be necessary to resort to an ashing preparative technique.

Viscous samples should be heated to 50–60°C, and all samples should be vigorously agitated prior to weighing.

An aliquot of the oil, say 10 g, is accurately weighed into a clean dry beaker. (For the estimations of barium and calcium in lubricating oils, potassium naphthenate solution is added as an ionization suppressant so that when diluted to analytical volumes the solution will contain about 1000 ppm K.) The oil is dissolved in the solvent, transferred to a graduated flask and adjusted to the mark with the solvent.

Estimation is made by aspirating the test solutions to the atomic absorption spectrophotometer noting the responses, and reading the metal contents from a calibrating curve obtained from the comparable standard solutions. One of the known standard solutions should be re-aspirated after every five samples to check that the original calibration has not changed.

The use of Mixed-solvent Systems. Organo-metallic compounds are more expensive than common inorganic salts, are of less reliable composition and less stable in solution. HOLDING and NOAR[198] proposed the use of a mixed-solvent system that allows standards prepared from inorganic salts to be used for oil analysis.

These workers established that a mixture of butanol (3 volumes), cyclohexanone (5 volumes) and industrial methylated spirit (2 volumes), could accommodate about 10 ml of water and 0·3 g of oil per 100 ml. The use of this solvent offers a convenient procedure for surmounting the difficulties, outlined above, that can occur when sodium is contained in a crude oil as droplets of brine.

The validity of this procedure has been confirmed in our own laboratories for the estimation of zinc in lubricating oils. For this estimation standards were prepared from an aqueous zinc master solution. Aliquots of this solution were transferred to 100 ml graduated flasks, deionized water added to bring the aqueous volumes up to 10 ml, and the solutions diluted to 100 ml with the above solvent.

Master test solutions were prepared by dispersing 0·2 g aliquots of the oils in the solvent and diluting to suitable known volumes with the organic solvent. Further dilutions to bring the zinc contents into optimum ranges for estimation were made by pipetting aliquots of this primary solution to 100 ml graduated flasks, adding 10 ml of deionized water and diluting to the marks with the solvent. Satisfactory results were obtained in this determination.

Less satisfactory results were obtained for the estimation of calcium when solutions containing 1000 ppm Na (to suppress calcium ionization) were prepared from an aqueous sodium chloride solution. In this case crystals of sodium chloride tended to deposit in the atomizer nozzle, causing complete blockage after about 10 min aspiration. Better results were obtained when the sodium was added in the form of sodium naphthenate.

Interferences. Of the metals normally determined in oils, very few are liable to chemical interferences from other metals. However standardization can cause problems because the sensitivity for a given metal can depend on the nature of the metal species present.[204–206]

The presence of aluminium depresses the absorption of calcium. Aluminium, though, is effectively removed in the refining process, so is absent as an impurity in common lubricating oils. It is used as an additive in only a few specialized oils.

Use of the nitrous oxide–acetylene flame system is essential for the estimation of barium and it is, therefore, convenient to utilize it also for calcium and zinc determinations. The use of this hotter flame, moreover, reduces the interference of aluminium upon calcium. It is of interest to note that the presence of phosphate in a non-ionic medium does not depress calcium absorption.

Perhaps the most exacting analysis required in this field is that of vanadium at the levels found in fuel oils. The trace (Fig. 5.33) illustrates the sensitivity of current standard commercial equipment to perform this estimation. However it should be stressed that calibration can be difficult.[204] Bis (1-phenyl, 1, 3 butanediano)-oxovanadium (IV) and vanadyl tetraphenylporphyrin give substantially different sensitivities with respect to vanadium. The former is often used as a vanadium standard whilst the latter is typical of the type of vanadium compounds present in naturally occurring oils. A similar effect has also been reported for nickel[205] and the use of tetraphenylporphyrin standards recommended. The addition of bromine or iodine to both test and sample solutions does not remedy the situation.[206] Flameless electrothermal atomization (see p. 233) can also be used to advantage for this type of determination and has even been used to determine phosphorus in edible oils and fats.[208] A R.F.E.D.L. source (see page 190) and the 213·6 nm phosphorus non-resonance line were used.

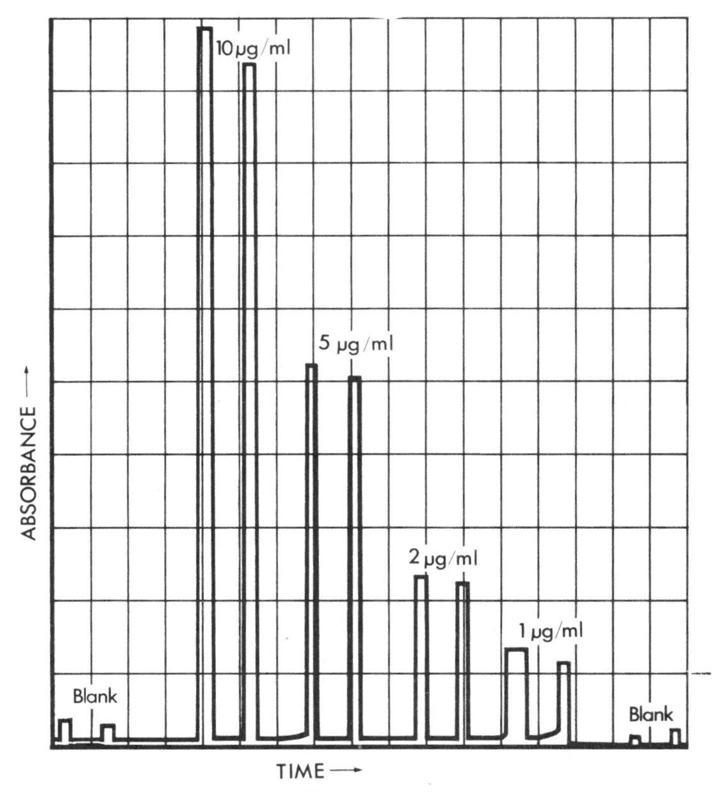

Fig. 5.33 Typical integrated responses for the estimation of vanadium in fuel oils diluted with propan-2-ol/white spirit mixture

WEAR METALS

The wear metals—iron, copper, silver, magnesium, chromium, tin, lead and nickel, may be determined in lubricating oils. The oils are diluted with a suitable solvent prior to analysis. Standard solutions are prepared from oil-soluble compounds as outlined above.

The estimation is only approximate since the wear metals will be present as particles suspended in the oil, and some of these particles may be too large to be converted into an atomic vapour in the optical path through the flame. It is possible to incorporate hydrochloric acid into the final solution if a mixed solvent system is used[(200)] and if left overnight this solvent will dissolve many types of wear metal particles. A similar technique that incorporates a mixture of hydrochloric and hydrofluoric acids has been used to determine titanium in wear metals.[(207)] 5 g oil is dissolved in 10 g M.I.B.K. and 0.15 ml of 1 + 3 hydrofluoric acid (48% m/m) + Hydrochloric acid (36% m/m) is then added and the solution is shaken for ten seconds and nebulized into a nitrous oxide-acetylene flame. It was found possible to prepare a master titanium standard (200 μg/g) by adding 325 mesh titanium powder to the blank oil and adding a small amount of the mixed acid reagent.

For more accurate determinations of wear metals it is necessary to ash the oil sample, dissolve the residue with acid, and carry out the measurement in aqueous medium. Even when this more prolonged sample preparation is resorted to, atomic absorption still offers the most versatile, rapid and frequently the most reliable procedure for final measurement of metal concentrations.

Flameless electrothermal atomization techniques (see p. 233) are applicable here.

Determination of Lead in Petrol

This is a determination that possesses peculiar characteristics, and for that reason it is discussed separately. The two compounds most commonly employed as anti-knock additives are tetraethyl and tetramethyl lead, but in order to apply the method directly it is necessary to know which of these, or what mixture of the two, has been included in the sample to be analysed. If the analyst possesses no information as to the nature of the lead additive, direct determination is unreliable. In such cases destruction of the lead alkyl followed by extraction of the lead into an aqueous phase is the most satisfactory procedure.

(A) INDIRECT DETERMINATION (after Shell United Kingdom Ltd)

Light petroleum distillates can contain lead at the μg/ml level, and the determination of this element is commonly effected by polarography, after extraction into an aqueous phase. This procedure requires about $2\frac{1}{2}$ h per determination. The following procedure, developed by Shell United Kingdom Ltd., and employing atomic absorption, requires only about 45 min.

The sample is treated with bromine to convert the lead to lead(II) bromide which is extracted with dilute nitric acid, and the lead content of the extract determined on the atomic absorption spectrophotometer. The following standard solutions and reagents are required.

Master Stock Solution (5000 μg/ml Pb). Dissolve 3·995 g of lead nitrate, $Pb(NO_3)_2$(A.R.), in approximately 200 ml of deionized water, transfer to a 500-ml graduated flask, and dilute to the mark with deionized water.

Intermediate Stock Solution (100 μg/ml). Pipette 10 ml of the master solution to a 500 ml graduated flask, and dilute to the mark with deionized water.

Working Standards. Prepare working standards containing 50, 25, 10 and 5 μg/ml by dilution from the intermediate solution.

Blank Solution. Dilute nitric acid (8 parts by volume nitric acid (70% m/m) (A.R.) + 992 parts deionized water).

Digestion Reagent. Bromine (A.R.).

Sample Preparation

1. 250 ml of sample are measured into a 600-ml beaker and bromine added dropwise until a deep wine colour persists for 2 min. The sample is then allowed to stand for a further 10 min to ensure complete bromination.
2. The solution is evaporated on a hotplate until the volume is reduced to approximately 50 ml.
3. The contents of the beaker are transferred to a 100 ml separating funnel and the beaker washed with a 10 ml portion of the dilute nitric acid which is then added to the separating funnel.
4. The separating funnel is shaken vigorously for 2 min, the two phases allowed to settle and the lower layer run into a 60 ml beaker.
5. The extraction is repeated with a second 10 ml portion of the dilute nitric acid, and the extracts combined and made up to 25 ml in a graduated flask with the dilute nitric acid.
6. This acid extract is now ready for analysis by atomic absorption spectroscopy.

Measurement. Measurement is made at 283·3 nm. The concentration of lead in the 25-ml extracts obtained from the samples is at most about 50 μg/ml. It is recommended that the dilute nitric acid should be aspirated to the spectrophotometer after each lead-containing solution, so as to clear the system.

An exhaustive series of investigations carried out by Shell United Kingdom

Ltd, indicated that both the procedure and instrument readings were highly repeatable, and that over the concentration range experienced with control samples the response was linear.

(B) Direct Determination

If the nature of the anti-knock additive is known it is possible to determine lead directly in petrol by atomic absorption.

Response curves for solutions of tetraethyl lead, tetramethyl lead and mixtures of these two compounds obtained at 283·3 nm using air–acetylene, are shown in Fig. 5.34.

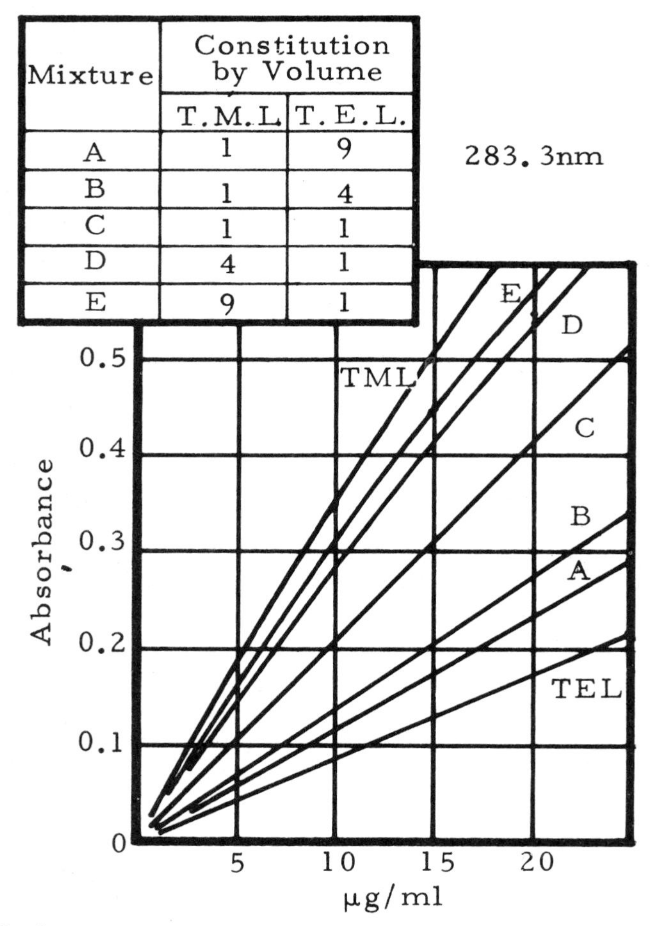

Mixture	Constitution by Volume	
	T.M.L.	T.E.L.
A	1	9
B	1	4
C	1	1
D	4	1
E	9	1

Fig. 5.34 Typical response curves for mixtures of T.E.L. and T.M.L.

As stated previously, besides pure T.E.L. and pure T.M.L., mixtures of these compounds are used, which will of course exhibit different response characteristics. These mixtures, though, are almost invariably prepared so that they have the same lead content as either the straight T.E.L. or T.M.L. compounds (112·25 g Pb per litre).

Since the master additives have identical lead contents it is possible for blenders to utilize atomic absorption to check that these master blends have been properly constituted; for example, response curves as shown in Fig. 5.34 could be constructed from dilutions of T.E.L. and T.M.L. mixtures whose proportions are known. Works batches could then be compared against these curves and prediction as to their relative contents of T.M.L. and T.E.L. made.

An investigation has been carried out (R.J.R.) to determine the effect of different hydrocarbons on the lead absorption. Identical response curves were obtained for T.E.L. originally diluted with xylene, iso-octane (2,2,4 trimethylpentane) and cyclohexane.

The effects of the presence of differing bromo- and chloro-compounds has also been investigated and found not to affect the absorption.

Preparation of Standard Solutions. Tetraethyl lead and tetramethyl lead are supplied by the Associated Octel Co. Ltd, in diluted form under the designations 'Dilute T.E.L.-B compound' and 'Dilute T.M.L.-CB compound'. Both contain 112·25 g lead/litre, in a medium of 70 per cent v/v xylene and 30 per cent v/v *n*-heptane. These compounds are highly toxic and Associated Octel's recommended safety precautions must be rigidly adhered to.

Prepare an intermediate standard (containing 560 μg/ml) by diluting 1 ml of either T.E.L.-B or T.M.L.-CB (according to which has been included in the sample to be analysed) to 200 ml with 'raw-petrol'.

Working Standards. Working standards are prepared from the intermediate solution by dilution with cyclohexanone or white spirit and isopropyl alcohol (propan-2-ol). It is wise to use an oxygenated compound for this operation, as further dilution with a hydrocarbon can give rise to a medium that produces a luminous flame at all flows of acetylene and air obtainable on the instrument.

The standards should cover the range 0–50 μg/ml if the tetraethyl compound is to be assessed, and 0–15 μg/ml lead if the tetramethyl compound is being estimated.

Blank Solution. Use the relevant diluent.

Sample Preparation. Dilute the petrol sample with cyclohexanone or white spirit and isopropyl alcohol, so that its estimated lead content lies in the relevant range stated above.

Measurement. Aspirate this sample and the standards to the atomic absorption spectrophotometer and by comparing the readings obtain the lead content of the petrol.

The successful direct determination of lead in petrol has been achieved by adding iodine and an organic liquid ion exchanger to the petrol in order to break down the lead alkyls followed by nebulization into the nitrous oxide-hydrogen flame.[202]

Chapter 5 References and further reading

In the space available it is only possible to give a selected number of references for this Chapter. During the period from 1971 to 1978 more than 700 papers dealing with applications of atomic absorption have been published each year. The interested reader can rapidly perform a literature survey, from 1971 onwards, for almost any specific field of analysis that utilizes atomic absorption, emission or fluorescence spectroscopy by consulting the Annual Reports on Analytical Atomic Spectroscopy (A.R.A.A.S.). This useful journal is published every July by the Chemical Society, Burlington House, London W1V 0BN, and gives a critical review of almost all published work and major lecture meetings in the preceding year on analytical atomic spectroscopy. The large methodology section is divided into eight sections—petroleum products; chemicals; metals; refractories, metal oxides, ceramics, slags and cements; minerals; soils, plants and agricultural products; environmental studies; and medicine. Each of these sections contains a comprehensive Table giving the basic essentials of each application paper listed by element as well as a critical review of the published work for that year.

TRACE METAL ESTIMATIONS, FOODSTUFFS AND GENERAL ORGANIC MATERIALS

1. FREY, S. W., The determination of copper, iron, calcium, sodium and potassium in beer by atomic absorption spectrophotometry. *Atomic Abs. Newsletter* 3 (1964) 127.
2. ISAAC, R. A., and JOHNSON, W. C., Collaborative study of wet and dry ashing techniques for the elemental analysis of plant tissue by AAS. *J. Ass. Off. Analyt. Chem.* 58 (1975) 436.
3. ADRIAN, W. J., and STEVENS, M. L., Effect of different sample preparation methods on the A.A.S. determination of calcium in plant material. *Analyst* 102 (1977) 446.
4. SLAVIN, W., Applications of atomic absorption spectroscopy in the food industry. *Atomic Abs. Newsletter* 4 (1965) 330.
5. SLAVIN, W., The determination of various metals in synthetic fibres using atomic absorption spectrophotometry. *Atomic Abs. Newsletter* 4 (1965) 192.
6. GUILLAUMIN, R., Determination of calcium and magnesium in vegetable oils and fats by atomic absorption spectrophotometry. *Atomic Abs. Newsletter* 5 (1966) 19.
7. Metallic impurities in organic matter sub-committee. The use of 50 per cent hydrogen peroxide for the destruction of organic matter. *Analyst* 92 (1967) 403, and *Analyst* 101 (1976) 62, parts i and ii.
8. ROACH, A. G., SANDERSON, P., and WILLIAMS, D. R., Determination of trace amounts of copper, zinc and magnesium in animal feeds by atomic absorption spectroscopy. *Analyst* 93 (1968) 42.
9. FRIEND, M. T., SMITH, C. A. and WISHART, D. Ashing and wet oxidation procedures for the determination of some volatile trace metals in foodstuffs and biological materials by A.A.S. *Atomic Abs. Newsletter* 16 (1977) 46.
10. CAPAR, S. G., A.A.S. determination of Pb, Cd, Zn and Cu in clams and oysters: collaborative study. *J. Ass. Off. Analyt. Chem.* 60 (1977) 1400.
11. HARTLEY, F. R., and INGLIS, A. S., The determination of metals in wool by atomic-absorption spectroscopy. *Analyst* 93 (1968) 394.
12. CAMERON, A. G., and HACKETT, D. R., Determination of Copper in Foods by Atomic Absorption Spectrophotometry. *J. Sci. Food Agr.* 21 (1970) 535.
13. PRICE, W. J., ROOS, J. T. H., and CLAY, A. F., Rapid determination of Nickel in Edible Fats by atomic absorption spectrophotometry. *Analyst* 95 (1970) 760.
14. FLETCHER, K., Direct Determination of Lead in plant materials by Atomic Absorption Spectrophotometry. *J. Sci. Food Agr.* 22 (1971) 260.

15. Metallic Impurities in Organic Matter Sub-Committee. The determination of small amounts of Copper in organic matter by atomic absorption spectroscopy. *Analyst* 96 (1971) 741.
16. HARTSTEIN, A. M., FREEDMAN, R. W., and PLATTER, D. W., Novel Wet-Digestion Procedure for Trace Metal Analysis of Coal by A.A. *Anal. Chem.* 45 (1973) 611.
17. ROSCHNIK, R. K., The Determination of Lead in Foods by A.A.S. *Analyst* 98 (1973) 596.
18. CROSBY, N. T., Determination of metals in foods – A review. *Analyst* 102 (1977) 225.
19. FOURIE, H. O., and PEISACH, M., Losses of trace elements during dehydration of marine zoological material. *Analyst* 102 (1977) 193.
20. SIROTA, G. R. and UTHE, J. F., Determination of tetraalkyllead compounds in biological materials. *Anal. Chem.* 49 (1977) 823.

FLAMELESS PROCEDURE FOR MERCURY

21. HATCH, W. R., and OTT, W. L., Determination of Sub-Microgram Quantities of Mercury by atomic absorption spectrophotometry. *Anal. Chem.* 40 (1968) 2085.
22. LINDSTEDT, G., A rapid method for the determination of Mercury in Urine. *Analyst* 95 (1970) 264.
23. FREELAND, G. N., and HOSKINSON, R. M., Non-aqueous atomic absorption spectrophotometric determination of Organo-metallic Biocides. *Analyst* 95 (1970) 579.
24. MANNING, D. C., Non-Flame methods for Mercury determination by atomic absorption. A review. *Atomic Abs. Newsletter* 9 (1970) 97.
25. APRIL, R. W., and HUME, D. N., Environmental Mercury. Rapid determination in Water at Nanogram Levels. *Science* 170 (1970) 849.
26. GARDNER, D., A rapid method for the determination of mercury in air. *Anal. Chim. Acta* 82 (1976) 321.
27. MAGOS, L., Selective Atomic Absorption determination of inorganic mercury and methyl mercury in undigested biological samples. *Analyst* 96 (1971) 847.
28. ANALYTICAL METHODS COMMITTEE. Determination of mercury and methyl mercury in fish. *Analyst* 102 (1977) 769.
29. OMANG, S. H., Determination of mercury in natural waters and effluents by flameless atomic absorption spectrophotometry. *Anal. Chim. Acta* 54 (1971) 415.
30. OMANG, S. H., and PAUS, P. E., Trace determination of Mercury in Geological Materials by flameless atomic absorption spectroscopy. *Anal. Chim. Acta* 56 (1971) 393.
31. MUNNS, R. K., and HOLLAND, D. C., Determination of Mercury in fish by flameless atomic absorption spectrophotometry. *Anal. Chim. Acta* 53 (1971) 415.
32. HOOVER, W. L., MELTON, J. R., and HOWARD, P. A., Determination of Trace Amounts of Mercury in Foods by flameless atomic absorption. *J. Ass. Off. Analyt. Chem.* 54 (1971) 860.
33. DUFFER, J. K., Determination of Mercury at the parts-per-million level. *J. Paint Technol.* 43 (1971) 67.
34. MELTON, J. R., HOOVER, W. L., and HOWARD, P. A., Determination of mercury in soils by flameless atomic absorption. *Soil Sci. Soc. Amer. Proc.* 35 (1971) 850.
35. BRAUN, R., and HUSBANDS, A. P., Determination of low levels of Mercury by Cold-Vapor Atomic Absorption. *Spectrovision* 26 (1971) 2.
36. ISKANDER, I. K., SYERS, J. K., JACOBS, L. W., KEENEY, D. R., and GILMOUR, J. T., Determination of Total Mercury in Sediments and Soils. *Analyst* 97 (1972) 388.
37. RAINS, T. C., and MENIS, O., Determination of Sub-Microgram amounts of Mercury in Standard Reference Materials by Flameless AAS. *J. Ass. Off. Analyt. Chem.* 55 (1972) 1339.
38. URE, A. M., and SHAND, C. A., The Determination of Mercury in soils and Related Materials by Cold Vapour AAS. *Anal. Chim. Acta* 72 (1974) 63.
39. MAGOS, L., and CLARKSON, T. W., Atomic Absorption Determination of Total, Inorganic and Organic Mercury in Blood. *J. Ass. Off. Analyt. Chem.* 55 (1972) 966.
40. OLAFSSON, J., Determination of Nanogram Quantities of Mercury in Sea Water. *Anal. Chim. Acta* 68 (1974) 207.
41. THOMPSON, K. C., and REYNOLDS, G. D., The Atomic-Fluorescence Determination of Mercury by the Cold Vapour Technique. *Analyst* 96 (1971) 771.
42. THOMPSON, K. C., and GODDEN, R. G., Improvements in the Atomic Fluorescence Determination of Mercury by the cold vapour technique. *Analyst* 100 (1975) 544.
43. LOPEZ-ESCOBAR, L., and HUME, D. N., Ozone as a Releasing Agent. The Determination of Trace Mercury in Organic Matrices by Cold Vapour Atomic Absorption. *Analyt. Letters* 6 (1973) 343.

44. FAREY, B. J., NELSON, L. A. and ROLFE, M. G., A rapid technique for the breakdown of organic mercury compounds in natural waters and effluents. *Analyst* 103 (1978) 656.
45. URE, A. M., The determination of mercury by non-flame atomic absorption and atomic fluorescence spectroscopy. *Anal. Chim. Acta* 76 (1975) 1. (A useful review with 442 references)
46. KOIRTYOHANN, S. R., and KHALIL, M., Variables in the determination of mercury by cold vapour atomic absorption. *Anal. Chem.* 48 (1976) 136.
47. CHILOV, S., Determination of small amounts of mercury. *Talanta* 22 (1975) 205. (A useful review with 339 references.)
48. MAHAN, K. I., and MAHAN, S. E., Mercury retention in untreated water samples at the ppb level. *Anal. Chem.* 49 (1977) 662.
49. CHRISTMAN, D. R., and INGLE, J. D., Problems with sub-pph mercury determinations. *Anal. Chim. Acta* 86 (1976) 53.
50. FELDMAN, C., Preservation of dilute mercury solutions. *Anal. Chem.* 46 (1974) 99.

WATER AND EFFLUENTS (see also references 20–22 Chapter 4)

51. DELAUGHTER, B., The determination of sub-p.p.m. concentrations of chromium and molybdenum in brines. *Atomic Abs. Newsletter* 4 (1965) 273.
52. PAUS, P. E., The application of atomic absorption spectroscopy to the analysis of natural waters. *Atomic Abs. Newsletter* 10 (1971) 69.
53. EDIGER, R. D., A review of water analysis by atomic absorption. *Atomic Abs. Newsletter* 12 (1973) 151.
54. BIECHLER, D. G., Determination of trace copper, lead, zinc, cadmium, nickel and iron in industrial waste waters by atomic absorption spectrometry after ion exchange concentration on Dowex A-1. *Anal. Chem.* 37 (1965) 1054.
55. FISHMAN, M. J., and ERDMANN, D. E., Water Analysis (General Review). *Anal. Chem.* 49 (1977) 139R.
56. FISHMAN, M. J., The use of atomic absorption for analysis of natural waters. *Atomic Abs. Newsletter* 5 (1966) 102.
57. JAYNES, T., and FINLEY, J. S., The determination of manganese and iron in sea water by atomic absorption spectrometry. *Atomic Abs. Newsletter* 5 (1966) 4.
58. BURRELL, D. C., and WOOD, G. G., Direct Determination of Zinc in sea water by atomic absorption spectrophotometry. *Anal. Chim. Acta* 48 (1969) 45.
59. NIX, J., and GOODWIN, T., The simultaneous extraction of iron, manganese, copper, cobalt, nickel, chromium, lead and zinc from natural water for determination by atomic absorption spectroscopy. *Atomic Abs. Newsletter* 9 (1970) 119.
60. FISHMAN, M. J., Determination of Aluminium in Water. *Atomic Abs. Newsletter* 11 (1972) 46.
61. SURLES, T., TUSCHALL, J. R., and COLLINS, T. T., Comparative A.A.S. study of trace metals in lake water. *Environmental Science and Tech.* 9 (1975) 1073.
62. DE VINE, J. C. and SUHR, N. H. Determination of silicon in water samples. *Atomic Abs. Newsletter* 16 (1977) 39.

SOIL, FERTILIZERS AND CHEMICALS

63. DAVID, D. J., The determination of exchangeable sodium, potassium, calcium and magnesium in soils by atomic absorption spectrophotometry. *Analyst* 85 (1960) 495.
64. ALLAN, J. E., The determination of zinc in agricultural materials by atomic absorption spectrophotometry. *Spectrochim. Acta* 17 (1961) 467.
65. DAVID, D. J., The determination of strontium in biological materials and exchangeable strontium in soils by atomic absorption spectrophotometry. *Analyst* 87 (1962) 576.
66. MCBRIDE, C. H., Determination of minor nutrients in fertilizers by atomic absorption spectrophotometry. *Atomic Abs. Newsletter* 3 (1964) 144.
67. NADIRSHAW, M., and CORNFIELD, A. H., Direct determination of manganese in soil extracts by atomic absorption spectroscopy. *Analyst* 93 (1968) 475.
68. MARSHALL, G. B., and WEST, T. S., Determination of Chromium in Aluminium Salts by Atomic Absorption Spectroscopy. *Analyst* 95 (1970) 343.
69. VINK, J. J., The determination of harmful trace elements in Manganese Dioxide for Dry Cell use. *Analyst* 95 (1970) 399.

70. KRISHAMURTY, K. V. SHPIRT, E. and REDDY, M. M., Trace metal extraction of soils and sediments by nitric acid-hydrogen peroxide. *Atomic Abs. Newsletter* 15 (1976) 68.
71. AGEMIAN, H. and CHAU, A. S. Y. Evaluation of extraction techniques for the determination of metals in aquatic sediments. *Analyst* 101 (1976) 761.

CLINICAL APPLICATIONS

72. BERMAN, E., The determination of lead in blood and urine by atomic absorption spectrophotometry. *Atomic Abs. Newsletter* 3 (1964) 111.
73. SPRAGUE, S., and SLAVIN, W., Determination of iron, copper and zinc in blood serum by an atomic absorption method requiring only dilution. *Atomic Abs. Newsletter* 4 (1965) 228.
74. CHANG, T. L., GOVER, T. A., and HARRISON, W. W., Determination of magnesium and zinc in human brain tissue by atomic absorption spectroscopy. *Anal. Chim. Acta* 34 (1966) 17.
75. KAHNKE, M. J., Atomic absorption spectrophotometry applied to the determination of zinc in formalinized human tissue. *Atomic Abs. Newsletter* 5 (1966) 7.
76. SPRAGUE, S., and SLAVIN, W., A simple method for the determination of lead in blood. *Atomic Abs. Newsletter* 5 (1966) 9.
77. KOIRTYUHANN, S. R., and HOPKINS, C. A., Losses of trace metals during the ashing of animal tissues. *Analyst* 101 (1976) 870.
78. HANSEN, JESSIE L., Measurement of serum and urine lithium by atomic absorption spectrophotometry. *Amer. J. Med. Tech.* 34 (1968) 625.
79. LORBER, A., COHEN, R. L., CHANG, C. C., and ANDERSON, H. E., Gold Determination in Biological Fluids by Atomic Absorption Spectrophotometry Application to Chrysotherapy in Rheumatoid Arthritis Patients. *Arthritis and Rheumatism*, Vol. XI, No. 2(1) (1968) 170.
80. KAHN, H. L., PETERSON, G. E., and SCHALLIS, JANE E., Atomic Absorption Micro-Sampling with the Sampling Boat Technique. *Atomic Abs. Newsletter* 7 (1968) 35.
81. PYBUS, J., Determination of Calcium and Magnesium in Serum and Urine by Atomic Absorption Spectrophotometry. *Clin. Chim. Acta* 23 (1968) 309.
82. SPIELHOLTZ, G. I., and TORALBALLA, G. C., The determination of Zinc in Crystalline Insulin and in Certain Insulin Preparations by atomic absorption spectroscopy. *Analyst* 94 (1969) 1072.
83. OLSON, A. D., and HAMLIN, W. D., A new method for serum iron and total iron-binding capacity by atomic absorption spectrophotometry. *Clin. Chem.* 15 (1969) 438.
84. SEGAL, R. J., Nonspecificity of Urinary Lead Measurements by Atomic Absorption Spectroscopy. A Spectrophotometric Method for Correction. *Clin. Chem.* 15 (1969) 1124.
85. EINARSSON, O., and LINDSTEDT, G., A non-extraction Atomic Absorption Method for the Determination of lead in blood. *Scan. J. Clin. Lab. Invest.* 23 (1969) 367.
86. WHITE, R. A., B.N.F.M.R.A. Research Report A1707, Sept. 1968 and Paper presented at International At. Abs. Spect. Conference, Sheffield, 1969.
87. DELVES, H. T., A Micro-Sampling Method for the Rapid Determination of Lead in Blood by Atomic Absorption Spectrophotometry. *Analyst* 95 (1970) 431.
88. ROSE, G. A., and WILLDEN, E. G., An Improved Method for Determination of Whole Blood Lead by using an AA Technique. *Analyst* 98 (1973) 243.
89. DELVES, H. T. A simple matrix modification to allow the direct determination of cadmium in blood by flame microsampling A.A.S. *Analyst* 102 (1977) 403.
90. SABET, S., OTTAWAY, J. M., and FELL, G. S., Comparison of the Delves Cup and carbon furnace atomization used in A.A.S. for the determination of lead in blood. *Proc. Analyt. Div. Chem. Soc.* 14 (1977) 300.
91. THOMPSON, K. C., and GODDEN, R. G., A simple method for monitoring excessive levels of lead in whole blood using atomic absorption spectrophotometry and a rapid, direct nebulisation technique. *Analyst* 101 (1976) 174.
92. KAHN, H. L., and SEBESTYEN, JANE, S., The Determination of Lead in Blood and Urine by Atomic Absorption Spectrophotometry, with the Sampling Boat System. *Atomic Abs. Newsletter* 9 (1970) 33.
93. SIDEMAN, L., MURPHY, J. J. JR., and WILSON, D. T., A collaborative study of the serum calcium determination by atomic absorption spectroscopy. *Clin. Chem.* 16 (1970) 597.
94. DELVES, H. T., SHEPHERD, G., and VINTER, P., Determination of Eleven Metals in Small Samples of Blood by sequential solvent extraction and atomic absorption spectrophotometry. *Analyst* 96 (1971) 260.

95. DUNCKLEY, J. V., Estimation of Gold in Serum by Atomic Absorption Spectroscopy. *Clin. Chem.* 17 (1971) 992.
96. MENDEN, E. E., BROCKMAN, D., CHOUDURY, H. and PETERING, H. G. Dry ashing of animal tissues for A.A.S. determination of Zn, Cu, Cd, Pb, Fe, Mn, Mg and Ca. *Anal. Chem.* 49 (1977) 1644.
97. BERMAN, E., Biochemical applications of flame emission and atomic absorption spectroscopy. *Appl. Spectroscopy* 29 (1975) 1.
98. CERNIK, A. A., and SAYERS, M. H. P., Determination of Lead in Capillary Blood using a paper punched disc atomic absorption technique. Application to the Supervision of lead workers. *Brit. J. Industr. Med.* 28 (1971) 392.
99. CERNIK, A. A. Lead in blood – the analyst's problem. *Chemistry in Britain* 10 (1974) 58.
100. CERNIK, A. A. and SAYERS, M. H. P. Application of blood calmium determination to industry using a punched disc technique. *Brit. J. Industr. Med.* 32 (1975) 155.
101. CERNIK, A. A. Determination of blood lead using a 4 mm paper punched disc – carbon sampling cup technique. *Brit. J. Industr. Med.* 31 (1974) 239.

METALLURGICAL ANALYSIS

102. ANDREW, T. R., and NICHOLS, P. N. R., The application of atomic absorption to the rapid determination of magnesium in electronic nickel and nickel alloys. *Analyst* 87 (1962) 25.
103. JOHNSON, H. N., KIRKBRIGHT, G. F., and WHITEHOUSE, R. J., Molecular and A.A.S. Methods for the Determination of Cerium utilising the Formation of Molybdocerophosphoric Acid. *Anal. Chem.* 45 (1973) 1603.
104. BELCHER, C. B., and BRAY, H. M., The determination of magnesium in iron by atomic absorption spectrophotometry. *Anal. Chim. Acta* 26 (1962) 322.
105. MCPHERSON, G. L., PRICE, J. W., and SCHAIFE, P. H., Application of atomic absorption spectroscopy to the determination of cobalt in steel, alloy steel and nickel. *Nature* 199 (1963) 371.
106. KINSON, K., HODGES, R. J., and BELCHER, C. B., The determination of chromium in low-alloy irons and steels by atomic absorption spectrophotometry. *Anal. Chim. Acta* 29 (1962) 134.
107. BELCHER, C. B., and KINSON, K., The determination of manganese in iron and steel by atomic absorption spectrophotometry. *Anal. Chim. Acta* 30 (1964) 483.
108. KINSON, K., and BELCHER, C. B., The determination of nickel in iron and steel by atomic absorption spectrophotometry. *Anal. Chim. Acta* 30 (1964) 64.
109. KINSON, K., and BELCHER, C. B., Determination of minor amounts of copper in iron and steel by atomic absorption spectrophotometry. *Anal. Chim. Acta* 31 (1964) 180.
110. SHAFTO, R. G., The determination of copper, iron, lead, and zinc in nickel plating solutions by atomic absorption. *Atomic Abs. Newsletter* 3 (1964) 115.
111. SPRAGUE, S., and SLAVIN, W., The determination of copper, nickel, cobalt, manganese and magnesium in iron and steels by atomic absorption spectrophotometry. *Atomic Abs. Newsletter* 3 (1964) 72.
112. WILSON, L., The determination of silver in aluminium alloys by atomic absorption spectroscopy. *Anal. Chim. Acta* 30 (1964) 377.
113. DYCK, R., The determination of chromium, magnesium and manganese in nickel alloys by atomic absorption spectrophotometry. *Atomic Abs. Newsletter* 4 (1965) 170.
114. MCPHERSON, G. L., Atomic Absorption spectrophotometry as an analytical tool in a metallurgical laboratory. *Atomic Abs. Newsletter* 4 (1965) 186.
115. BEYER, M., Determination of manganese, copper, chromium, nickel and magnesium in cast iron and steel. *Atomic Abs. Newsletter* 4 (1965) 212.
116. FARRAR, B., Determination of copper and zinc in ore samples and lead-base alloys. *Atomic Abs. Newsletter* 4 (1965) 325.
117. HUMPHREY, J. R., Determination of magnesium in uranium by atomic absorption. *Anal. Chem.* 37 (1965) 1604.
118. BELL, G. F., The analysis of aluminium alloys by means of atomic absorption spectrophotometry. *Atomic Abs. Newsletter* 5 (1966) 73.
119. CAPACHO-DELGADO, L., and MANNING, D. C., Determination of vanadium in steels and gas oils. *Atomic Abs. Newsletter* 5 (1966) 1.
120. KIRKBRIGHT, G. F., SMITH, A. M., and WEST, T. S., Rapid determination of molybdenum

in alloy steels by atomic absorption spectroscopy in a nitrous oxide–acetylene flame. *Analyst* 91 (1966) 700.
121. MOSTYN, R. A., and CUNNINGHAM, A. F., Determination of molybdenum in ferrous alloys by atomic absorption spectrometry. *Anal. Chem.* 38 (1966) 121.
122. MANSELL, R. E., EMMEL, H. W., and MCLAUGHLIN, E. L., Analysis of magnesium and aluminium alloys by atomic absorption spectroscopy. *Appl. Spectroscopy* 20 (1966) 231.
123. KIRKBRIGHT, G. F., PETERS, M. K., and WEST, T. S., Determination of small amounts of molybdenum in niobium and tantalum by atomic absorption spectroscopy in a nitrous oxide–acetylene flame. *Analyst* 91 (1966) 705.
124. BARNES, L. JR., Determination of chromium in low alloy steels by atomic absorption spectrometry. *Anal. Chem.* 38 (1966) 1083.
125. OTTAWAY, J. M., and PRADHAN, N. K., Determination of Chromium in Steel by A.A.S. with an Air–Acetylene Flame. *Talanta* 20 (1973) 927.
126. DAGNALL, R. M., WEST, T. S., and YOUNG, P., Determination of trace amounts of lead in steels, brass and bronze alloys by atomic absorption spectrometry. *Anal. Chem.* 38 (1966) 358.
127. KIRKBRIGHT, G. F., PETERS, M. K., and WEST, T. S., Determination of trace amounts of copper in niobium and tantalum by atomic absorption spectroscopy. *Analyst* 91 (1966) 411.
128. WILSON, L., The determination of cadmium in stainless steel by atomic absorption spectroscopy. *Anal. Chim. Acta* 35 (1966) 123.
129. BELL, G. F., On the effect of copper on the determination of zinc in aluminium. *Atomic Abs. Newsletter* 6 (1967) 18.
130. MCAULIFFE, J. J., A method for determination of silicon in cast iron and steel by atomic absorption spectrometry. *Atomic Abs. Newsletter* 6 (1967) 69.
131. JURSIK, M. L., Application of atomic absorption to the determination of Al, Fe and Ni in the same uranium-base sample. *Atomic Abs. Newsletter* 6 (1967) 21.
132. SCOTT, T. C., ROBERTS, E. D., and CAIN, D. A., Determination of minor constituents in ferrous materials by atomic absorption spectrophotometry. *Atomic Abs. Newsletter* 6 (1967) 1.
133. SCHOLES, P. H., The application of atomic absorption spectrophotometry to the analysis of iron and steel. *Analyst* 93 (1968) 197.
134. PRICE, W. J., and ROOS, J. T. H., The determination of silicon by atomic absorption spectrophotometry, with particular reference to steel, cast iron, aluminium alloys and cement. *Analyst* 93 (1968) 709.
135. CAMPBELL, D. E., Determination of Silicon in Aluminium Alloys by Atomic Absorption Spectroscopy. *Anal. Chim. Acta* 46 (1969) 31.
136. HEADRIDGE, J. B., and RICHARDSON, J., The determination of trace amounts of cadmium in stainless steel by solvent extraction followed by atomic absorption spectrophotometry. *Analyst* 94 (1969) 968.
137. SCHNEPFE, M. M., and GRIMALDI, F. S., Atomic Absorption determination of rhodium in chromite concentrates. *Talanta* 16 (1969) 1461.
138. HEADRIDGE, J. B., and RICHARDSON, J., Determination of Trace amounts of bismuth in ferrous alloys by solvent extraction followed by atomic absorption spectrophotometry. *Analyst* 95 (1970) 930.
139. MICHAILOVA, T. P., and REZEPINA, V. A., The determination of gold and silver in plant liquors and electrolytes by atomic absorption spectrophotometry. *Analyst* 95 (1970) 769.
140. JANSSEN, A., and UMLAND, F., Determination of the platinum metals in the presence of each other by atomic absorption flame photometry. Determination of small amounts of rhodium, palladium, iridium, platinum using copper (2+) and sodium (+) as a spectroscopic buffer. *Z. Anal. Chem.* 251 (1970) 101.
141. REYNOLDS, R. J., and LAGDEN, D. S., Potential hazards by formation of silver acetylide upon aspirating solutions containing high concentrations of silver to an atomic absorption spectrophotometer when acetylene is used as fuel. *Analyst* 96 (1971) 319.
142. THOMERSON, D. R., and PRICE, W. J., The determination of chromium and molybdenum in a complete range of steels by atomic absorption spectrometry with a nitrous oxide–acetylene flame. *Analyst* 96 (1971) 321.
143. BURKE, K. E., Determination of microgram amounts of antimony, bismuth, lead and tin in aluminium, iron and nickel-base alloys by non-aqueous atomic absorption spectroscopy.

Analyst 97 (1972) 19.
144. MARTIN, MARGARET J., The determination of niobium in steel by atomic absorption spectrophotometry. *Analyst* 97 (1972) 394.
145. ROONEY, R. C., and PRATT, C. G., The determination of tungsten and silicon in highly alloyed materials by atomic absorption spectroscopy. *Analyst* 97 (1972) 400.
146. HEADRIDGE, J. B., and SOWERBUTTS, A., The determination of tin in steels by solvent extraction followed by atomic absorption spectrophotometry. *Analyst* 97 (1972) 442.
147. JOHNSON, H. N., KIRKBRIGHT, G. F., and WEST, T. S., An indirect amplification procedure for the determination of vanadium in aluminium alloys by atomic absorption spectroscopy. *Analyst* 97 (1972) 696.
148. MALLETT, R. C., PEARTON, D. C. G., RING, E. J., and STEELE, T. W., Interferences and their elimination in the determination of the noble metals by atomic absorption spectrophotometry. *Talanta* 19 (1972) 181.
149. ASHY, M. A., and HEADRIDGE, J. B., The Determination of Iridium and Ruthenium in Rhodium Sponge by Solvent Extraction followed by A.A.S. *Analyst* 99 (1974) 285.
150. WALTON, G., A Method for the Determination of Silver in Ores and Mineral Products by A.A.S. *Analyst* 98 (1973) 335.
151. QUARRELL, T. M., POWELL, R. J. W., and CLULEY, H. J., Determination of Tin and Antimony in Lead Alloy for Cable Sheathing by A.A.S. *Analyst* 98 (1973) 443.
152. BAILEY, N. T., and WOOD, S. J., A Comparison of Two Rapid Methods for the Analysis of Copper Smelting Slags by A.A.S. *Anal. Chim. Acta* 69 (1974) 19.
153. NAKAHARA, T., MUNEMORI, M., and MUSHA, S., A.A.S. Determination of tin in premixed inert gas (entrained air) hydrogen flames. *Anal. Chim. Acta* 62 (1972) 267.
154. SCHWEINSBERG, D. P., and HEFFERMAN, B. J., Decomposition of Antimony Concentrates with Ammonium Iodide. *Anal. Chim. Acta* 67 (1973) 213.
155. JONES, M. H., and WOODCOCK, T. T., On-Stream AA determinations of Mn and Zn in Flotation Liquors containing Calcium Sulphate. *Anal. Chim. Acta* 69 (1974) 275.
156. GHIGLIONE, M., ELJURI, E., and CUEVAS, C., A new method for preparing metallic samples for A.A.S. *Appl. Spectroscopy* 30 (1976) 320.
157. PETERSON, G. E. and KERBER, J. D., The application of A.A.S. to the analysis of ferrous alloys. *Atomic Abs. Newsletter*, 15 (1976) 134. (A useful review, 51 references).
157. (a) PETERSON, G. E., Application of A.A.S. to the analysis of nonferrous alloys. *Atomic Abs. Newsletter* 16 (1977) 133. (A useful review, 29 references.)

SILICATE ANALYSIS
158. BILLINGS, G. K., and ADAMS, J. A. S., The analysis of geological materials by atomic absorption spectrometry. *Atomic Abs. Newsletter* 3 (1964) 65.
159. TRENT, D. J., and SLAVIN, W., Determination of the major metals in granitic and diabasic rocks by atomic absorption spectrophotometry. *Atomic Abs. Newsletter* 3 (1964) 17.
160. TRENT, D. J., and SLAVIN, W., Determination of various metals in silicate samples by atomic absorption spectrophotometry. *Atomic Abs. Newsletter* 3 (1964) 118.
161. BURRELL, D. C., An atomic absorption method for the determination of cobalt iron and nickel in the asphaltic fraction of recent sediments. *Atomic Abs. Newsletter* 4 (1965) 328.
162. BURDO, R. A., and WISE, W. M., Determination of silicon in glasses and minerals by A.A.S. *Anal. Chem.* 47 (1975) 2360.
163. HANNAKER, P. and HUGHES, T. C. Multi-element trace analysis of geological materials with solvent extraction and flame A.A.S. *Anal. Chem.* 49 (1977) 1485.
164. JONES, A. H., Analysis of glass and ceramic frits by atomic absorption spectrophotometry. *Anal. Chem.* 37 (1965) 1761.
165. BERNAS, B., A new method for decomposition and comprehensive analysis of silicates by atomic absorption spectrometry. *Anal. Chem.* 40 (1968) 1682.
166. HENDEL, Y., Replacement of platinum vessels with a pressure device for acid dissolution in the rapid analysis of glass by A.A.S. *Analyst* 98 (1973) 450.
167. PRICE, W. J. and WHITESIDE, P. J., General method for the analysis of siliceous materials by A.A.S. and its application to macro and micro samples. *Analyst* 102 (1977) 664.
168. SIDEROPOULOS, N., The determination of calcium in chrome refractories by atomic absorption spectroscopy. *Analyst* 94 (1969) 389.
169. VAN LOON, J. C., and PARISSIS, C. M., Scheme of silicate analysis based on the lithium

metaborate fusion followed by atomic absorption spectrophotometry. *Analyst* 94 (1969) 1057.
170. INGAMELLS, C. O., Lithium metaborate flux in silicate analysis. *Anal. Chim. Acta*, 52 (1970) 323.
171. STONE, M., and CHESTER, S. E., Determination of lithium oxide in silicate rocks by atomic absorption spectrophotometry. *Analyst* 94 (1969) 1063.
172. BOAR, P. L., and INGRAM, L. K., The comprehensive analysis of coal ash and silicate rocks by atomic absorption spectrophotometry by a fusion technique. *Analyst* 95 (1970) 124.
173. FERRIS, A. P., JEPSON, W. B., and SHAPLAND, R. C., Evaluation and correction of interference between aluminium, silicon and iron in atomic absorption spectrophotometry. *Analyst* 95 (1970) 574.
174. SIMONSEN, A., Determination of platinum in basic rocks by solvent extraction and atomic absorption spectroscopy. *Anal. Chim. Acta* 49 (1970) 368.
175. O'GORMAN, J. V., and SUHR, N. H., A rapid atomic absorption technique for the determination of lithium in silicate materials. *Analyst* 96 (1971) 335.
176. COBB, W. D., and HARRISON, T. S., The determination of alumina in iron ores, slags and refractory materials by atomic absorption spectroscopy. *Analyst* 96 (1971) 764.
177. HUTCHISON, D., The determination of molybdenum in geological materials by a combined solvent-extraction atomic absorption procedure. *Analyst* 97 (1972) 118.
178. KILROY, W. P., and MOYNIHAN, C. T., A.A. analysis of borosilicate glasses. *Anal. Chim. Acta* 83 (1976) 389.
179. ARMANNSSON, H., The use of dithizone extraction and A.A.S. for the determination of Cd, Zn, Cu, Ni and Co in rocks and sediments. *Anal. Chim. Acta* 88 (1977) 89.

CEMENT ANALYSIS
180. SPRAGUE, S., Cement Analysis. *Atomic Abs. Newsletter* 2 (1963) 43.
181. TAKEUCHI, T., and SUZUKI, M., The determination of sodium, potassium, magnesium, manganese and calcium in cement by atomic absorption spectrophotometry. *Talanta* 11 (1964) 1391.
182. CAPACHO-DELGADO, L., and MANNING, D. C., The determination by atomic absorption spectroscopy of several elements including silicon, aluminium and titanium in cement. *Analyst* 92 (1967) 553.
183. PRICE, W. J., and ROOS, J. T. H., The determination of silicon by atomic absorption spectrophotometry, with particular reference to steel, cast iron, aluminium alloys and cement. *Analyst* 93 (1968) 709.
184. ROOS, J. T. H., and PRICE, W. J., A comprehensive scheme for the analysis of cement by atomic absorption spectrophotometry. *Analyst* 94 (1969) 89.
185. CROW, R. F., and CONNOLLY, J. D., A.A. analyses of portland cement and raw mix using a lithium metaborate fusion. *J. Test. Eval.* 1 (1973) 382.
186. MARUTA, T., and SUDOH, G., Determination of calcium in cements by A.A.S. based on germanium-calcium spectral overlap. *Anal. Chim. Acta* 86 (1976) 277.

PETROLEUM ANALYSIS
187. ROBINSON, J. W., Determination of lead in gasoline by atomic absorption spectroscopy. *Anal. Chim. Acta* 24 (1961) 451.
188. SPRAGUE, S., and SLAVIN, W., Determination of the metal content of lubricating oils by atomic absorption spectrophotometry. *Atomic Abs. Newsletter* 2 (1963) 20.
189. TRENT, D. J., and SLAVIN, W., The direct determination of trace quantities of nickel in catalytic cracking feedstocks by atomic absorption spectrophotometry. *Atomic Abs. Newsletter* 3 (1964) 131.
190. BURROWS, J. A., HEERDT, J. C., and WILLIS, J. B., The determination of wear metals in used lubricating oils by atomic absorption spectroscopy. *Anal. Chem.* 37 (1965) 579.
191. MEANS, E. A., and RATCLIFF, D., Determination of wear metals in lubricating oils by atomic absorption spectroscopy. *Atomic Abs. Newsletter* 4 (1965) 174.
192. SPRAGUE, S., and SLAVIN, W., A rapid method for the determination of trace metals in used aircraft lubricating oils. *Atomic Abs. Newsletter* 4 (1965) 367.
193. TRENT, D. J., The determination of lead in gasoline by atomic absorption spectroscopy. *Atomic Abs. Newsletter* 4 (1965) 348.

194. KERBER, J. D., The direct determination of nickel in catalytic-cracking feedstocks by atomic absorption spectrophotometry. *Appl. Spectroscopy* 20 (1966) 212.
195. MOORE, E. J., MILNER, O. I., and GLASS, J. R., Application of atomic absorption spectroscopy to trace analyses of petroleum. *Microchem. J.* 10 (1966) 148.
196. WILSON, H. W., Note on the determination of lead in gasoline by atomic absorption spectrometry. *Anal. Chim. Acta* 38 (1966) 921.
197. MOSTYN, R. A., and CUNNINGHAM, A. F., Some applications of atomic absorption spectroscopy to the analysis of fuels and lubricants. *J. Inst. Petrol* 53 (1967) 101.
198. HOLDING, S. T., and NOAR, J. W., Mixed-solvent systems for the flame analysis of petroleum materials. *Analyst* 95 (1970) 1041.
199. KASHIKI, M., YAMAZOE, S., and OSHIMA, S., Determination of lead in gasoline by atomic absorption spectroscopy. *Anal. Chim. Acta* 53 (1971) 95.
200. HOLDING, S. T., and MATTHEWS, P. H. D., The use of a mixed-solvent system for the determination of calcium and zinc in petroleum products by atomic absorption spectroscopy. *Analyst* 97 (1972) 189.
201. INSTITUTE OF PETROLEUM STANDARD FOR PETROLEUM AND ITS PRODUCTS. Part I. Methods for analysis and testing. 34th edition. Applied Science Publishers Ltd., England, 1975.
202. LUKASIEWICZ, R. J., BERENS, P. H., and BUELL, B. E., Rapid determination of lead in gasoline by AAS in the nitrous oxide–hydrogen flame. *Anal. Chem.* 47 (1975) 1045.
203. HOLDING, S. T., and ROWSON, J. J., The determination of barium in unused lubricating oil by means of AAS. *Analyst* 100 (1975) 465.
204. SEBOR, G., LANG, I. VAVRECKA, P., SYCHRA, V., and WEISSER, O. The determination of metals in petroleum samples by A.A.S. Part 1. Determination of vanadium. *Anal. Chim. Acta* 78 (1975) 99.
205. LANG, I., SEBOR, G., SYCHRA, V., KOLIHOVA, D., and WEISSER, O. The determination of metals in petroleum samples by A.A.S. Part 2. Determination of nickel. *Anal. Chim. Acta*, 84 (1976) 299.
206. SEBOR, G., and LANG, I., The determination of metals in petroleum samples by A.A.S. Part 4. The effect of halogens on the determination of vanadium and nickel in xylene solutions. *Anal. Chim. Acta*, 89 (1977) 221.
207. SABA, C. S., and EISENTRAUT, K. J., Determination of titanium in aircraft lubricating oils by A.A.S. *Anal. Chem.* 49 (1977) 454.
208. PRÉVÔT, A., and GENTE-JAUNIAUX, M., Rapid determination of phosphorus in oils by flameless AA. *Atomic Abs. Newsletter* 17 (1978) 1.

6 *Characteristics of Standard Equipment*

One of the more time-consuming, exacting, and sometimes frustrating duties that fall to the lot of a chief chemist is that of purchasing expensive analytical instruments. This responsibility is not always facilitated by the sales publications of the manufacturers. For this reason it is felt that a basic description of the main features of commercial equipment, and the points a chemist should consider in choosing a unit for his own uses, may be of interest.

There are several ways of choosing an atomic absorption spectrophotometer. One method, not infrequently employed, is that of selecting the most expensive unit within the laboratory budget and hoping that it will automatically possess a capability in excess of that required. This approach can often produce unlooked-for complications. The equipment, even if the senior analyst is willing to allow junior staff to use it, can be so complicated to operate that only specially trained personnel can do so. It can also be slower in operation than less expensive equipment!

Desirable Features of an Atomic Absorption Spectrophotometer

In attempting to assess the relative merits of various commercial units for use in a particular laboratory the chemist should consider what features he requires the instrument to possess. It is advisable firstly to consider the characteristics and concentration levels of the elements to be determined, together with the accuracy required.

The dramatic improvements that have been made since the end of 1966 in the manufacture of hollow-cathode lamps have given instruments built around simple optics a performance comparable to their more elaborate counterparts.

Secondly, it is important to take account of the ability and experience of the operators who are to use the equipment. Obviously if the unit is to be used by junior staff for routine work, or by personnel not specifically trained in the disciplines of analytical chemistry and instrument engineering, it should be simple and reliable in operation.

Speed of operation is always of importance, and generally the simpler the equipment the faster will it be to operate.

Versatility is a final important feature. Most commercial units can be operated as atomic absorption spectrophotometers and flame emission units. Some can be adapted also for atomic fluorescence. It is when a unit is to be used for high-temperature flame emission or atomic fluorescence that it is essential that it should incorporate a high-resolution monochromator, possessing high light-gathering capacity and low stray light characteristics.

Basic Features of Standard Instruments

All atomic absorption spectrophotometers consist of:

1. a nebulizer system
2. a flame system
3. an optical system, comprising lamps, single or double beam light passage and monochromator
4. a photomultiplier
5. an electronic readout system.

It is in the components and assembly of these essential units (Fig. 6.1) that differences in performance and price (especially the latter) occur.

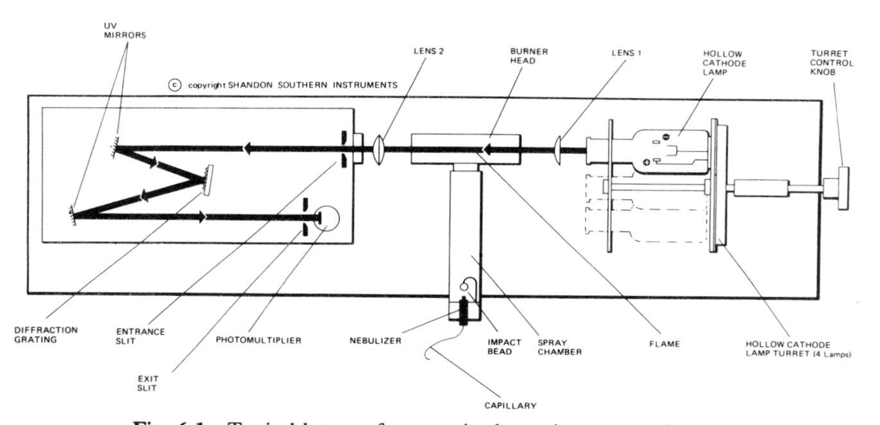

Fig. 6.1 Typical layout of an atomic absorption spectrophotometer

Nebulizers

Nebulizers for Premix Burners

The ideal nebulizer would convert all the sample into a fine mist of uniform droplet size at a constant rate, so that a steady absorption measurement would be obtained. In doing so it would use exactly the amount of nebulizing gas required by the burner arrangement. It would be of simple construction, not easily blocked by particulate matter, made of corrosion-resistant materials, and would remain in adjustment for long periods. The performance should not be dependent on viscosity, surface tension, density, or volatility of the liquid.

The break-up into drops of a liquid emerging from an orifice has received much study. It is well known that a neck appears in the liquid jet, which finally breaks, to form a drop just in front of the neck, with the formation of an extra tiny drop (Plateau's spherule) at the neck itself. With a laminar flow burner the larger drops settle out in the expansion chamber, and only the fine mist enters the burner.

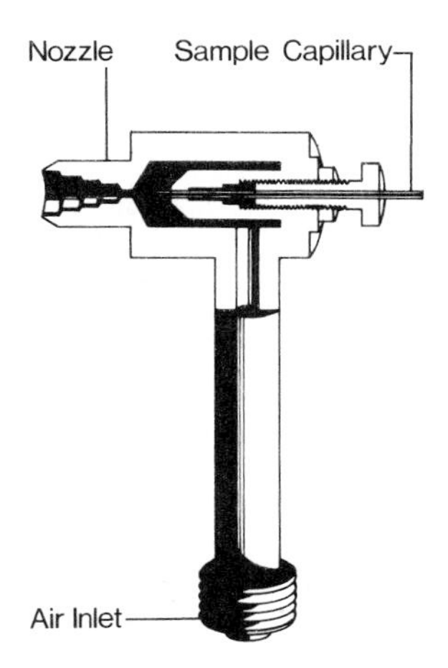

Fig. 6.2 Simple pneumatic nebulizer

A simple pneumatic nebulizer is shown diagramatically in Fig. 6.2. All nebulizers of this type operate on the same principle. A liquid stream emerging from an orifice is broken up into droplets by the flow of the nebulizing gas. The droplet size distribution (i.e., efficiency of nebulization) depends on the flow rate of the nebulizing gas and the relative positions of the end of the capillary and the nozzle.

Nebulizers of this type are therefore operated at a set optimum flow rate for the oxidizing gas, and in some units the position of the capillary can be adjusted by the operator while in others it is preset by the manufacturer. The maximum efficiency (production of Plateau's spherules) attainable with nebulizers of this type is about 10–15 per cent of the liquid taken up. This usually corresponds to a solution uptake rate of between 3 and 6 ml per minute.

Nebulizers are commonly made of stainless steel. This material is sensibly resistant to attack by a large number of reagents but is susceptible to some corrosion from solutions containing hydrochloric acid especially in the presence of iron(III) salts.

Most manufacturers now offer 'corrosion resistant' nebulizers that incorporate platinum–iridium capillaries and titanium, tantalum or 'Kel-F' nozzles.

Impact Bead Nebulizers

These devices, shown diagrammatically in Fig. 6.3, are designed to increase the percentage of small droplets produced and thereby increase the efficiency of nebulization.

A bead of corrosion-resistant material is placed close to the nozzle, so that the droplets (which are moving at almost sonic speed at this point) impinge upon it. The larger droplets are thus comminuted by impact, and the

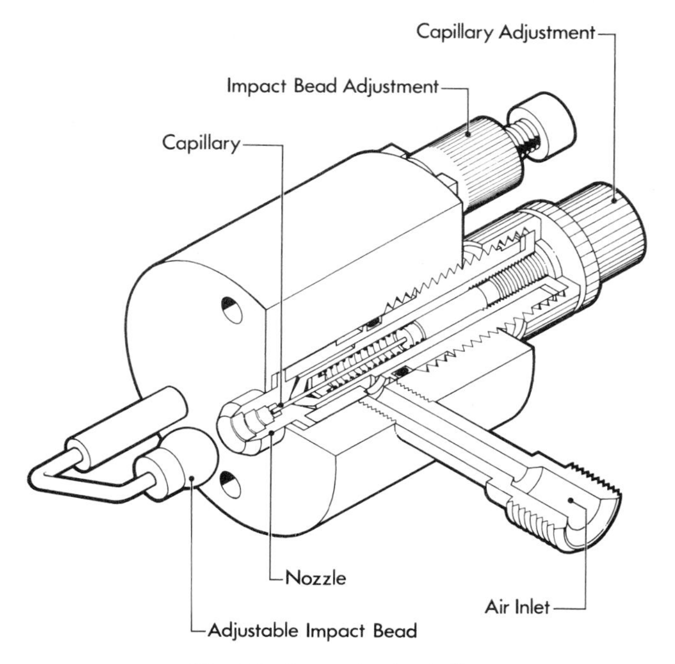

Fig. 6.3 Impact bead nebulizer

proportion of small droplets that are carried to the flame can be increased by a factor of 1·5 to 4 times.

Flame System

There are two types: turbulent flow or total consumption burners and laminar flow or premix burners. The former are seldom used nowadays.

Turbulent Flow Burners

In construction, see Fig. 6.4, this type of burner consists of three concentric tubes, the inner one being a fine capillary which carries the sample into the flame (by means of the suction created by the passage of gases in the surrounding tubes). The sample is injected into the flame at a rate which is dependent upon the flow of gas and also the viscosity of the sample.

Thus, in general, this type of burner is characterized by the fuel gas and supporting gas being unmixed until they reach the base of the flame. The solution to be nebulized is also introduced at this point and thus the burner and nebulization system are built as a single unit. These burners are often known as 'direct injection' or 'total consumption' burners, since all the sample liquid enters the flame and is converted to spray at the point of entry. At first sight

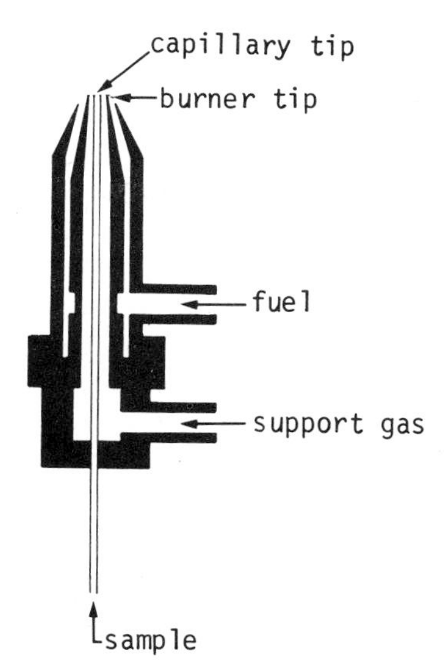

Fig. 6.4 Turbulent flow burner in section

such a system might appear to offer 100 per cent efficiency, and cause one to question why premix burners are used at all.

In practice it is found that total consumption burners must be constructed in a manner similar to an oxy-acetylene blow torch, i.e., they can only have a very short path length, through which the light beam can interact with the atomized species.

Secondly, the droplets aspirated to the flame are of widely varying sizes, so that the processes of evaporation and thermal decomposition occur at virtually all heights in the flame; i.e., it is impossible to select a particular height at which an overwhelming majority of the absorbing species will be found.

The turbulent nature of the flame, in which the gases do not mix until they start to burn, gives rise to a high noise level. The shortness of the light path through the flame, and the non-uniformity of droplet size, reduce sensitivity, and further aggravate the poor signal-to-noise ratio.

Attempts have been made in some instruments to overcome these disadvantages by arranging for the light beam to make multiple passes through the flame. Apart from complicating the optical system and adding to the expense of the unit, such a set-up is intrinsically inefficient. The carrier gas/fuel ratio is widely different at varying heights in the flame, so that the absorption, and in extreme cases interferences, vary at each of the different light paths. Also a sensible percentage of the light is dissipated at the reflecting surfaces so that wider monochromator slit widths and higher photomultiplier gain have to be used, thereby increasing the background noise and possibly reducing sensitivity.

Laminar Flow Burners or Premix Burners

In this type of combustion system the supporting gas nebulizes the sample solution into an expansion chamber, where the larger droplets (constituting 85–90 per cent total aerosol) settle out and go to waste. The remaining portion of the aspirated solution, which is in the form of a mist, of small droplets of reasonably uniform size, is carried into the burner barrel where it is pre-mixed with fuel gas (in some designs the fuel gas is added in the expansion chamber). The pre-mixed gases travel up the burner barrel and are burnt either on an array of holes or a slot. A slot is preferable for high-burning-velocity gas mixtures, e.g., air–acetylene or nitrous oxide–acetylene.

With air–acetylene, air–hydrogen and air–propane, most commercial burners have flame paths 100–120 mm long. The slits of nitrous oxide–acetylene burners, because of the higher burning velocity of this gas mixture, are usually 50 mm in length.

Premix burners of this construction, by virtue of the length of the flame path give good sensitivity, and also a more stable flame which results in much improved signal-to-noise ratios.

Finally, because the droplets are of more uniform size, it is possible to select a definite portion of the flame where the optimum response for the absorbing species will be found. This allows much more reproducible, sensitive and in some cases interference-free determinations to be performed than with a total-consumption system.

A typical air–acetylene burner and a nitrous oxide burner of the grooved-head type are shown in Figs 6.5 and 6.6.

Nitrous oxide–acetylene burners are susceptible to two sources of inconvenience to the analyst. Firstly, carbonization of the slot due to pre-decomposition of the acetylene at the hot edges, and secondly a dropping off of sensitivity as the burner heats up.

Carbonization can be greatly minimized by resorting to the sloping head design. With burners of this design, air entrainment around the base of the flame tends to oxidize any carbon formed on the burner jaws.

This second inconvenience can sometimes be reduced by the addition of cooling fins to the burner grid.

Optical System

Lamps

Hollow-Cathode Lamps are the main spectral sources. There has been a steady increase during the last ten years in the number of elements for which reliable hollow-cathode lamps can be fabricated, and in the light-output intensity from such devices. The production of the so-called 'high-spectral-output' lamps (not to be confused with high-intensity lamps, see later) incorporating shielded cathodes, improved electrode assemblies and better

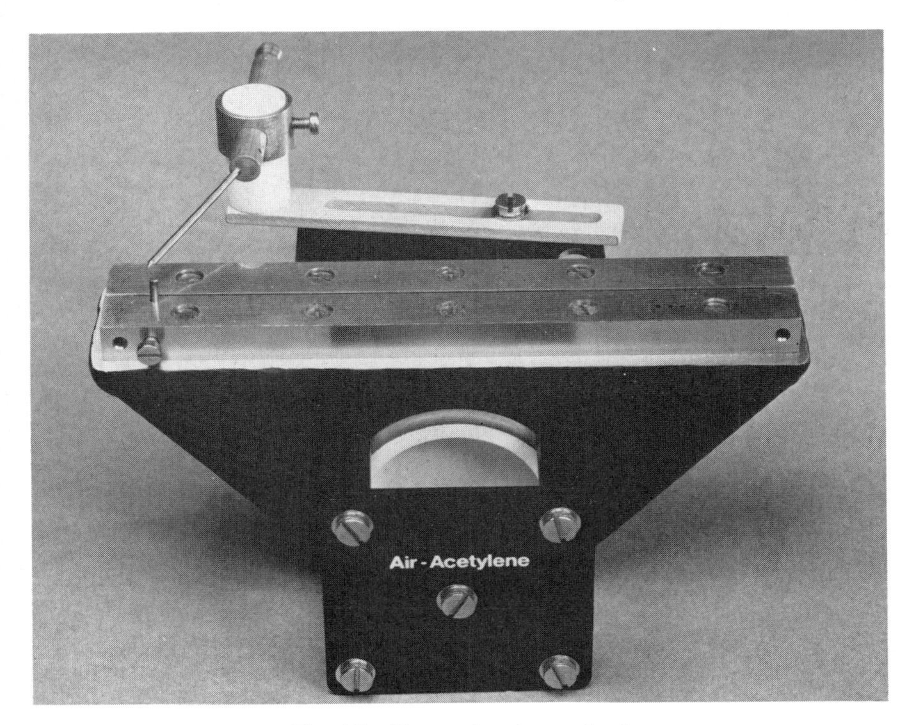

Fig. 6.5 Air–acetylene burner head

choices of filler gases, now permits even the simplest atomic absorption spectrophotometers to select more absorbing wavelengths and gives them a performance comparable to more elaborate and expensive units.

For example, the initial type of argon-filled iron lamp gave a very low output intensity. This necessitated the use of fairly wide monochromator slit settings so that radiation not only of the most absorbing line at 248·3 nm but also that from the less absorbing line at 248·8 nm together with the non-absorbing background was focused on to the photomultiplier. An insensitive and grossly non-linear response curve was thereby produced. An instrument that had superior light-gathering power and resolution would obviously perform analyses better, with such a low-power spectral device, than a unit that had a less efficient monochromator.

The spectra from the currently available neon-filled lamps is not only more powerful but the relative intensity of the most absorbing line at 248·3 nm to the less absorbing line at 248·8 nm has been improved. The improved output from this newer type of source allows narrow monochromator slit settings to be used, so that an instrument having a lower resolution monochromator approaches the analytical capability of one designed around more powerful optics.

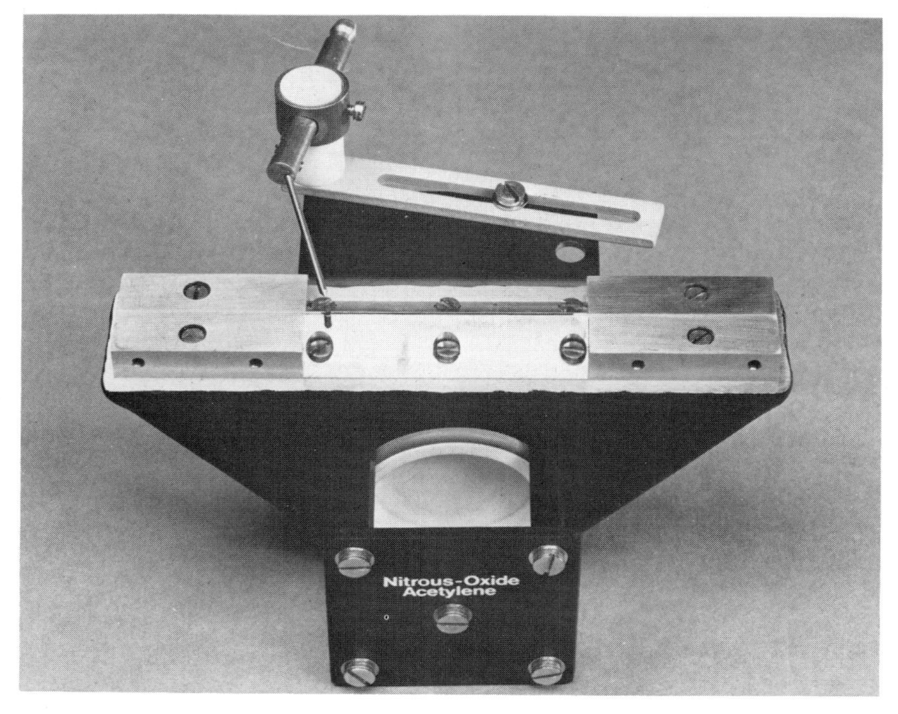

Fig. 6.6 Nitrous oxide–acetylene burner head

For nickel (see below) the output intensity of lamps was firstly improved by substituting neon for argon as filler gas, and the intensity of the unwanted ionic line at 231·6 nm was considerably reduced by restricting the filler gas to a low pressure.

Perhaps the most valuable advance in this series of improvements was the production of hollow-cathode lead lamps that permit reliable use of the very sensitive 217·0 nm line with simple standard atomic absorption spectrophotometers.

High Intensity Hollow-Cathode Lamps (also called 'High Brightness'). These devices were developed by SULLIVAN and WALSH[(1)] and are essentially an open-structure hollow-cathode lamp containing two auxiliary electrodes capable of emitting a sensible population of electrons when electrically heated (see Fig. 6.8). The emitted electrons are designed to pass as a booster current across the hollow cathode. The energy available from the electrons of the booster current is low, so that the easily excited ground-state lines of the metal atoms sputtered from the cathode are preferentially enhanced, and the non-absorbing ion lines are reduced in intensity. Because of this enhancement of the ground-state atomic resonance lines, and suppression of unwanted ion

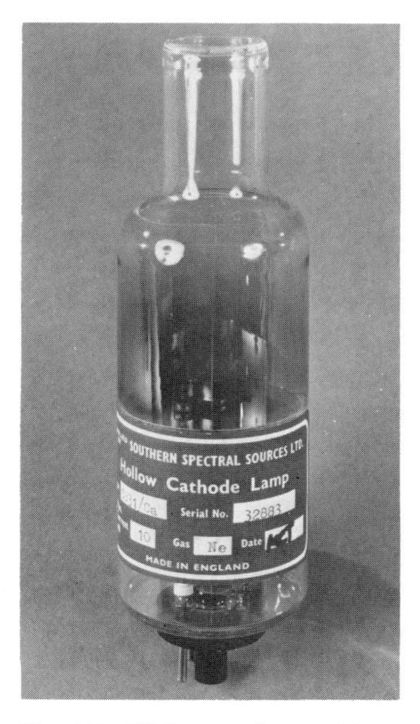

Fig. 6.7 'High spectral output' lamp

Fig. 6.8 High intensity lamp showing open structure and booster electrodes

and filler gas lines, it was anticipated that these high-intensity devices would permit more linear, sensitive and stable atomic absorption measurements to be made. For some elements indeed, notably nickel, this aim was realized.

High-intensity lamps, though, have a number of serious practical disadvantages, and of the two companies that originally produced them, both have now abandoned their manufacture. Briefly, these lamps require an extra power supply to provide the high auxiliary current, they are costly, their lifetimes are short, and they can only be produced for certain elements.

Finally, the drastic improvements in performance that have been made in standard hollow-cathode designs have all but rendered the high-intensity types superfluous.

These points are more specifically illustrated by the data for nickel reproduced in Figs 6.9 to 6.11.

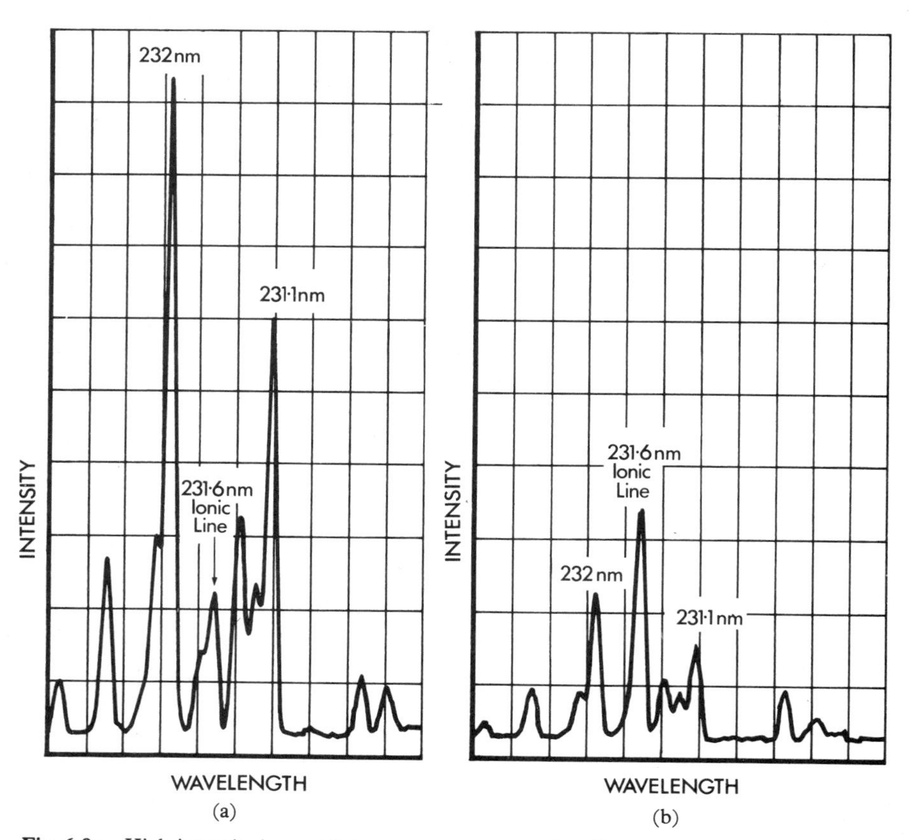

Fig. 6.9a High-intensity lamp with booster current operating. Note reduction in the intensity of the ionic line and the increase in output from the resonance line.

Fig. 6.9b High-intensity lamp operated as a normal hollow cathode, i.e., without the booster current. Note high relative intensity of the ionic line compared to the resonance line

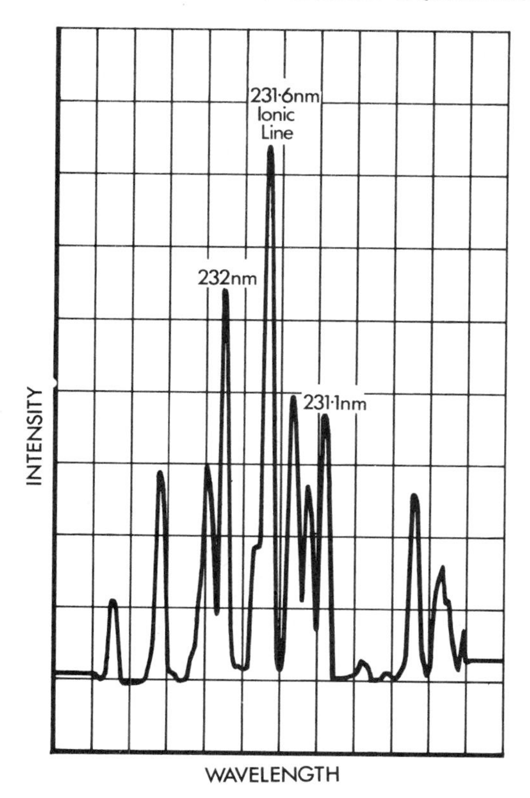

Fig. 6.10 Spectral trace from a hollow-cathode lamp, containing too high a pressure of filler gas. Note the very high relative intensity of the ionic line

Figure 6.9 shows the spectrum obtained from a high-intensity lamp operated with and without the booster current. Without this auxiliary current, i.e., with the source operated as a normal hollow-cathode lamp, not only is the output much less intense than when used in the true high-intensity mode, but the non-absorbing ionic line at 231·6 nm is more prominent than the resonance line at 232·0 nm.

The ionic radiation is due to collision of filler gas atoms with nickel atoms, and this fact is illustrated in Fig. 6.10 and 6.11. Figure 6.10 shows the spectrum for a normal hollow-cathode lamp containing too high a pressure of filler gas. Figure 6.11 shows a spectral trace for a second standard hollow-cathode lamp, that contains neon at a lower pressure than that in the lamp for which the previous trace was produced. It illustrates the fact that undesirable ionic radiation can in some cases be reduced to an insignificant level by controlling filler gas pressure, thereby enabling a normal hollow-cathode lamp to be constructed for nickel that possesses all the essential benefits of a high-intensity lamp. With such a device an instrument designed around a low-resolution monochromator provides an analytical performance similar to that of a unit having high-resolution optics.

Microwave Excited Electrodeless Discharge Tubes. (E.D.T.s*) The production and operation of these devices have been described in detail by WEST *et al.*[2,3] Essentially, they consist of a sealed quartz vial, about 0·8 cm internal diameter, 4 cm long and wall thickness 1 mm, containing a small quantity of the required element, usually as the iodide, under a pressure of between 1 and 6 Torr of argon. When placed in a high-frequency electromagnetic field, produced by a microwave generator, they emit energy of the characteristic wavelengths of the material they contain (see Figs 6.12 and 6.13).

The general advantages and disadvantages of electrodeless discharge tubes with respect to hollow-cathode lamps are listed below.

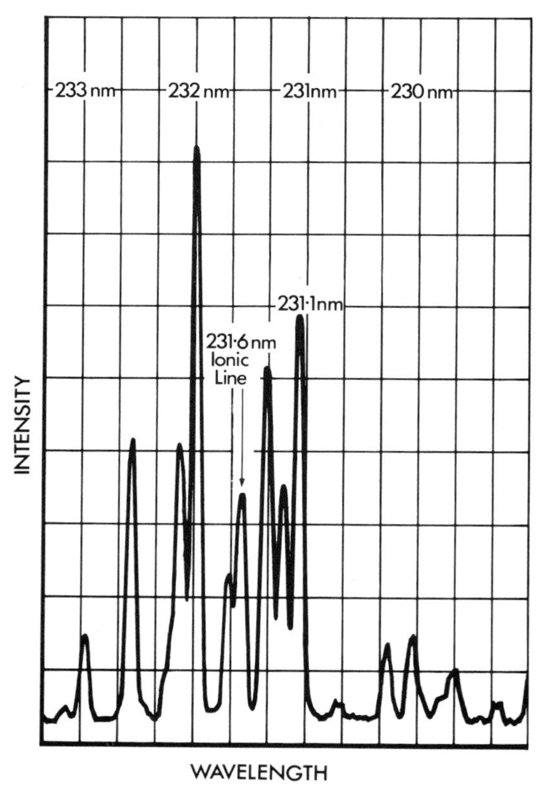

Fig. 6.11 Spectral trace of output from a hollow cathode lamp of shielded design, containing a low pressure of neon. Note the drastic reduction in the relative intensity of the ionic line

*These are also referred to as microwave excited electrodeless discharge lamps (E.D.L.s.).

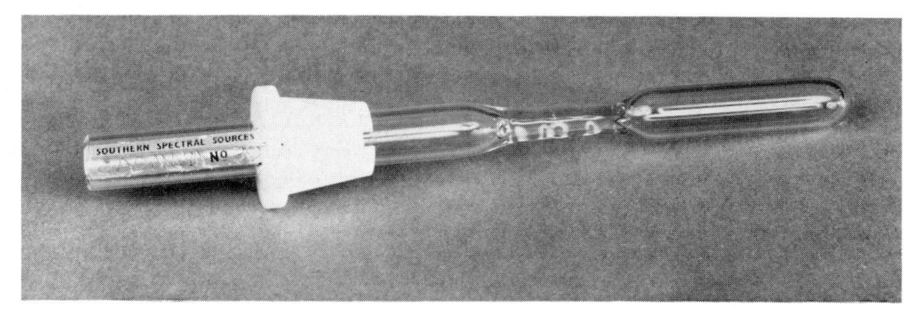

Fig. 6.12
Microwave excited electrodeless discharge tube

Microwave-excited electrodeless discharge tubes retail for about half the price of hollow-cathode lamps and possess exceedingly long lifetimes. They can even be manufactured by a competent technician in a laboratory equipped with a high vacuum and glass blowing facilities. They emit very sharp spectral lines of much higher intensity than can be obtained from a hollow-cathode lamp.

Electrodeless discharge tubes suffer from the disadvantage that they require for their operation a special microwave generator, which is a fairly costly piece of equipment. More seriously, at their present state of development the tubes

Fig. 6.13 Microwave cavity

vary rather widely in characteristics and are subject to more operating parameters than are hollow-cathode lamps. To be specific, the performance obtainable from an electrodeless discharge tube depends upon the position of the tube in the microwave cavity, the extent of cooling to which the cavity is subject, the cavity tuning and the operating power. Most require a very long warm-up period to reach stability of output, at least 30 min, and in extreme cases over 2 h. The variations in output that occur over this period are usually towards higher intensity in the early stages, but can later show a drop-off in energy. This phenomenon, it is thought, is due to the substance which gives rise to the discharge physically moving around inside the tube.

The hollow-cathode lamps that are currently being produced require only about 7 to 15 min to reach stability of output. They also emit sharp spectral lines, and are subject to only one operating parameter, lamp current.

BROWNER *et al.*[7,8] have succeeded in improving the stability and warm-up times of electrodeless discharge lamps by passing a stream of heated air over a microwave lamp positioned above an 'A'-type antenna. (This gives a uniform microwave field around the lamp.) The lamp is maintained at the optimum operating temperature by the heated airstream and is excited by a relatively low microwave power. With this system the tuning of the lamp is no longer critical. The temperature of the airstream can easily be adjusted to suit each type of lamp. Thus, the microwave energy is not required to heat the lamp, but only to excite it. With conventional operation of microwave lamps, a large proportion of the microwave energy is used to heat the lamp. This heating is the main reason for the instabilities observed. As the lamp temperature changes, the cavity tuning and vapour pressure of the lamp contents radically alter, resulting in a positive feedback type of system. The heated airstream method effectively thermostats the lamp, with the microwave energy only being used to excite the lamp.

With this type of system the warm-up time seldom exceeded 7 min. Typical long-term stabilities observed were 0·2–2% per hour. Typical peak-to-peak noise at a time constant of 0·5 s was 0·1–0·3%.

It must be admitted, at the present time, though, that microwave excited electrodeless discharge lamps cannot generally be recommended as a substitute for hollow-cathode lamps in the routine laboratory, where reliability, speed, and convenience of operation are important. For further information, the interested reader is referred to reference 9 that gives a critical review on the preparation and operation of these lamps.

Radiofrequency excited electrodeless discharge lamps. The use of simple radiofrequency excited electrodeless discharge lamps (R.F.E.D.L.s) in place of hollow-cathode lamps has been reported.[10] These lamps are excited from a low-power (up to about 30 Watts) 27 MHz radiofrequency generator and are 5–100 times more intense than the equivalent hollow-cathode lamp. The advantages of the R.F.E.D.L.s compared to their microwave-excited equivalents are that the radiofrequency lamp coil is permanently attached to the lamp during manufacture so that tuning and lamp alignment problems are eliminated, and also that the radiofrequency power supply costs somewhat less than a microwave power supply. The main disadvantages of R.F.E.D.L.s are that satisfactory lamps have only been prepared for a limited number of elements (this can also be said of microwave-excited lamps) and in general R.F.E.D.L.s are not as intense as microwave-excited lamps. This is an important consideration for atomic fluorescence studies. The lamps that have been prepared are for relatively volatile elements or elements that form relatively volatile halides (e.g., As, Bi, Cd, Cs, Ge, Hg, P, Pb, Rb, Sb, Se, Sn, Te, Tl and Zn).

The atomic absorption detection limits obtained when using these radiofrequency sources with a double beam instrument were found to be generally slightly better than those obtained using a hollow-cathode lamp source.[10]

The Single and Double Beam Systems

Some commercial atomic absorption spectrophotometers use a single beam optical system. With this system, as its name implies, the light from the lamp is simply passed through the flame, as a single beam, to the monochromator, and focused on to the photomultiplier. Units incorporating this principle are simple in construction and easy and quick to operate.

With the double-beam system the light from the source is split, usually by means of a rotating sector mirror into two beams, one of which passes through the flame, while the second (the reference beam) is deviated around it. At a point beyond the flame the two beams are recombined and their ratio is electronically compared (Fig. 6.14).

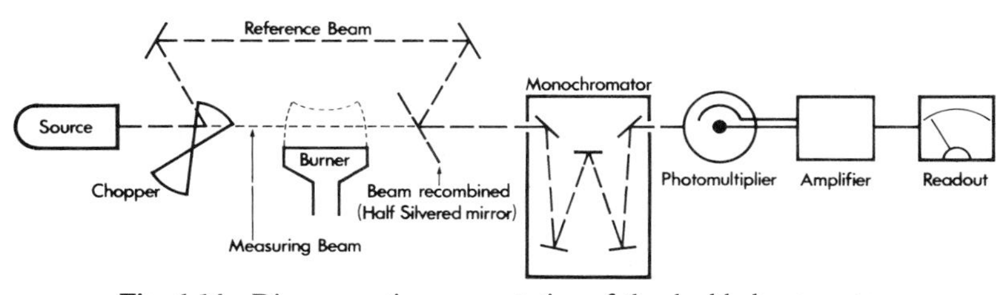

Fig. 6.14 Diagrammatic representation of the double beam system

The advantages claimed for such a system are:

(i) That it is possible to take readings directly a lamp is switched on, instead of waiting for it to reach stable maximum emission. As this warming period is only about 5 to 15 minutes and virtually all standard instruments are now offered with multilamp turrets, this is not a very important advantage. Also, the cathode temperature changes during the warm-up period, causing slight changes in the slope of the calibration curve.

(ii) The second advantage quoted is that the system corrects for small variations in intensity of light output during a run, so that a steady base line is obtained. This may have been of some significance in the early days of atomic absorption, but most lamps now produced are so stable that this advantage is not very apparent for most determinations.

For very prolonged runs using a continuous sampling procedure, for example, the determination of calcium in blood serum in say 100 or more samples that are selected, prepared, diluted and fed automatically to an atomic absorption spectrophotometer, the ability of a double-beam instrument to correct for zero drift could be advantageous. It so happens, however, that calcium lamps are more prone to variations in output over long running periods than lamps for most of the other elements. The use of auto zero can overcome this (see p. 196) when using single-beam instruments.

The double-beam system, though, possesses some shortcomings.

(i) The reference beam corrects only for the variations in lamp output, which are often virtually negligible. It does nothing to overcome the most serious sources of instability, fluctuations in the flame and nebulizer systems.

(ii) The utilization of the light from the lamp is much less efficient with a double-beam system than with a single beam, because a high percentage of the energy available is lost by sharing between the two beams, and on the more numerous optical surfaces required. This means that wider slit settings have to be employed so that unless a high-resolution monochromator of good light-gathering capacity is used, a decrease in sensitivity can result. This effect is particularly pronounced for arsenic and selenium estimations.

The Monochromator

A monochromator disperses polychromatic light into its constituent wavelengths by using the properties of a prism or a grating. Prisms are usually constructed from fused silica and vary in performance according to their size and quality of manufacture. Grating performances vary mainly according to their size, quality of manufacture and the number of lines per unit width. The typical monochromator consists of:

1. An entrance slit—a long narrow adjustable slit (~0·02–1 mm).
2. A collimating mirror—a concave mirror to render all rays from the entry slit parallel before reaching the dispersing element.
3. A dispersing element—a prism or grating (i.e., a device that alters the intensity of light in a way dependent upon wavelength).
4. A focusing mirror—a concave mirror which receives the parallel dispersed radiation and images it upon the exit slit.
5. An exit slit—a long narrow adjustable slit (~0·02–1 mm) usually mechanically linked to the entry slit.
6. A detector (photomultiplier).

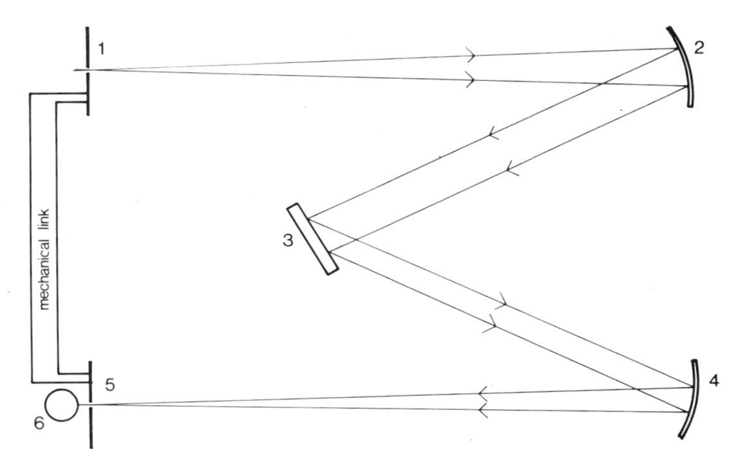

Fig. 6.15 Typical Czerny–Turner monochromator layout

This is depicted in Fig. 6.15 which shows the optical layout of a Czerny–Turner grating monochromator.

The two optical characteristics that influence the performance of an atomic absorption spectrophotometer are (a) the light intensity that the monochromator can focus on to the photomultiplier; and (b) the ability of the monochromator to separate lines that are close together in the spectrum.

The factors that affect the intensity of light focused on the photomultiplier are f-number and efficiency of light utilization.

f-number. The brightness of an image formed by an optical instrument, or in the case under consideration, the light intensity that can be focused upon the photomultiplier, is governed by the f-number, which is given by the ratio f/a, where f is the focal length of the instrument and a is the diameter of the effective entrance pupil. The lower the f-number the brighter is the image, so that, in theory, for a given focal length, it is advantageous if the value of a is made as large as possible. This can be accomplished by increasing the physical size of the internal optics of the monochromator, i.e., a unit containing large mirrors and a large grating, or prism, will transmit more light than one of equal focal length, containing smaller components. In practice the requirements in atomic absorption are for efficient collimation of the light from the hollow-cathode through a long (12 cm) thin flame. Very low f-numbers are not required and can, in certain circumstances, cause an increase in stray light level.

Efficiency of light utilization (prism versus grating). Light dispersed into its component wavelengths by a prism is all of one order whereas with a diffraction grating it is shared between the several orders produced. From this point of view, therefore, a prism should be capable of transmitting a higher intensity of light to the detector. In practice the angles at which gratings are blazed can be adjusted so as to concentrate about 70% of the light into a particular order.

The factors that govern the separation of spectral lines are the linear dispersion and resolution.

Linear Dispersion. This is defined as $dl/d\lambda$ (where λ = wavelength and l = the distance along the focal plane). The focal plane passes through the exit slit and hence the linear dispersion is a measure of the spread of the spectrum in mm per nm at the exit slit of the monochromator.

The reciprocal of the linear dispersion is called the reciprocal dispersion, and gives the spread of the spectrum in nm per mm at the focal plane of the system. From this figure the bandpass, i.e., width of the spectrum in nm that will be focused upon the detector at various slit settings, can be computed.

With a grating the reciprocal dispersion is effectively independent of wavelength whilst with a prism the reciprocal dispersion increases with increasing wavelength.

Resolution. The resolution, or resolving power of an optical instrument, defines its ability to produce separate images of objects that are very close together. For a monochromator it is given by the ratio $\lambda/d\lambda$, whereas $d\lambda$ is the smallest wavelength difference that produces resolved images. With a grating the diffraction pattern sets a theoretical upper limit to this characteristic, and it can be shown that, in a given order, it is proportional to the total number of rulings, but is independent of spacing. For a prism it increases with the refractive index and the length of the base.

With a grating monochromator, a spectrum formed in the first order of diffraction is always used, so as to avoid complications due to overlapping orders. To obtain adequate dispersion, the grating space must be made as small as possible. At the same time the resolution will improve as the number of rulings is increased.

A 50 mm width grating ruled at 600 lines per mm has a theoretical resolution given by

$$R = \frac{\lambda}{d\lambda} = mN$$

where m is the diffraction order and N is the total number of rulings on the grating, i.e.

$$R = 50 \times 600 = 30\ 000 \text{ (for the first order).}$$

Thus, at 240 nm, the smallest wavelength interval that can be resolved by the grating will be

$$d\lambda = \frac{240}{30\ 000} = 0{\cdot}008 \text{ nm.}$$

Most grating monochromators do not achieve this theoretical resolution. The resolution is usually limited by the smallest usable slit width (0·02 mm). For example, with a quarter-metre focal length monochromator with a 50 mm grating ruled at 600 lines per mm with a minimum slit width of 0·02 mm the maximum obtainable resolution is approximately 0·13 nm.

Summary of factors affecting characteristics of a grating monochromator

	Characteristic				
Factor Increased	**Light gathering**	**Linear dispersion**	**Resolution as measured at fixed slit settings**	**Overall size of instrument**	**Price**
Focal Length	**Reduced**	**Improved**	**Improved**	**Increased**	**Increased**
Size of grating and mirrors	**Improved**	—	—	—	**Increased**
Lines per mm on grating	—	**Improved**	**Improved**	—	**Increased**

To summarize, an instrument using a grating monochromator possesses the characteristics of uniform dispersion over the whole spectrum. It will almost certainly have superior dispersion at wavelengths above 250 nm compared with a similar-priced prism monochromator.

Photomultiplier

Some photomultipliers used in commercial atomic absorption spectrophotometers, together with their nominal characteristics, are tabulated below:

E.M.I.	9783R	Quartz window	Range 165–750 nm
E.M.I.	9781R	u.v. glass window	Range 185–750 nm
E.M.I.	9785B	u.v. glass window	Range 185–850 nm
H.T.V.	R382	Quartz window	Range 165–750 nm
H.T.V.	R446	u.v. window	Range 185–850 nm
H.T.V.	R456	Quartz window	Range 165–850 nm
R.C.A.	I.P.28R	u.v. glass window	Range 185–750 nm

There can be considerable variation in performance between tubes of the same nominal characteristics. For example, the H.T.V. R446 tubes, which are widely used in atomic absorption spectrophotometers, are all specially selected, firstly by the component manufacturer and then again by the instrument manufacturer.

A quartz window tube will transmit light down to 165 nm, but below 185 nm air and standard premix flames totally absorb all radiation (see p. 212).

Electronic Readout System

Modulation

In general, not only will there be light of a particular wavelength originating from the lamp, falling upon the photodetector, but also light of the same wavelength arising from the flame. It is necessary to distinguish between these two sources of radiation, since it is the measurement of that from the lamp only which is required. This requirement is attained by modulating the light from the lamp, with either a mechanical chopper, or by using pulsed power supply and tuning the electronics of the detector to this particular frequency. The d.c. signals from the flame and other extraneous sources are thus eliminated.

The main advantage of electronic modulation over a mechanical chopper is that no moving parts are required. However, if a double beam system using 'time sharing', i.e., one in which the light from the measuring and reference beams is passed onto the same photomultiplier, is used, then a mechanical chopper is essential.

The random background noise that originates from the flame is always more noticeable at low frequencies. High-frequency photomultiplier noise increases above about 1000 Hz. The optimum modulation frequency is thus round about 300 to 400 Hz. Harmonics of mains frequency should be avoided.

Auto-Zero

The object of this facility is to correct automatically for instrumental drift, on single beam instruments, by allowing the response for a blank solution to be reset to read zero by the operation of a button. The auto-zero facility can also be remotely operated when using an automatic sampler, so that when a selected blank solution is sampled the auto-zero facility operates and brings the baseline back to the zero setting.

Auto-Concentration

This facility permits a standard solution to be set to a pre-selected reading, and allows recalibration to this value during a series of determinations to be achieved at will by operation of a push-button, or by remote operation.

The combined facilities will permit an operator to re-zero and re-calibrate an instrument even when subject to fairly considerable levels of drift and fluctuation. The user, though, should be advised that incorrect results can nevertheless be obtained if these facilities are used indiscriminately.

Specifically, deviations due to fluctuations in the intensity of the light source, slight variations in the nebulization efficiency and instability of the electronic circuitry are validly corrected for by auto-zero and auto-concentration facilities.

Deviations due to wavelength drift, partial blockage of the nebulizer, and contamination of the burner system, although they may be electronically corrected for on the blank and standards, can, if excessive, lead to erroneous results; because under these conditions the shape and slope of the response curve are sometimes drastically changed.

Curvature Correction

The relationship between concentration of an absorbing species in solution and absorbance is seldom absolutely linear, particularly at absorbances above about 0·5. Incorporation of facilities that permit a good degree of correction for this effect has become fairly standard practice in commercial equipment.

The calibration curve is plotted and in general is sensibly linear up to absorbances of 0·2–0·3. Above this level the curvature increases with increasing concentration. Some of the correction systems employed allow the analyst to select the absorbance where non-linearity commences and then apply a variable amount of non-linear correction for concentrations above this point, the degree of correction increasing with absorbance. Other systems have a preset absorbance level above which the non-linear correction is applied. It should be emphasized that although grossly non-linear curves can often be 'corrected' the ultimate precision obtainable using a grossly non-linear calibration curve will be poorer than that using a linear calibration curve.

Before applying curve correction the analyst should ensure that the lamp and burner position are optimized. Non-linearity of calibration curves is increased if the light beam from the hollow-cathode lamp does not pass through the centre of the flame.

Integration of the Signal

The two parameters that control limits of detection, and hence the precision, with which a determination can be performed are sensitivity and background noise. Diminution of background noise will improve the precision of all estimations, and will noticeably enhance the performance of an instrument for low-level determinations. Fluctuations in readings can very simply be observed and averaged by feeding the signal to a chart recorder and drawing a line midway between the maximum height of the peaks and minimum of the valleys.

A neat and valuable capability on all modern instruments is that provided by the inclusion of an electronic integrating device. This functions by allowing the total fluctuating current, over a period of 3–15 s, to charge a capacitor. The voltage attained by this unit is then applied to the read-out device. The reading is usually held until a 're-set' button is operated.

The integrated readings so obtained are thus absolutely steady so that there is no uncertainty in taking a single reading. Scale expanded readings for tin

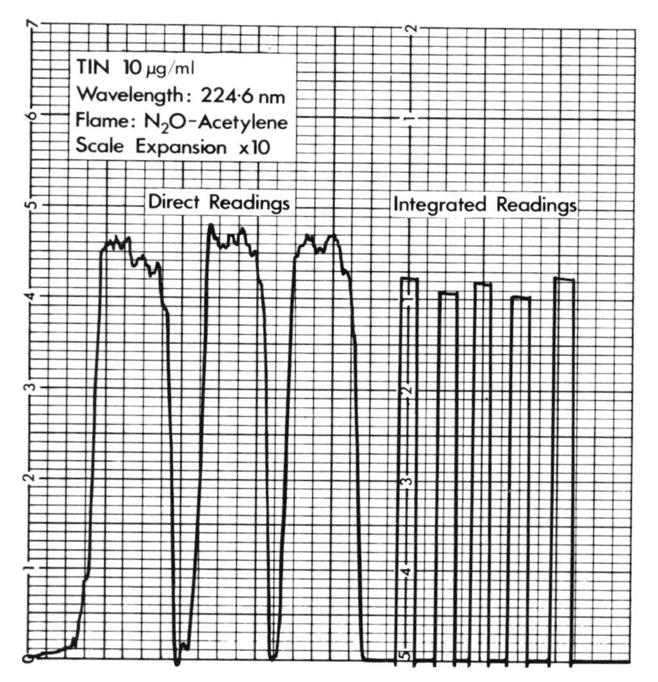

Fig. 6.16 Scale expanded readings for tin in the direct and integrated modes

taken in the direct and integrated modes are shown in Fig. 6.16 which illustrate this point.

Figure 6.16 also illustrates how the inclusion of integration can sometimes improve the limit of detection. Although the noise level in the direct mode is high and there is a wide area of uncertainty in taking a single reading, it is also fairly uniform. Thus, when successive signals are integrated over constant periods of 3–15 s the average values obtained show only small differences.

Scale Expansion and Zero Suppression

Since 1966 all atomic absorption spectrophotometers have been provided with built-in facilities that permit scale expansion. Such facilities are, of course, essential for low-level estimations. In almost all instruments the expansion is continuous up to a maximum of twenty or fifty times.

An ability to suppress a particular response to zero and then to apply scale expansion over a small selected range is a very valuable facility that all modern atomic absorption spectrophotometers possess.

Alternatively, the use of a $3\frac{1}{2}$-digit digital voltmeter (D.V.M.) readout unit allows non-zero suppressed integrated readings to be taken with sufficient precision in most cases. For example, in the determination of calcium, a

4 μg/ml calcium solution may give an integrated absorbance meter reading of 0·35. This would be registered as a D.V.M. reading of 350. By suitable use of the scale expansion this D.V.M. reading could be increased to 400 (or 800) to obtain precise direct (or direct multiple) concentration readout. In this case the readout precision would be 0·25 (0·125)% and in most determinations this would not be the limiting factor in the ultimate precision attainable. For non-linear calibration curves some form of curve correction would be necessary. Many modern instruments now have facilities to automatically correct non-linear calibration curves (see p. 197).

Chap. 6 References and further reading

1. SULLIVAN, J. V., and WALSH, A., High intensity hollow cathode lamps. *Spectrochim. Acta* 21 (1965) 721.
2. DAGNALL, R. M., THOMPSON, K. C., and WEST, T. S., Microwave-excited, electrodeless discharge tubes as spectral sources for atomic fluorescence and atomic absorption spectroscopy. *Talanta* 14 (1967) 551.
3. DAGNALL, R. M., and WEST, T. S., Some applications of microwave-excited, electrodeless discharge tubes in atomic spectroscopy. *Applied Optics* 7 (1968) 1287.
4. HEADRIDGE, J. B., and RICHARDSON, J., Comparison of electrodeless discharge tubes and hollow cathode lamps in atomic absorption spectroscopy. *Laboratory Practice* 19 (1970) 372.
5. ALDOUS, K. M., ALGER, D., DAGNALL, R. M., and WEST, T. S., Preparation and operation of microwave excited electrodeless discharge tubes for use in atomic spectroscopy. *Laboratory Practice* 19 (1970) 587.
6. COOKE, D. O., DAGNALL, R. M., and WEST, T. S., Optimisation of some experimental parameters in the preparation and operation of microwave-excited electrodeless discharge lamps. *Anal. Chim. Acta* 54 (1971) 381.
7. BROWNER, R. F., PATEL, B. M., GLENN, T. H., RIETTA, M. E., and WINEFORDNER, J. D., A device for precise temperature control of electrodeless discharge lamps. *Spectroscopy Letters* 5 (1972) 311.
8. BROWNER, R. F., PATEL, B. M., and WINEFORDNER, J. D., Design and Operation of Temperature Controlled Multiple Element Electrodeless Discharge Lamps for Atomic Fluorescence Spectrometry. *Anal. Chem.* 44 (1972) 2272.
9. HAARSMA, J. P. S., DE JONG, G. J., and AGTERDENBOS, J., The preparation and operation of electrodeless discharge lamps. A critical review. *Spectrochim. Acta* 29B (1974) 1.
10. BARNETT, W. B., VOLLMER, J. W., and DE NUZZO, S. M., The application of RF electrodeless discharge lamps in atomic absorption. *Atomic Abs. Newsletter* 15 (1976) 33.
11. BENTLEY, G. E., and PARSONS, M. L., Preparation of E.D.L.'s for elements forming gaseous covalent hydrides. *Anal. Chem.* 49 (1977) 551.

7 *Some Further Techniques*

The information in the preceding chapters, it is hoped, will enable an analyst to develop and perform reliable estimations by atomic absorption spectroscopy. The subject, though, is one that tends to interest scientists to an extent that they enquire beyond the bounds of knowledge essentially required for their purposes. It is at this point that the interested newcomer to the field can start to feel confused about the hardware and procedures referred to at lectures and in original articles. This chapter is therefore devoted to a description of these developments, of which some are of mere historical interest, some may be incorporated into standard instruments in the future, and others have already found specialized practical application.

1. Nebulization

(A) Ultrasonic Nebulization

When a beam of ultrasonic waves is passed through a liquid and directed at a gas interface, atomization of the liquid occurs. Liquid particles are ejected from the surface into the surrounding gas. Under proper conditions a very fine dense fog is produced.

The technique of ultrasonic nebulization has at least one advantage over conventional nebulization techniques; the fog particle size and fog density can be independently controlled.

With the pneumatic nebulizers normally employed in atomic absorption systems, reduction in particle size can only be effected at the expense of fog density, because to attain it the gas flow must be increased. In ultrasonic nebulization the fog density can be varied simply by adjusting the gas flow past the liquid surface. The amount of liquid suspended in gas is limited only by the rate at which it condenses out. The size of fog particles can be controlled by varying the frequency of the ultrasound. The higher the frequency the smaller the droplet. Lang[1] experimentally found the size of droplets to be

$$D = 0{\cdot}34\,\lambda$$

where D = median drop diameter
λ = capillary wavelength produced by ultrasound.

$$\lambda = \left(\frac{8\pi\sigma}{\rho f^2}\right)^{1/3}$$

where σ = surface tension
ρ = liquid density
f = ultrasonic frequency

The minimum gas flow required to transport the droplets must be in excess of their terminal velocity, which can be calculated from Stokes' Law.

Ultrasonic nebulizers were first used in conjunction with R.F.[2,3,4] plasmas, where low gas flow rates make the use of pneumatic nebulizers impracticable. More recently attention has been given to their use with conventional flames in atomic absorption spectroscopy.[5–10] The high efficiency of liquid sample to mist conversion, which is independent of gas flow rate, makes the system ideal in atomic spectroscopy. Also the range of droplet sizes produced is much narrower than with the normal mechanical nebulizer.

In the early design of ultrasonic nebulizers[3] sample changing was inconvenient. The sample was placed inside a tank, on to the base of which an ultrasonic transducer was bonded. The ultrasonic waves, produced by applying an 800 kHz signal, were focused by a plano-concave leucite lens on to the liquid surface of the sample. As the R.F. output was increased to about 3 watts/cm^2 at the transducer, a stable fog was produced above the liquid surface, which could be transported over appreciable distances with little condensation of droplets. The average droplet size was about 5 microns.

With a later design the sample was drip-fed on to the transducer at a fixed rate and total nebulization was achieved.[9] A much quicker sample changeover and wash through could be obtained with this system.

(B) The Pulsed Nebulization Technique

The maximum concentration of an iron or steel solution that can be reliably nebulized into a nitrous oxide–acetylene flame is generally accepted to be 2–3% m/v (2–3 g sample per 100 ml). If higher concentrations are nebulized the burner slot rapidly blocks. A method of overcoming this limitation is to pipette 25–200 μl of the sample into a suitable container and totally nebulize this volume whilst recording the transient output pulse on a pen recorder.[11–15]

This procedure permits results to be obtained for solutions containing up to 10% m/v iron or steel, or 2·5% m/v aluminium on a routine basis using the nitrous oxide–acetylene flame. The method is not very satisfactory with the cooler air–acetylene flame because of severe background absorption effects (see p. 239). Figure 7.1 shows some results obtained in the author's (K.C.T.) laboratory using a specially modified nitrous oxide–acetylene burner, with a slot width of 0·60 mm and a 200-μl sample volume.[12] A solution of 0·1 M hydrochloric acid was nebulized between each discrete sample. The risk of explosive flashback using this wide slot burner was effectively eliminated by adding nitrogen at a flow rate of 2 l/min to the flame.

The pulse nebulization technique should work satisfactorily with most modern nitrous oxide–acetylene burners, but partial blockage of the burner slot will sometimes be observed after 15–30 min operation when pulse nebulizing 10% m/v steel solutions. It should be stressed that for optimum precision the optimum setting of the damping control should be ascertained for

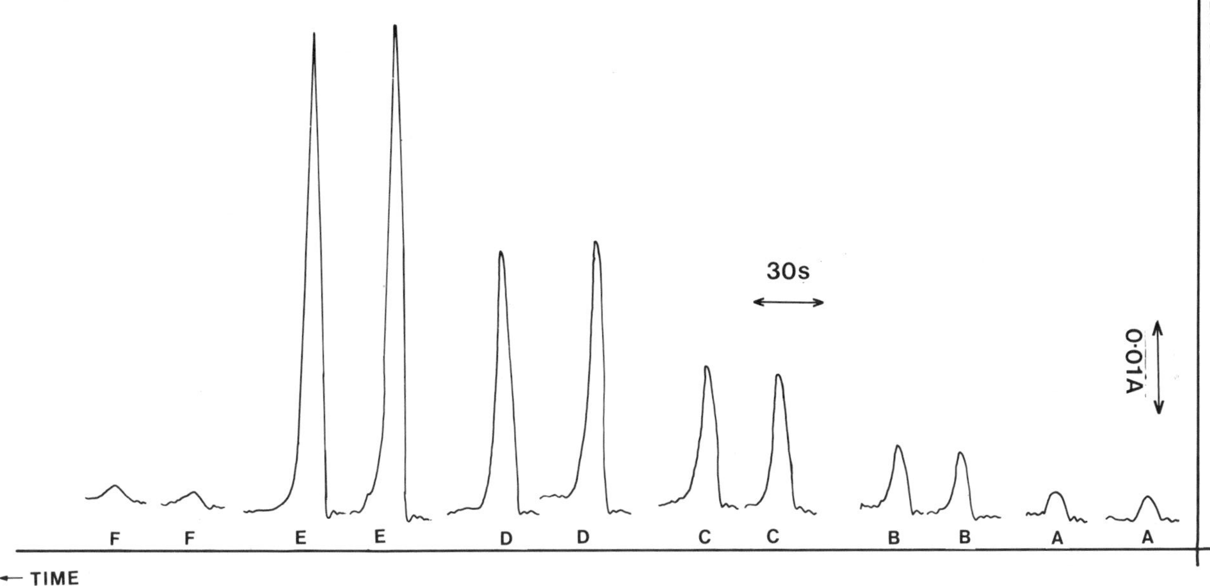

Fig. 7.1 Typical traces obtained using the pulse nebulization technique. Flame, nitrous oxide–acetylene (diluted with 2 l/min nitrogen and maintained on a wide-slot burner). Element, aluminium. Wavelength, 396·2 nm. Sample volume 200 μl. Damping (time constant) 2 sec.

A 100 000 μg/ml iron solution
B As A + 1 μg/ml aluminium
C As A + 2·5 μg/ml aluminium
D As A + 5 μg/ml aluminium
E As A + 10 μg/ml aluminium
F As A using palladium 395·9 nm non-resonance line for background correction.

each new type of determination. Typically, a time constant of 1–2 sec has been found to be optimum.

If conventional background correction using a deuterium hollow-cathode lamp is used for this type of analysis, where 5–10% m/v steel solutions are pulse-nebulized, erroneous results can be obtained. For example, if a 10% m/v high-purity iron solution is nebulized the background absorption signal as registered using a deuterium lamp is significantly greater than that observed using the aluminium 309·3 nm or tin 286·3 nm lines.(12) This anomalous behaviour was attributed to atomic iron lines falling within the monochromator spectral bandpass. There are about 6000 iron lines between 200 and 850 nm. The high concentration of iron atoms present in the flame will absorb radiation from the deuterium lamp over the wavelength region corresponding to the absorption profile of some of these iron lines that fall within the monochromator spectral bandpass. For absorption to occur the lower energy level of a given iron line should be sufficiently populated at the temperature of the nitrous oxide–acetylene flame (see Chapter 8). Although this absorption is small (less than 0·01 absolute absorbance units), it can be significant. For example, there are five iron lines within ±0·15 nm of the 309·3 nm aluminium line and there are three iron lines within ±0·15 nm of the 286·3 nm tin line. (One of these latter lines has a lower energy level of only 0·09 eV.)

In order to correct for the very small amount of genuine background absorption a non-resonance line from an element that is unlikely to be encountered in the samples should be used. Ideally, the chosen background correction line and analyte resonance line should not overlap any atomic line profiles of the major elements in the sample. Table 7.1 shows some detection limits (95% confidence) for the determination of arsenic, tin and soluble

Table 7.1 Results obtained using pulsed nebulization technique(12)

200 μl 10% m/v steel solution pulse nebulized. 0·3 nm spectral bandpass

Element	Wavelength nm	Detection limit. % analyte in steel	Background correction wavelength and lamp	Background absorption from 10^5 μg/ml^{-1} steel solution. % analyte in steel
Al	396·2	0·00025	395·9 Pd	0·00025
Al	309·3	0·00025	311·1 Pd	0·00025
			309·3 D_2	0·0012
Sn	286·3	0·0008	285·5 Pd	0·001
			286·3 D_2	0·005
As	193·7	0·002	193·7 D_2	0·002

aluminium in steel and the degree of background absorption. For arsenic, the deuterium lamp could be satisfactorily used for background correction because of the absence of iron lines in the vicinity of 193·7 nm.

The pulse nebulization technique can also be successfully applied to the analysis of blood and other biological fluids (see p. 135), and it can also be easily automated.[15]

(C) Branched Input Nebulization

The capillary uptake tube to an atomic absorption spectrophotometer can be branched by means of a T-piece. The T-piece is connected to the nebulizer via a short length (about 20 mm) of plastic capillary tubing, and two lengths of plastic capillary tubing (each about 200 mm in length) are connected to the arms of the T-piece.[16–19] The T-piece can be conveniently adapted from an automatic analyser glass T-piece[18] or drilled out from a small P.T.F.E. block.[16,17] The advantages that accrue from the use of this simple device are listed below.

(1) *Addition of buffer or ionization suppressant*

It is possible to nebulize the samples through one uptake capillary tube and the buffer solution (e.g., lanthanum chloride solution) or an ionization suppressant solution through the other. This procedure saves the time-consuming procedure of accurately adding lanthanum chloride or ionization suppressant solution to all sample and standard solutions prior to nebulization. It also considerably reduces the quantity of reagents (e.g., $LaCl_3$) required. However, it must be stressed that the analytical sensitivity is halved when using this technique.

A good example of this is the direct determination of calcium in natural waters by continuously nebulizing a 2000μg ml^{-1} sodium solution through one uptake capillary tube and the blank, standards and undiluted samples through the other into a nitrous oxide–acetylene flame maintained perpendicular to the optical axis.[17] Alternatively, the air–acetylene flame can be used and a 4000μg ml^{-1} lanthanum solution used in place of the sodium solution.[19] Magnesium can also be determined.

(2) *Standard additions calibration*[18]

The two arms of the uptake capillary tube are adjusted by altering their relative lengths for equal sensitivity, then the sample is continuously nebulized through one uptake capillary tube and various standards and water (or the solvent blank) are nebulized in turn through the other. The various readings are then noted. Finally, the blank for the procedure is measured by replacing the sample with the blank and nebulizing water (or the solvent blank) through the other uptake tube. A normal standard additions calibration graph is then plotted.

(3) *Interference studies*[18]

By nebulizing a fixed concentration of analyte through one arm of the T-piece and various potential interferants and water (or the solvent blank) through the other, it is possible to rapidly carry out an interelement study with minimum sample preparation. It would appear from the results[18] that adequate mixing of the solutions takes place.

(D) BABINGTON NEBULIZATION

Conventional pneumatic nebulizers require the sample to pass through a small capillary tube (typical internal diameter 0·35–0·45 mm) and are not very suitable for the direct nebulization of viscous samples and samples with a high total dissolved solids content, such as blood, some urine samples, seawater, evaporated milk or concentrated wet digests of various materials.

The BABINGTON nebulization technique[20] overcomes this disadvantage and basically consists of a hollow sphere into which the nebulizing gas is passed via a small inlet tube. A small exit hole is drilled on the side surface of the sphere where the nebulizing gas escapes. The sample is simply allowed to flow down from the top over the surface of the sphere. As the sample passes over the exit hole region, efficient nebulization occurs. There are no small orifices that will restrict the sample flow or that can block up.

FRY and DENTON[21] have described a modified BABINGTON nebulizer suitable for flame atomic absorption work. It consisted of a polished 6 mm o.d., 5·5 mm i.d. closed round-ended metal tube (mounted vertically) with a 0·61 mm diameter orifice located 3 mm below the rounded end of the tube. The sample (1 ml volume) was pumped through a 2·38 mm i.d. sample delivery tube using a peristaltic pump (flow rate 20–25 ml/min), the end of the tube was positioned 1 mm above the rounded end of the metal tube. An impact bead (see page 180) was positioned opposite the exit hole to produce additional fragmentation. The sensitivity obtained for cadmium was similar to that obtained using a wide range of conventional pneumatic nebulizers.

The modified BABINGTON nebulizer was used in conjunction with a wide slot, 'high solids' air-acetylene burner and gave satisfactory results for the determination of copper and zinc in a range of matrices including haemolysed whole blood, urine, seawater and a wet digestion of NBS orchard leaves.[21] It was essential to calibrate by the method of standard additions. With whole, unhaemolysed blood the results for zinc and iron, which are mainly present in the red blood cells, were low. This was explained by the fact that the intact red blood cells have a diameter of 5–9 microns (5000–9000 nm) which is considerably greater than the median mass diameter, 3·6 microns (3600 nm) of solution droplets produced by the modified BABINGTON nebulizer. Thus the relative number of red blood cells that will be carried through to the flame will be less than in the original blood.

2. Types of Flame

Low Temperature Flames

Air–coal gas and air–propane flames have been used for the determination by A.A.S. of certain elements, where the samples are in virtually pure aqueous media and chemical interferences are not present. The introduction of the air–acetylene flame in about 1958 to A.A.S. improved determinations generally, and gave useful sensitivity for about 35 elements.[22] Even with this flame, though, many elements are still incompletely atomized, and chemical interferences are still severe for calcium, in the presence of phosphates or aluminium.

Atomization, however, is not only a function of temperature, as shown by the absorptions of molybdenum and tin which exhibit higher values in the cooler fuel-rich air–acetylene flame than the hotter stoichiometric one.

High Temperature Flames

Premixed oxygen–hydrogen or oxygen–acetylene mixtures have such high burning velocities that they have rarely been used for A.A.S. Also the turbulent oxygen–hydrogen flame used for flame emission spectroscopy is considerably reduced in temperature when aqueous samples are nebulized into it, and may only have a temperature of 2300°C[24] under these conditions. For this reason results obtained using this flame are less sensitive than might be expected, and chemical interferences still occur.

Table 7.2 shows the burning velocity and flame temperature for a number of gas mixtures that have been used in atomic absorption studies. The burning velocity values quoted refer to a burner grid operating at room temperature

Table 7.2 Burning velocities and flame temperatures

	Maximum burning velocity (cm s^{-1})	Maximum temperature (°C)
Air–acetylene	150	2300
Air–coal gas	55	1840
Air–propane	43	1925
Air–hydrogen	320	2100
50% Oxygen–50% nitrogen–acetylene	640	2815
Oxygen–acetylene	1130	3060
Oxygen–propane	380	2800
Oxygen–cyanogen	140	4640
Nitrous oxide–acetylene	285	2750
Nitric oxide–acetylene	90	3095
Nitrogen dioxide–hydrogen	150	2660
Nitrous oxide–hydrogen	390	2650

(e.g., a water-cooled burner). The burning velocities increase with increasing temperature of the unburnt flame gases.

The Oxygen—Cyanogen Flame

The first published attempt to use this flame was made by ROBINSON,[25] using a total-consumption burner in which the sample was aspirated using pre-mixed oxy–cyanogen. For several metals he achieved sensitivities slightly better than those obtained with a turbulent-flow oxygen–hydrogen flame, but he was unable to detect any absorption for tin, tantalum, tungsten, and aluminium. Vanadium was only detectable with poor sensitivity. Since emission was obtained for these metals at the same wavelengths, in the former flame, lack of absorption signal may be explained by the emission being due to chemiluminescence rather than thermal excitation.

Although the high temperature and low burning velocity of the oxygen–cyanogen flame looks promising for A.A.S., it suffers from the drawbacks that cyanogen is highly toxic and explosive, so it is unlikely to be accepted for routine use. Again, because of the poor diffusion of cyanogen into oxygen it is difficult to maintain a steady flame, and even small quantities of liquids aspirated can lower the temperature by as much as 2000°C.[26]

Oxygen–Acetylene Flame

Because of its high burning velocity, work on atomic absorption was first reported using a turbulent burner,[27] with a multi-pass arrangement, and a xenon arc as a source. SLAVIN and MANNING,[28] employing hollow-cathode lamps, demonstrated strong absorption for aluminium, vanadium, titanium and beryllium.

The first published absorption for the lanthanides was reported in the oxygen–acetylene flame by SKOGERBOE and WOODRIFF.[29] Atoms produced in this flame were highly localized, and the replacement of water for the organic solvent originally used reduced the absorption to almost zero. Because of the inherent risk in using this flame little routine use has been found for it.

Nitrogen–Oxygen–Acetylene Flame

AMOS and THOMAS[30] investigated the atomic absorption of aluminium in a pre-mixed air–acetylene flame, in which the air was enriched with 50 per cent oxygen, burning on a stainless steel block 32 mm thick, with a 30 mm × 0·45 mm slot. This flame, however, has some disadvantages for routine work, since it requires the provision of either a commercial O_2/N_2 mixture or a fairly complex gas mixing unit designed to prevent inadvertent use of oxygen-rich mixtures. Further, the high burning velocity of the gas mixtures required to produce atomization of metals forming refractory oxides limits the burner slit length to 30 mm.

Nitrous Oxide–Hydrogen Flame

For most applications there is little advantage gained in using this flame, although for the alkaline earth elements, which are hardly ionized in it, useful sensitivity and limits of detection have been found.[31] Molybdenum, germanium, beryllium and aluminium show negligible absorption in the nitrous oxide–hydrogen flame, although aluminium gives some weak atomic emission at 394·2 and 396·1 nm.[32] This flame has been applied to boron analysis in fertilizer using the 518 nm BO_2 band emission (see Ref. 15, Chapter 4). It has also been applied to boron analysis in steels using both atomic absorption and flame emission at 249·7 nm line and a considerable improvement in detection limit compared to the nitrous oxide–acetylene flame was observed.[33]

This flame is very useful for the nebulization of a wide range of organic solvents and has been successfully applied to the determination of lead in petrol (see Ref. 202, Chapter 5).

Nitric Oxide/Nitrogen Dioxide–Acetylene Flames

The lower burning velocity and higher temperatures attainable by burning acetylene with nitric oxide or nitrogen dioxide suggest the flames may be useful, but the corrosive and very toxic nature of these oxides of nitrogen prevents their routine use.

SLAVIN *et al.*[34] have found that the sensitivity for several metals is slightly greater in the nitric oxide–acetylene flame than in the nitrous oxide–acetylene system, but the stability of the former flame is so poor that the detection limits are usually worse.

Nitrous Oxide–Acetylene Flame

This is a commonly used flame for atomic absorption work. Here the burning velocity is not excessive, and the high temperature of the flame is due in part to the energy liberated by the decomposition of the nitrous oxide. This flame can be burned on a 100 mm slot, although optimum sensitivity for most metals is obtained using a 50 mm slot.

AMOS and WILLIS[23] demonstrated that the flame could be used with commercial equipment, and found good sensitivity for most of the metals which form refractory oxides and had proved difficult or impossible to atomize with the air–acetylene flame.

3. Separated Flames

Premixed hydrocarbon–air flames consist of two separate reaction zones: the primary reaction zone, where the combustible gas mixture burns principally to carbon monoxide, hydrogen and water, and the outer mantle, or secondary diffusion flame, where the hot gases burn with the atmospheric oxygen to carbon dioxide and water. The hottest part of these premixed flames is normally in the centre of the flame just above the primary reaction zone.

In 1891 TECLU, and SMITHELS and INGLE[35] independently demonstrated the existence of the two zones, by mechanically operating the flame with a glass tube. The lower end of the tube was given an airtight fitting around the burner stem, and this prevented atmospheric oxygen reacting with the combustion products from the primary zone until they reached the top of the glass tube, where the secondary diffusion zone burned as before. In this way the space between the reaction zones, the interconal zone, is extended in length, and can be viewed without interference from the radiation of the secondary zone, which normally surrounds it in all unseparated flames.

It is the interconal zone which is normally used for analytical flame spectroscopy, and the potentiality of separated flames for flame emission and atomic absorption spectroscopy has been investigated by WEST, KIRKBRIGHT *et al.*[36–43]

SMITHELS and INGLE[35] reported the separation of the low-burning-velocity air-supported flames, coal-gas, ethylene, methane, etc., on an open Bunsen-type burner port. KIRKBRIGHT, SEMB and WEST[36] reported the mechanical separation of an air–acetylene flame burning on a Meker-type head. A silica mechanical separator was used as this gave transmission of emission lines down to 200 nm.

The air–acetylene flame was separated (see Fig. 7.2) using a 120 mm long, silica tube of 20 mm internal bore, which was placed round a Méker-type burner, with an airtight sleeve enabling various separating distances to be obtained. Between 60 to 80 mm was found to be the most convenient distance from the top of the separator to the stainless steel burner head. The Méker head used had 13 holes of about 1·2 mm diameter arranged in a 9-mm-square pattern.

With this arrangement the flame is easily separated into a primary zone, consisting of 13 separate cones on the burner head, inside the silica tube, and a secondary diffusion zone burning at the top of the silica tube. Stable separated flames are obtained over a wide range of fuel/air ratios. The flame becomes turbulent with very lean mixtures and the use of very rich (luminous) mixtures leads to carbon forming on the inside walls of the separator. Turbulent noise occurs if a wide tube is used for separation, or if a separating distance greater than 100 mm is used with the 20-mm bore separator.

Emission Characteristics of Separated Air–Acetylene and Nitrous Oxide–Acetylene Flames

The primary reaction zone of premixed air–acetylene flames radiates strongly over much of the visible and near-u.v. region of the spectrum, notably the molecular band emissions of CH, C_2, OH, and the strong CO continuum. This intense radiation, covering the larger part of the spectrum, makes the primary cone of little general use in flame spectroscopy. The secondary reaction zone radiates much less strongly. The emission from this region is due to the CO continuum, generated by the reaction $2CO + O_2 \rightarrow 2CO_2 + h\nu$,

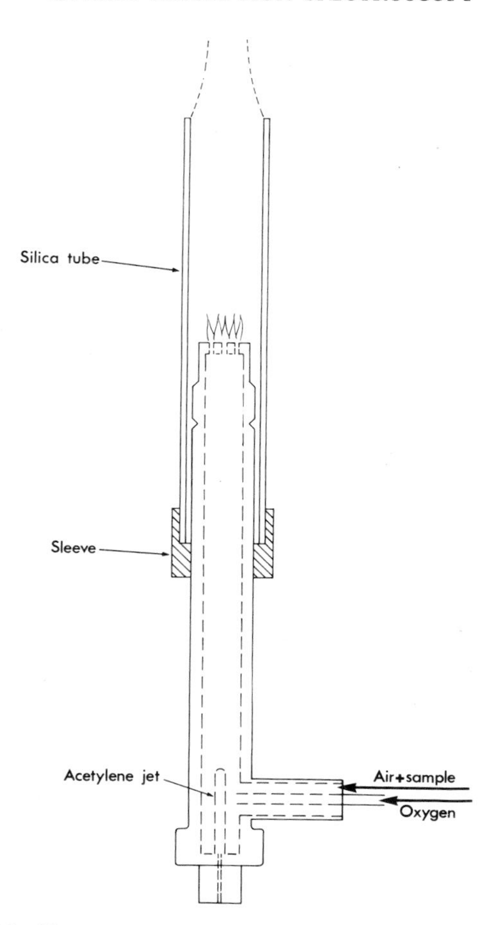

Fig. 7.2 Flame separation with a silica tube mechanical separator[36]

and also the strong OH band system in the near-u.v. with strong peaks around 282 nm and 307 nm. The flame background due to CO and OH, that is observed when an unseparated flame is viewed, can be reduced by 2 orders of magnitude in the separated flame.

The noise level is also reduced considerably, and in emission studies some advantage is gained–particularly for elements whose emission lines lie in regions of very high background in the normal unseparated flame, e.g., bismuth (306·8 nm).

A further effect of separation is to cool the flame because the heat normally provided by the reaction

$$CO + O \rightarrow CO_2 + 67{\cdot}6 \text{ kcal}$$

that occurs in the surrounding secondary diffusion zone has been removed from the interconal region. This tends to reduce emission intensities.

In a later communication[37] KIRKBRIGHT, SEMB, and WEST describe the mechanically separated nitrous oxide–acetylene flame for emission work. This also possessed a reduced CO and OH background. The nitrous oxide–acetylene flame possesses in the interconal region a red zone, due to the CN emission, which contributes to the highly reducing nature of the flame. When this flame is separated the CN emission is protected from the atmospheric oxygen and is thus extended. The burner used was in the form of a circular annulus 0·5 mm in width and 11 mm in diameter. This produced an intense primary reaction zone and red CN zone inside the separator, with the displaced secondary zone burning as a diffusion flame at the top of the silica tube.

The separated nitrous oxide–acetylene flame gave improved determinations of the refractory elements aluminium, molybdenum, and beryllium[41,43] by emission and fluorescence spectroscopy. The use of a long-path mechanically separated flame has been reported[38] for use in atomic absorption spectroscopy. This involved the use of the normal separated air–acetylene flame in conjunction with an electrically heated furnace. The interconal region was extended into a tube so as to obtain a long absorption path. Highly sensitive atomic absorption estimations of zinc, iron, copper, mercury, magnesium, and arsenic were reported.

The mechanical separator, however, eventually became corroded, and the transmission of the silica became progressively worse. The use of replaceable silica windows cemented to ground glass sockets gave the separator an increased life, but the advent of gas sheathing made it redundant.

The gas-sheathed flame was reported by HOBBS, KIRKBRIGHT, SARGENT, and WEST,[40] the mechanical silica separator being replaced by a protective wall of nitrogen, flowing in a laminar annulus around the flame (see Fig. 7.3). This has the effect of 'lifting-off' the secondary zone, by preventing the access of atmospheric oxygen to support a diffusion zone in the lower parts of the flame.

This alternative method of separating the flame possesses the same advantages as the silica separator described earlier. The interconal zone exhibits low radiative background, so that improved analytical signal to flame background ratios are obtained from aspirated metal ion solutions.

Several advantages are gained by gas sheathing over the mechanical silica separator. A wider range of fuel/air ratios may be used with safety and the analytical emission need not be viewed through the silica tube, which at low wavelengths becomes absorbing. The problem of etching or sooting of the inner surface of the separator is also eliminated. Again, the flame temperature above the primary cones is slightly reduced by separating, and from the electronic excitation temperature for iron using the two-line method, was found to fall from 2420 ± 20K to 2320 ± 20K for the air–acetylene flame.[40]

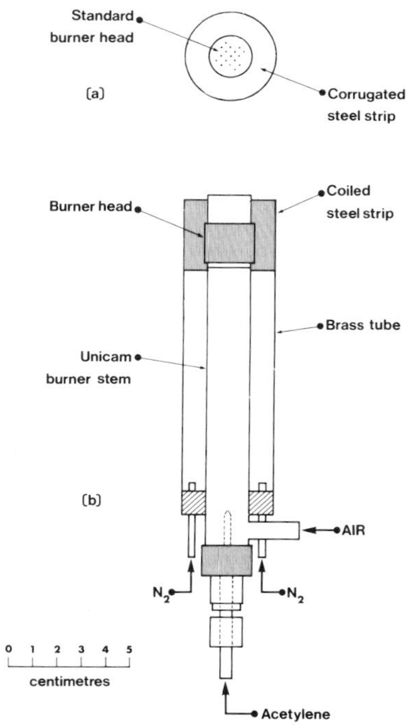

Fig. 7.3 Burner arrangement for nitrogen-separated air-acetylene flame[40]

The laminar nitrogen sheath was produced by passing nitrogen gas through a matrix of steel strip wound round the emission burner head in a spiral of alternate corrugated and flat 20 mm wide strip (0·1 mm in thickness).

Although silica separators could not easily be applied to long-path flames the gas-sheathed arrangement can be. 50 mm path length nitrous oxide–acetylene and 120 mm path length air–acetylene flames have been gas sheathed in a similar manner, and improved detection limits obtained for some atomic absorption determinations.

The red CN zone in the nitrous oxide–acetylene flame is increased, due to less attack by atmospheric oxygen, and detection limits for refractory elements can generally be improved in the shielded flame.

The other property of the nitrogen-separated nitrous oxide–acetylene flame is its transparency to radiation below 190 nm, where most conventional flames absorb radiation strongly.

THE DIRECT ATOMIC ABSORPTION DETERMINATION OF IODINE, PHOSPHORUS AND SULPHUR

Atomic absorption techniques can be applied to determine over 65 elements. The elements that cannot be satisfactorily determined either form very refractory oxides (e.g., Ce and Th) that cannot be appreciably broken down to atoms in the hottest usable flame (nitrous oxide–acetylene) or have their resonance lines in the vacuum U.V. below 190 nm, where oxygen (in the air) and conventional premixed flames absorb radiation strongly. Table 7.3 gives a list of elements which have their main resonance lines in the vacuum ultraviolet region of the spectrum.

Table 7.3 Elements which have their main resonance lines in the vacuum ultraviolet region of the spectrum

Element	Main resonance line wavelengths (nm)		
Inert gases	He (58·4) to Xe (149·1)		
Fluorine	95·5		
Nitrogen	120·0		
Hydrogen	121·6		
Oxygen	130·2		
Chlorine	138·0		
Bromine	154·1		
Carbon	156·0		
Phosphorus	177·5	178·3	178·8
Iodine	178·3	183·0	
Sulphur	180·7		
Mercury	185·0	(253·7)	

Oxygen and conventional air–fuel gas flames absorb radiation strongly below 190 nm, thus making conventional flame atomic absorption measurements almost impossible. KIRKBRIGHT *et al.*[48–53] observed that the fuel-rich nitrogen-sheathed nitrous oxide–acetylene flame is remarkably transparent down to 175 nm, and could be used to determine iodine, phosphorus and sulphur with a slightly modified atomic absorption spectrophotometer (see Fig. 7.4). Microwave-excited electrodeless-discharge lamps were used as sources. (Demountable sulphur and iodine hollow-cathode lamps have also

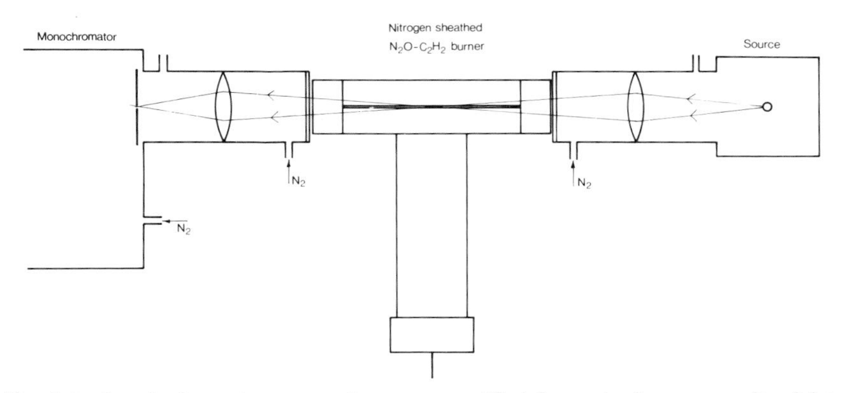

Fig. 7.4 Atomic absorption spectrophotometer modified for use in the vacuum ultraviolet

Table 7.4 Summary of iodine, phosphorus and sulphur results

Element	Wave-length	Sensitivity for 1% absorption μg/ml	Limit of detection μg/ml	Reference	Comments
I	183·0	12	25	48	Aqueous solution NH_4I
I	183·0	4·2	8	48	CHI_3 in M.I.B.K.
I	183·0	0·8	—	48	CH_3I in M.I.B.K.
I	183·0	0·32	—	49	Leipert amplification $I^- \xrightarrow{Br_2} IO_3^-$. $IO_3 + 5\ I^- + 6\ H^+ \rightarrow 3\ I_2 + 3\ H_2O$. I_2 extracted into M.I.B.K.
P	177·5	4·8	29	50	Aqueous orthophosphate
P	178·3	5·4	21	50	Aqueous orthophosphate
S	180·7	9	30	51	Aqueous solution K_2SO_4
S	180·7	2·7	—	52	Di-Benzyl Sulphide in M.I.B.K.
S	180·7	0·6	—	52	Thiophene in M.I.B.K.

been developed.[53]) The monochromator and external optical path was purged with dry nitrogen and a photomultiplier with a silica window was used. The results obtained by KIRKBRIGHT *et al.* are summarized in Table 7.4.

The limits of detection do not appear very impressive, but with further instrumental refinements should be capable of some improvement. As can be seen from Table 7.4, volatile compounds give appreciably better sensitivity figures for 1% absorption compared to involatile compounds. This is presumably due to appreciable volatilization of the volatile element species occurring within the spray chamber. The direct determination of sulphur in oils was accomplished by using di-benzyl sulphide as a standard and assuming volatile sulphur compounds to be absent.[52] The results showed reasonable agreement with an X-ray fluorescence method. The direct determination of phosphorus[50] in foodstuffs showed good agreement with a wet ashing–colorimetric method. Sulphur has also been determined in a graphite tube furnace using flameless electrothermal atomization.[54]

4. Modified Burner and Nebulizer Systems

THE FUWA-VALLEE LONG TUBE BURNER[55–57]

This is a device for increasing the optical path of a total consumption burner system. The burner itself is inclined almost horizontally, and the flame directed into the end of a long ceramic tube. This tube is designed to confine the flame to the optical axis of the spectrophotometer, and thereby to combine the benefits of one hundred per cent sample transfer to the flame and a long light-absorbing path. Improved detection limits, over those of other turbulent burners, are reported for metals (e.g., Pb, Zn, Cd) having easily dissociated oxides, but others such as Mg, Ca, Cr, Mo are converted to the stable oxide in the upper portions of the flame and, therefore, exhibit no improvement in sensitivity. The use of this device is rather untidy and of somewhat doubtful advantage, and is unlikely to be incorporated routinely into standard production instruments.

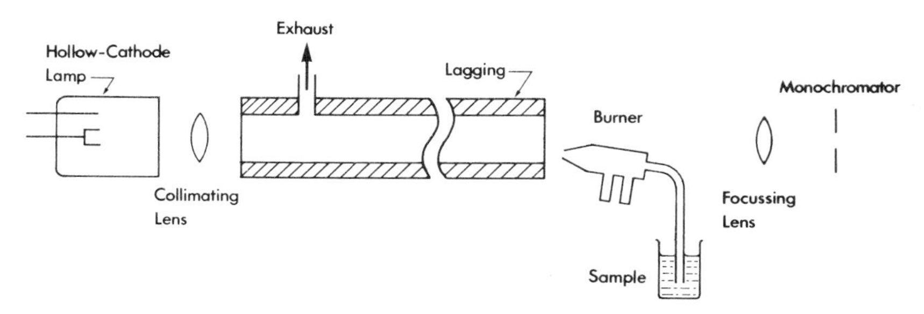

Fig. 7.5 Long tube burner arrangement

The Boling Three-slot Burner[58,59]

Several advantages are claimed for this burner, which is designed to provide a flame that will more completely enclose the optical beam than a normal single-slot burner. It is stated that because the outer sheath of the flame is outside the light path, the background noise is diminished, while the sensitivity is increased for elements that form refractory oxides (e.g., Cr, Mo). It is further asserted that the fuel/air ratio is less critical. The sensitivity is decreased for elements that are easily atomized (e.g. Cd, Cu, Pb, etc.) [60]

Force-Feed Burner[61]

This is a modification of the total consumption turbulent flow burner. The sample, instead of depending on the suction produced by the flame gases, is mechanically fed into the base of the flame. This means that the viscosity of the sample does not control the aspiration rate, and different solvents may be fed into the flame at constant rates of flow. The liquid sample is broken up by the shearing action of the high-pressure gases issuing from the annular space round the capillary.

The Heated Spray Chamber

More complete vaporization of the nebulized sample can be obtained if the fuel and carrier gases are preheated. The nebulized sample will then be flash dried into a fine aerosol of solid particles on entering the spray chamber. Nearly all of the nebulized sample will reach the flame and this will result in an increased sensitivity. Rawson[62] reported an increase of about six times in the sensitivity of magnesium by preheating the air and the fuel (coal gas). In some further work[63] the technique was improved by using a heated spray chamber maintained at 600°C. This was followed by a condenser which removed water vapour but allowed the sample to reach the flame. It was possible to safely introduce acetylene after the condenser stage. Without the condenser, the large amount of water vapour reaching the flame would cause a marked drop in temperature and this would result in an increase in condensed phase interference effects. Increases in sensitivity of 30–40 times for Co, Cu, Mg, Mn, Ni, Pb and Zn reported. Other workers[64–66] using an infra-red heated spray chamber followed by a condenser have reported increases of 5–30 times for a range of elements.

ISSAQ and MORGENTHALER[10] have made a comprehensive study of the combination of an ultrasonic nebulizer connected to a heated tube (145 mm length by 10 mm i.d.) followed by a condenser and a mixing chamber. The optimum temperature of the heated tube was said to vary from element to element and the response was also markedly dependent on the anionic form of the element.

In practice heated spray chambers and condensers are rather cumbersome and can take a long time to come to thermal equilibrium. Although the technique has been applied in one commercial unit it is not, at present, attractive to the majority of manufacturers. The procedure could possibly be dangerous if the nitrous oxide-acetylene flame is employed.

5 Flameless Electrothermal Atomic Absorption Spectroscopy

INTRODUCTION

The original and most commonly used system of atomization in atomic absorption spectroscopy is the flame; in particular the air–acetylene or nitrous oxide–acetylene systems. The advantages and disadvantages of flame systems are listed below.

Advantages

(i) The analyses of a large number of elements can conveniently be performed with either the air–acetylene or nitrous oxide–acetylene flames. Optimization of flame conditions is relatively easy.
(ii) A steady (non-transient) output signal is obtained.
(iii) Conventional burners are compact, durable and inexpensive.
(iv) Good signal reproducibility (0·4–2% R.S.D.) can be readily obtained.
(v) The detection limits obtained for many elements are adequate for many types of analysis.
(vi) Automatic sampling and measurement using flames is considerably easier than with flameless devices.
(vii) A rapid rate of analysis can be achieved using flames.
(viii) Interelement effects in flames are reasonably well documented.

Disadvantages

(i) The maximum attainable atom population in flames is restricted due to dilution by the flame gases (see p. 300). For many elements the flameless technique is 10–100 times more sensitive than the corresponding flame technique.

(ii) The available sample volume may in some applications (clinical, pathological, etc.) be insufficient to give a reliable reading by aspiration to a flame through a conventional pneumatic nebulizer.

(iii) Samples such as oils, blood and serum must either be diluted with a suitable solvent or suitably ashed before the sample can be satisfactorily nebulized.

Historical

It was realized by L'Vov,[67–69] West[70] and Massmann[71] that non-flame electrothermal atomization offered the possibilities of very high absolute sensitivities, performing analyses on samples of small volume and of handling many complex samples without pretreatment.

The analytical sensitivity of the procedures described by these original workers was very good but, when applied to practical analytical samples, very severe non-specific background absorption and chemical effects were often encountered. It should be noted that the first graphite tube furnace used for spectroscopic measurements was reported by King in 1905 who used it mainly for physical studies.

Since 1970, though, a number of commercial devices have appeared, based upon the Massmann and West procedures and designed to offer the potential advantages of non-flame atomization to routine laboratories. The main systems offered are the heated graphite tube atomizer (Massmann Furnace), the graphite rod with transverse hole or tube (Mini-Massmann), and various shapes of graphite rods and tantalum ribbons.

The basic operating procedure when using these devices is as follows. An aliquot (typically 1–20 μl) of the liquid sample is placed in or on the device using a suitable micropipette. A small current is then passed through the device for a preset time in order to evaporate the solvent in a controlled manner. This is known as the solvent evaporation stage. A larger current is then passed through the device for a preset period in order to dry ash the sample matrix and vaporize any volatile matrix components. This is known as the dry-ashing stage. The temperature used for the dry-ashing stage must be carefully selected so that the analyte is not vaporized. Finally, a much larger current is passed through the device for a shorter preset time in order to vaporize and atomize the analyte. This is known as the atomization stage and during this stage the transient absorbance signal is monitored. The signal can be directly monitored on a pen recorder and with some units the peak height and/or the peak area of the transient absorbance signal can also be monitored.

Description of Flameless A.A.S. Systems

(a) *The Massmann Furnace or Heated Graphite Tube Atomizer*

In its most widely used form the work head of this device consists of a graphite tube about 50 mm long, 8·6 mm i.d. and 10·6 mm o.d. through which the light beam passes. The cylinder is encased in a water-cooled metal housing and the system is purged with an inert gas, normally nitrogen or argon.

The sample (normally 5–100 μl in volume) is introduced into the tube through a small hole (2 mm diameter) midway along its length. The tube is electrically heated in a programmed sequence so that evaporation, dry-ashing and atomization occur as distinct regulated stages. Some form of temperature readout is incorporated into most commercial units.

The sensitivity can be increased by automatically stopping the flow of the purge gas during the atomization stage, thus retaining the atomized sample in the light beam for a longer period of time.

There is a temperature gradient from the centre to the ends of the tube which can in certain instances (e.g., sea-water analysis) lead to memory effects and severe background absorption due to condensation of the matrix in the cooler ends of the tube. This can be minimized by introducing the inert gas into both ends of the graphite tubes and allowing it to escape through the sample introduction hole.[(72)]

The graphite tube furnace is the most common type of commercially available electrothermal atomizer. It is also the largest of the commercial type devices and utilizes the heaviest power supply (approximately 3–5 KVA).

It is also possible to monitor atomic emission signals from the optical axis of a graphite tube furnace.[(73–77)] For this mode of operation the atomic absorption instrument is used in the emission mode and the hollow-cathode lamp of the analyte element is used to set the monochromator wavelength and then switched off. Elaborate precautions must be taken to reduce the monochromator aperture (i.e., increase the *f* number, see p. 193) so that the hot graphite surface is not viewed by the detector. The sample is introduced into the furnace, dried, ashed and then the emission emanating from the optical axis of the furnace is monitored during the atomization stage. The procedure is then repeated using the blank solution. Some impressive detection limits have been reported for Ba, Ga, In, Li, Na, K and Tl. Using a wavelength modulation technique, barium has been determined in calcium carbonate rocks.[(75)] The detection limit was equivalent to 0·036 μg/g of barium in the rock sample. Using commercially available equipment, this technique is unlikely, in the immediate future, to supersede conventional use of the graphite tube furnace in the absorption mode.

(b) *The Graphite Rod Atomizer with Transverse Hole or Tube (Mini-Massmann)*

The design of the original work head of this device[(78,79)] comprises a graphite rod 50 mm long and 5 mm in diameter (Fig. 7.6). The sample cavity consists of a hole (1·5 mm dia.) bored radially through the mid-portion of the rod, and provided with a small sampling port. Samples between 0·5 and 2 μl volume are normally used.

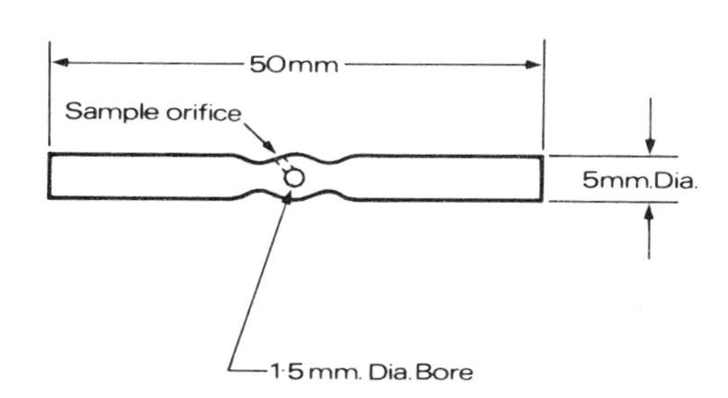

Fig. 7.6 Mini-Massmann graphite rod furnace with transverse hole. Original design of graphite heating element

The system only requires a moderate-sized power supply and is designed to satisfy the important criterion that the atomic vapour should be observed within an environment of uniform high temperature.

A later modified design is shown in Fig. 7.7 which comprises a hollow-graphite cylinder (9 mm long, 3 mm internal diameter) which is held in the optical beam by spring-loaded graphite rods. One end of each rod is machined to the external diameter of the graphite cylinder and presses against the wall of the cylinder rather than the ends (as in the larger Massmann furnaces). This modified design allows the use of a larger sample volume (up to 10 μl) with standard tubes and up to 20 μl with tubes with an internal thread cut into them. It also transmits more light by reason of the larger internal diameter.

The design ensures that the heating is localized and that conduction of heat away from the tube is minimized. Power utilization is also improved so that the larger tube can still be heated with a moderate-sized power supply (2–3 kVA). The power supply has been designed so that during the atomization stage the graphite cylinder heats up to a preselected temperature, up to 3000°C, and then remains at this temperature for a preselected time. The rate at which the graphite cylinder attains this preselected temperature can also be varied with a maximum heating rate of 800°C per second. The same work head can also be used with graphite cups. These will accept certain types of solid samples as well as liquid samples (Fig. 7.8).

The complete assembly consisting of electrical connections with water-cooled terminals and gas sheathing is shown in Fig. 7.8 and is mounted in place of the standard atomic absorption burner. If required, the graphite tube or cup

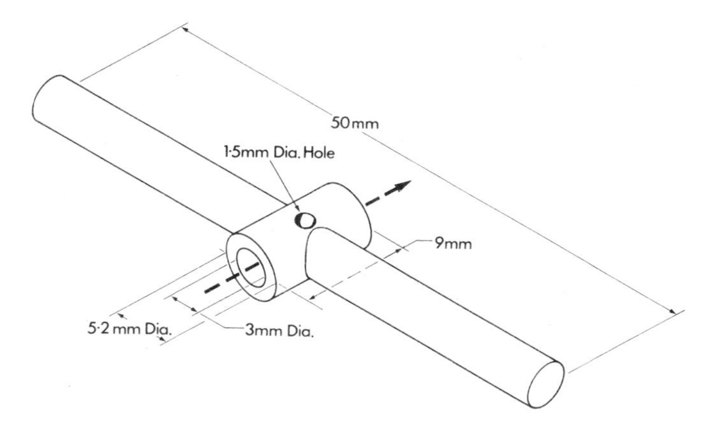

Fig. 7.7 Modified Mini-Massmann graphite rod work head (with transverse graphite tube)

can be surrounded by an argon–hydrogen flame with the object of minimizing interference effects.

Provisions for programmed heating to preselected temperatures for preselected times during the solvent evaporation and dry ashing stages have been made.

(c) *The Graphite Rod Atomizer*

WEST and co-workers[70,80–86] at Imperial College have pioneered the use of small-diameter (2 mm) graphite rods for use in flameless atomization. Devices

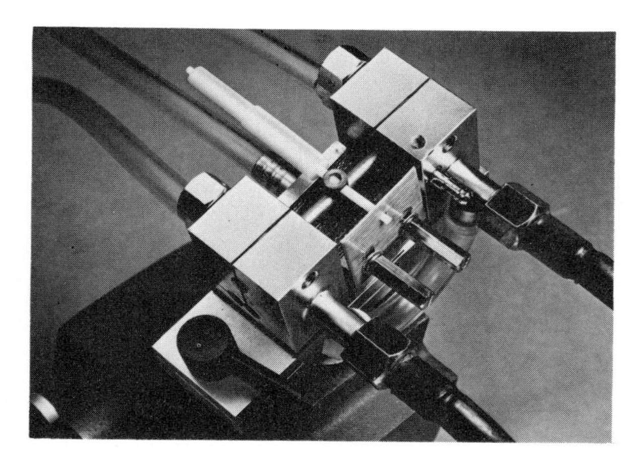

Fig. 7.8 Modified Mini-Massmann graphite rod work head (with graphite cup). The light path is illustrated by means of the alignment tool. (By courtesy of Varian Associates Ltd.)

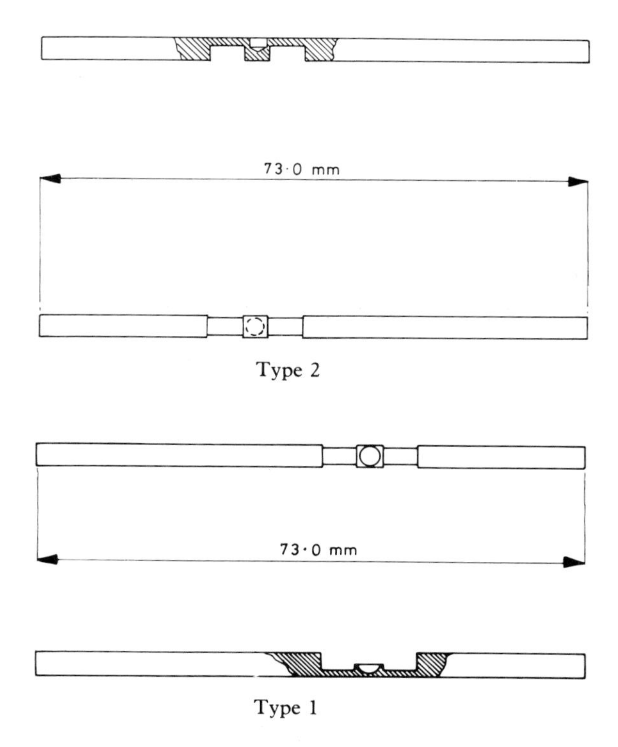

Fig 7.9a Shandon Southern Instruments (Baird Atomic) A3470 Graphite Rods

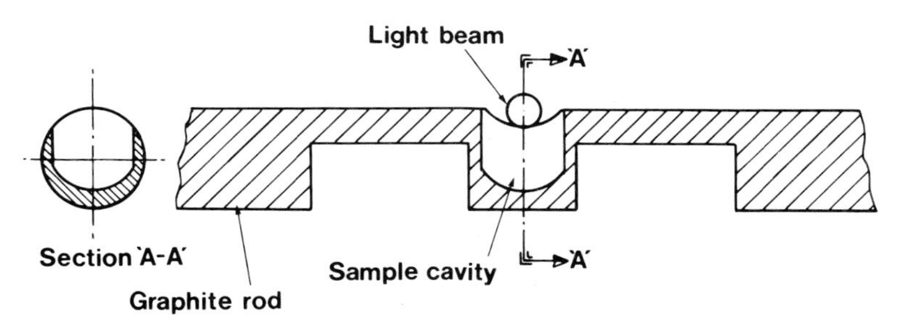

Fig. 7.9b Section through type 2 graphite rod

of this type using a slightly larger diameter (3·05 mm) graphite rod, specially shaped to ensure localized heating around the sample cavity, are now commercially available. Typical rods are shown in Fig. 7.9a. The Type 1 rod has a hemispherical sample cavity that will accept sample volumes up to 5 μl. The Type 2 rod has a vertical cylindrical sample cavity that will accept sample

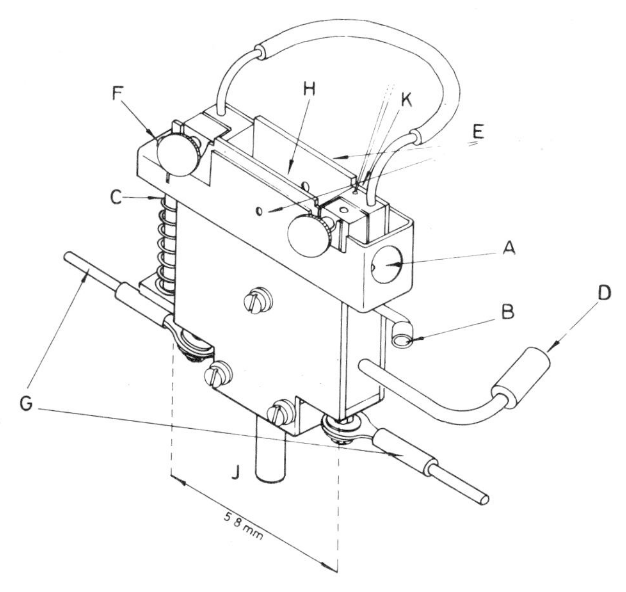

Fig. 7.10a Shandon Southern Instruments (Baird Atomic) A3470 graphite rod head
A. Access to graphite rod.
B. Air cooling in.
C. Air cooling out.
D. Purge gas in.
E. Aperture.
F. Aperture height adjusting screw.
G. Transformer connections.
H. Graphite rod.
J. Mounting pillar.
K. Thermistor.

volumes up to 10 μl. Measurements are made with a 1·8-mm-diameter light beam set to grazing incidence above the sample cavity (Fig. 7.9b). The efficient collimation of the light beam with respect to the top of the sample cavity minimizes interelement effects. If measurements are taken some distance above the rod, interelement and background absorption effects become very pronounced and poor sensitivity is observed.

The design of the graphite rod ensures that maximum heat occurs at the sampling point whilst the clamped ends of the rod remain relatively cool. The rod is held between two metal pillars which are free to move with the expansion of the rod, the viewing apertures being rigidly fixed with respect to the optical shelf. The complete assembly is shown in Fig. 7.10a and is mounted in place of the collimating lens situated between the burner and monochromator. This allows flame measurements to be taken with the graphite rod head in situ. The power requirements are relatively modest (1·5 kVA) (Fig. 7.10b and c). The pillars are air-cooled, water-cooling not being required. The temperature

Fig. 7.10b A3470 graphite rod head in actual operating position

Fig. 7.10c The A3470 flameless atomizer unit. (Positioned above left-hand chassis of A3400 atomic absorption spectrophotometer)

of one of the air-cooled pillars is monitored by a thermistor probe. For good precision it is essential to add the sample to the rod at a constant preselected pillar temperature.[87,88] For most work a pillar temperature of 55°C is used. The pillar temperature increases to 75–80°C immediately after the atomization stage and cools down to 55°C within a period of 20 sec. The system can be used for the determination of a large range of elements including Mo, Ti and V, which have boiling points in excess of 3000°C. Provision for programmed heating on four independent channels (evaporation, dry ash 1, dry ash 2, atomization) is provided. Three forms of readout are incorporated: peak height, integrated or direct readout.

Table 7.5 Limits of Detection using a Shandon Southern Instruments (Baird Atomic) A3470 Graphite Rod Atomizer coupled with an A3400 Spectrophotometer

Rod Sheath Gas Argon 2·6 l/min Methane, 15 ml/min 1·8 mm diameter aperture set to grazing incidence with respect to the rod sample cavity

Element	Wavelength nm	Detection limit ($\times 10^{-12}$ g)	Detection limit concentration using a 5 μl sample (μg/ml)
Al	309·3	20	0·004[1]
Ba	553·6	20	0·004[2]
Cd	228·8	0·08	0·000016
Cr	357·9	8	0·0016
Cu	324·7	5	0·001
Fe	248·3	10	0·002
Mg	285·2	<0·1	<0·00002
Mn	279·5	2·5	0·0005
Mo	313·3	20	0·004
Ni	232·0	10	0·002
Pb	283·3	4	0·0008
Si	251·6	100	0·02
Ti	364·3	75	0·015
V	318·4	50	0·010
Zn	213·9	0·08	0·000016

[1] In a 1000 μg/ml calcium matrix[88]
[2] In a 10 000 μg/ml calcium matrix aperture height raised 0·2 mm[100]

Some typical detection limits obtained in the author's (K.C.T.) laboratory are given in Table 7.5. Figure 7.11 shows some typical traces for all three readout modes. The small peak (C) following the peak of the refractory oxide forming element (Ti) shown in Fig. 7.11 is due to absorption by volatilized carbon from the graphite rod. The presence of this small carbon peak ensures that there is no memory effect. This procedure does not appreciably shorten the rod lifetime which is usually between 100 and 200 'firings' for an element such as titanium.

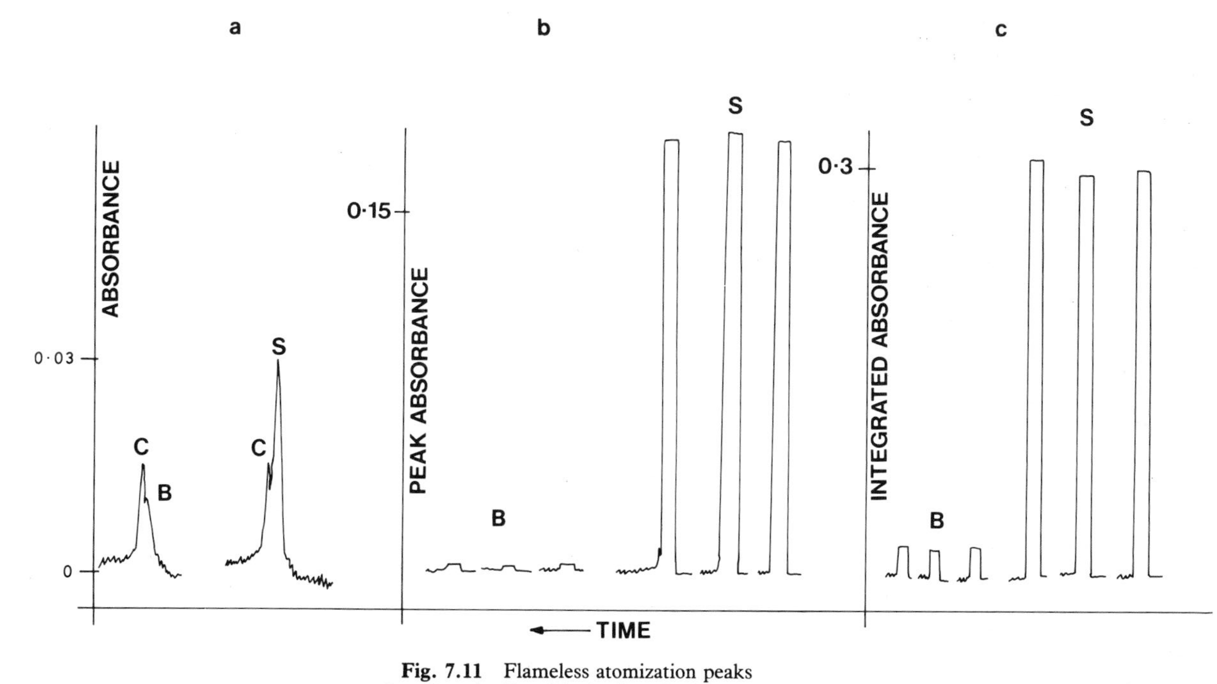

Fig. 7.11 Flameless atomization peaks
S = Sample; B = Blank; C = Carbon peak
a. 2 μl 0·2 μg/ml titanium 364·3 nm direct readout mode.
b. 5 μl 0·1 μg/ml lead, 283·3 nm peak height readout mode.
c. 7·5 μl 0·05 μg/ml chromium 357·9 nm integrated readout mode.
Sheath gas Argon 2·6 l/min, Methane 15 ml/min.

A type 3 rod has recently been developed; it is similar to the type 2 rod (Fig. 7.9) except that the sample cavity region has been extended along the rod to form a 11-mm-length slot. Absorption measurements are then made along the rod length rather than transversely across the rod using a slightly modified graphite rod head.

This type of rod will accept sample volumes of up to 20 μl and is especially useful for trace analyses of biological fluids, steels and oils. The relative non-specific background absorption signal is significantly less than that observed with a type 1 or 2 rod.

Note on the formation of pyrolytic graphite coatings

It is possible to coat graphite rods and tubes with a layer of pyrolytic graphite during manufacture. This hard impervious coating gives improved sensitivity for carbide-forming elements such as Al, Ti and V and can improve the precision for many analyses. The main disadvantages of these precoated rods (or tubes) is that the pyrolytic coating increases the unit cost and slowly degrades with each firing. This can result in a gradual change in sensitivity during the rod (or tube) lifetime especially for carbide-forming elements. The constant addition of a small volume of propane[87,88] (0·3% v/v), ethylene[88] (0·4% v/v) or methane[88,89] (0·3–0·8% v/v) to the argon (or nitrogen) purge gas has been found to result in the continual formation of a pyrolytic coating on the graphite rod or tube surface during the atomization stage. A standard unpyrolysed graphite rod or tube can then be used. The rod or tube is pyrolytically coated during each firing by the thermal decomposition of the hydrocarbon on the graphite surface. The resulting impervious pyrolytic coating results in an increased rod or tube lifetime, increased sensitivity for elements such as Al, Ti and V and improved precision for most elements. At the optimum hydrocarbon gas flow rates, non-specific background absorption is negligible. The rod or tube lifetime can be extended by as much as five times that of an untreated one.

For certain elements (e.g., silicon) a severe decrease in sensitivity is observed in the presence of a hydrocarbon gas. This could be due to carbide formation in the gas phase. In this case the rod or tube can be initially pyrolytically coated by prolonged controlled heating in the presence of a small amount of methane in the purge gas.[90] The device is then used without the addition of methane to the purge gas and when the sensitivity starts to decrease the device is then recoated with pyrolytic graphite.

(d) *The Tantalum Ribbon Atomizer*

These utilize a strip of tantalum foil[92–96] mounted in a totally or partially enclosed chamber. Sample volumes of 1–25 μl are suitable. The power requirements are relatively modest, 1·5 kVA. Facilities for maintenance of an inert argon atmosphere and for programmed heating are provided. Argon must be used as nitrogen reacts with tantalum at high temperatures. (It is quite permissible to use nitrogen with graphite rods or tubes.)

Tungsten and molybdenum[96] can also be used in place of tantalum, but show no significant advantages for the majority of elements. A problem with the ribbon devices is that they tend to become brittle with prolonged use and also, with repeated use at near the maximum operating temperature, changes of the ribbon shape can occur. Attack of the metal surface by certain acid media used for the dissolution of the sample can be serious. In the authors' opinion these devices are not as versatile as those based on graphite.

Response of Amplifier–Recorder System

All flameless A.A.S. devices give rise to transient signals. The rate of vaporization is governed by the rate at which the temperature increases. As the rate increases the absorption signal duration decreases and the degree of absorption increases until the median residence time of the atoms in the optical path is somewhat greater than the median time of atomization. An essential requirement, therefore, is that the response time of the electronic circuitry should be fast enough to permit accurate recording of the signal, particularly for higher readings. An instrumental time constant of 0·3 s or less is necessary for most applications. A faster response is required for a rod or ribbon device than for tube devices. This is because the vaporized sample is contained in the optical path for a longer time within the confines of a tube than in the open atmosphere above a rod or ribbon.

If direct pen recorder readout is used, the response time of the pen recorder (0·5–1 s F.S.D.) is usually the limiting factor. Efficient automatic peak height retrieval (no pen recorder being used) on a rod or ribbon necessitates an instrumental time constant of less than 0·1 s.

In general, flameless calibration curves tend to bend towards the concentration axis at lower absolute absorbance values than with conventional flames. This is probably due to the smaller Doppler half-width of the absorption line (see p. 280), and also to the overall speed of response of the electronics and recorder of most commercially available units. It has been found by the author (K.C.T.) that automatic peak height retrieval (using a Shandon Southern (Baird Atomic) A3470 flameless atomizer and an A3400 atomic absorption spectrophotometer with an overall signal output time constant of 40 ms) gave calibration curve linearity to 30–60% higher absorbance values than direct readout to a conventional (0·8 s F.S.D.) pen recorder. This effect was more pronounced with volatile elements (e.g., Cd and Zn).

Interference Effects

(a) *Non-Specific Background Absorption*

The sample matrix may vaporize as a molecular species which can absorb radiation over specific wavelength regions. The alkali metal and alkaline earth halides NaCl, KCl, KBr, KI, NaI and $CaCl_2$ exhibit well-defined molecular absorption spectra when atomized in a graphite tube furnace,[97–99] the background absorption being markedly dependent upon the wavelength. Condensation of the vaporized samples to a particulate smoke in the light path results in scattering of the source radiation. In this case the background absorption is not so wavelength dependent. This can be observed when trying to determine trace amounts of volatile elements such as cadmium and lead in a very large excess of other substances (e.g., seawater) where the matrix volatilizes at a similar temperature to the analyte.

Whenever a new type of analysis is performed, a check on the magnitude of the non-specific background absorption signal using a deuterium hollow-cathode lamp or a suitable non-resonance line within 2 nm of the analyte resonance line should always be made. If an excessive amount of background absorption is observed relative to the expected analyte signal, efforts should be made to minimize the level of this background absorption (see p. 229). If background correction is required, automatic simultaneous background correction rather than manual sequential background correction is the preferred technique when using electrothermal methods of atomization. This is because the background absorption signal for a given sample is much more likely to vary from firing to firing of an electrothermal device than when repeatedly nebulizing typical samples into a flame. Also background absorption effects tend to be much less pronounced in flames than in electrothermal atomization. For wavelengths above 300 nm the relatively low intensity of hydrogen or deuterium lamps can cause problems and then sequential background correction using a suitable non-resonance line should be used (see p. 241).

(b) *Chemical Interferences*

These are observed when the free atom population is altered by the presence of matrix components and can arise in several ways.

(i) The matrix may contain a high chloride (or other halide) concentration and the analyte can then partially vaporize as relatively volatile stable molecules (e.g., $CaCl_2$, $PbCl_2$) at a temperature below that at which efficient dissociation to atoms occurs.

(ii) The analyte and the matrix can react on the hot graphite surface and retard or enhance the rate of dissociation and vaporization of the analyte. This will result in a change in the overall shape of the transient peak. The effect can be minimized by measuring the peak area using a signal integration technique. These effects can also vary as the heating element ages. For example, as a graphite rod or tube ages it tends to become more porous so that the sample will penetrate further into the graphite before the solvent evaporation stage is completed.

(iii) The analyte atoms can react with matrix components after leaving the heated graphite surface. This results in a decrease in the measured transient absorbance. This type of interference is often observed when the boiling point of the analyte is similar to that of the matrix.

(iv) Reaction can occur between the analyte and the graphite resulting in involatile carbide formation. The degree of this carbide formation can be dependent upon the nature of the matrix. It has been found[(88,100)] that the addition of calcium (as the nitrate) to solutions of Al, Ba, Be, Si and Sn gave a considerable signal enhancement. This was attributed mainly to preferential carbide formation by the calcium. The age of the heating element can also influence this type of interference effect (see ii).

Methods of minimizing interferences

The elements that are most reliably estimated with a standard A.A.S. flame cell (e.g., Cd, Pb and Zn) are often subject to interelement effects using flameless systems. Zinc[(82,101)] has been found to be more interelement prone than molybdenum. On the other hand, interference effects due to the presence or formation of refractory compounds appear to be significantly less with the flameless atomizer than with the flame. For elements with high boiling points (e.g., Mo, Ti and V) it is often possible to completely remove the sample matrix during the dry-ashing stage(s) prior to the final atomization.

Ideally, after solvent evaporation and dry-ashing is completed a metallic oxide or metallic species should be present (e.g., Au, Al_2O_3, NiO, PbO, ZnO, etc.). For some elements the presence of hydrochloric acid results in metallic halides being present after dry-ashing. This can cause volatilization of the undissociated halide without atomization or can cause molecular absorption or light scatter by halide molecules of the matrix. If possible the dissolution of solid samples should be carried out in nitric acid rather than hydrochloric acid. If hydrochloric acid must be used, excess sulphuric acid, nitric acid or preferably ammonium nitrate should be added to the final solution. This should result in the removal of hydrochloric acid during the dry-ashing phase.

The anion(s) present can have a significant effect on electrothermal atomization signals. This has been demonstrated by FULLER,[(102–105)] who has shown that for some elements pronounced differences in signal can result by simply adding 0·1% v/v nitric, sulphuric or perchloric acid to pure aqueous solutions of certain elements. His results are shown in Table 7.5.

This Table demonstrates that the anion has far more effect on the signal in electrothermal atomization than in conventional flame atomization.

Background absorption and chemical interference effects can often be reduced by using various refinements, e.g., the use of limited field viewing with grazing incidence of the light beam across the sample cavity.[(81)] (This obviously does not apply to tube devices); alternative sample dissolution methods to avoid the formation of molecular halides (see above); shielding the Mini-Massmann device with an argon–hydrogen diffusion flame;[(79)] using the maximum possible dry-ashing temperature to remove as much as possible of

Table 7.6 Effect of 0·1% v/v acids on atomic absorption signals for various elements using a graphite tube furnace[102]

Element	Aqueous solution signal	Relative signal in the presence of 0·1% v/v acid		
		HNO_3	H_2SO_4	$HClO_4$
Cr	100	94	100	132
Sb	100	121	137	120
Se	100	97	53	80
Sn	100	150	6	1420
Te	100	159	202	205
Tl	100	100	100	<5

the sample matrix without volatilization of the analyte; setting the atomization voltage and time so that the analyte element is volatilized prior to the matrix.

EDIGER[106] has proposed various chemical methods for the minimization of non-specific background absorption. The addition of excess nickel ions to a solution lessens the volatility of arsenic, selenium and tellurium during the dry-ashing phase and allows much higher dry-ashing temperature to be used (up to 1200°C for selenium and tellurium, and up to 1400°C for arsenic). The addition of excess ammonium nitrate to samples containing large amounts of sodium chloride results in the volatilization of ammonium chloride during the dry-ashing phase (300°C) and leaves the sodium behind, mainly as sodium nitrate. This procedure considerably reduces the background absorption because sodium present as nitrate gives a much lower background absorption than when present as chloride. The addition of a large excess of ammonium fluoride or phosphate to solutions of cadmium allows the dry-ashing temperature to be increased from 500°C to 800–900°C without loss of cadmium.

The addition of orthophosphoric acid or ammonium dihydrogen phosphate has been shown to minimize chemical interferences when determining lead in a number of matrices. However, when this is attempted with urine samples a large background absorption signal is observed, this being caused by the decomposition of sodium and calcium phosphate salts simultaneously with the atomization of lead in the sample. This has been overcome by either adding ammonium molybdate to the samples or prior coating of the graphite tube with a 1% m/v ammonium molybdate solution.[107] It was postulated that the sodium phosphate was reduced to the phosphide and the reaction catalysed by either molybdenum or molybdenum carbide.

A particularly awkward problem is the determination of lead in non-saline waters and effluents. Although the matrix is not particularly complex, it can result in suppressions of the lead signal of up to 90% compared to pure aqueous lead standards. This means that it is necessary to make standard additions to every sample to ensure that a valid result is obtained. The effect does not appear to be linked to just one matrix constituent, but appears to be dependent upon the interaction of a number of matrix constituents (e.g., Ca, Cl, Mg, Na, SO_4). Various procedural modifications have been suggested in order to minimize this effect. The author (K.C.T.) has found that the addition of oxalic acid so that the sample, standard and blank solutions contained 3000 μg/ml oxalic acid almost eliminated the suppression of the signal for many samples. It was necessary to shake the samples prior to injection because calcium oxalate slowly precipitates out. Other methods that have been proposed are the addition of E.D.T.A. (diammonium salt) or ascorbic acid to all samples, standards and blank solutions so that the final solutions contained 0·004% m/v E.D.T.A.[108] or 1% m/v ascorbic acid.[109,204] This latter method using ascorbic acid gave an enhanced lead response compared to pure aqueous standards and eliminated the suppression of the lead signal for all the samples that were tested. It has been found possible to virtually eliminate the suppression of both lead and cadmium signals in a wide range of filtered non-saline water samples by coating of the graphite tube with lanthanum compound(s).[110] The coating was achieved by injecting 15 μl of a 20% m/v lanthanum nitrate hexahydrate solution into a modified Mini-Massmann graphite tube (Fig. 7.7) and carefully evaporating the solution to dryness. The graphite tube was then passed through a dry-ashing stage with a temperature setting of 550°C and finally through an atomization stage with a setting of 2000°C. This coating procedure was then repeated twice more. The coated tube could then be used with the same dry ash and atomization temperature for the determination of lead. For cadmium the dry ash temperature was reduced to 475°C. The only sample pretreatment of the filtered samples was the addition of nitric acid (1 ml nitric acid (70% m/m) per 100 ml sample). The lanthanum coating (probably present as oxide and/or carbide) is thought to act by causing the lead atoms in the sample to volatilize at a different time from the interfering matrix components that normally react with the liberated lead atoms and reduce the free lead atom population. Further work indicates that technique will work for manganese and zinc, but that for copper and nickel the lanthanum coating results in a large loss of sensitivity.

If, when developing a method, it is impossible to eliminate chemical interference effects, the method of standard additions should be used for calibration.

Comparison of the Massmann Tube, Mini-Massmann Rod and Rod-Type Systems

When atomization is effected with a bare rod or ribbon the vapour produced passes immediately into an unconfined space in which there is no further energy available to prevent condensation of the atoms with each other or with other species. The effective lifetime of the atomic vapour is therefore shorter for a rod or ribbon than for a furnace-type device. Hence, the likelihood of non-specific background absorption or chemical interference is somewhat greater unless the light beam has a small diameter (less than 2 mm) and is restricted to a grazing incidence over the sample cavity of the rod or ribbon (see Fig. 7.9b).[80,81] At the same time it might be anticipated that memory effects should be less for a graphite rod than for a graphite tube furnace. For elements such as molybdenum, titanium and vanadium this is observed in practice.

An important practical point in favour of the rod is that it requires the lowest power consumption of the available devices and takes less time to cool between analyses than a furnace. Furnace-type devices (Massmann and Mini-Massmann) are able to contain the atomic species in a restricted space at a high temperature for a longer time and this can often result in better precision. Sensitivity increases as the furnace cross-section decreases and the length increases. The continuum emission from the incandescent graphite can give rise to 'shot noise' which is usually manifested as an increase in the noise level on the readout trace as the graphite temperature increases. In extreme cases overload of the demodulator occurs and a change in the baseline level or a spurious peak then occurs. The black body continuum emission rapidly increases in intensity at wavelengths greater than 350 nm. It is most apparent when determining refractory oxide and carbide forming elements with resonance lines above 350 nm (e.g., Ba, Ca, Sr and Ti). Since the surface area of a tube is greater than that of a rod or ribbon such noise is more likely to be observed with a tube.

In practice, the amount of background radiation sensed by the photomultiplier can be constrained to a low level by careful collimation of the light beam,[100,111] running the hollow-cathode lamp at 90–100% of the maximum current rating, and by using a narrow spectral bandpass.

Applications

Flameless electrothermal atomization has been applied to the trace element analysis of a large variety of materials with considerable success. The precision of the technique is dependent upon many factors including the ability of the operator to accurately pipette small sample volumes, the analyte concentration and the matrix. Typical precisions of 2–7% R.S.D. can routinely be expected. More operator skill is required for electrothermal atomization than for conventional flame techniques.

The direct estimation of trace metals in sea-water might reasonably be considered an ideal application, but direct determination of the more volatile elements (e.g., Cd and Pb) is difficult because of the high sodium chloride content which gives rise to gross spectral interference. For these elements a solvent extraction method has been recommended.[112–114] However OTTAWAY *et. al*[115] have directly determined zinc and cadmium in seawater by using a low atomizing temperature (1492°C) in order to separate the atomic absorption from the background absorption signal. For elements that volatilize at higher temperature than sodium chloride (e.g., Cu, Fe, Mn, etc.) it is possible to determine these elements in sea-water but a distinct loss in sensitivity compared to pure solutions of the elements is observed which is attributed to chemical interference with the less volatile components of the sea-water matrix.[72] Thus, a standard addition method of calibration has to be used. A study of results for the direct estimations of a number of metals in river-water also indicated[108,119] that the standard addition method of calibration was essential. The determination of lead in cast-iron or steel using a graphite rod[116] or tube[117,118] is an example where with correct dry-ashing temperatures it is possible to obtain very sensitive results with negligible background absorption. The direct determination of phosphorus in steel using the phosphorus 213·6 nm non-resonance line has also been reported.[140]

In the space available it is impossible to give a comprehensive account of the large number of applications of this technique. A brief reference index has been included at the end of this section in order to assist the reader to obtain some relevant information on most aspects of electrothermal atomization. The technique has been especially successful for the analysis of a wide range of trace elements in a large variety of clinical samples (blood, bone, hair, kidney, liver, serum, skin, teeth, urine and various other tissues and organs) where the amount of sample available is often limited. The determination of lead and cadmium using the punched disc technique has already been described (see page 134). Other successful applications include trace element analysis in chemicals, foodstuffs, metals, minerals, oils and petroleum products, refractories, and soils and plants.

Some of the elements that at present cannot be determined using flameless electrothermal atomization are Ce, Hf, La, Nb, Ta, Th, U, W and Zr. This inability is thought to be due to involatility of the metals and carbide formation. Although it is possible to detect appreciable boron absorption from both graphite rod and tube atomizers, there is a memory effect.

Index of References to Various Aspects of Flameless Electrothermal Atomization

Books 69, 104
Chemicals and miscellaneous 85–87, 100, 105, 111, 120, 172
Clinical 78, 79, 92, 107, 119, 121–130, 161, 165–167
Emission measurements 73–77
Interference effects
(a) General 80–82, 94–96, 101, 103, 105–111, 131, 171
(b) Non-specific background absorption 72, 97–99, 106, 107, 114, 193, 194, 196–200
Metals 116–118, 133–140
Minerals 85, 100, 119, 132, 141, 142, 146, 203
Oils and petroleum 83, 84, 143–145, 201
Pyrolytic coating techniques 87–91
Refractory materials 85, 132, 141, 142, 146–148
Reviews 95, 104, 149–152
Soils and plants 153–161

CONCLUSIONS

It is highly unlikely that flameless atomization techniques will ever replace flame techniques for the majority of analyses. However, for certain specific types of analyses the flameless technique is invaluable and should be regarded as a complementary technique to the flame.

Graphite tube flameless atomization devices are rather bulky in construction, due mainly to the equipment required to provide the heavy current essential for their operation. Water cooling is essential. They can cost about as much as a medium-priced atomic absorption spectrophotometer. Graphite rod and tantalum ribbon devices are somewhat simpler, less bulky and can cost as little as one-half of the price of a medium-priced atomic absorption spectrophotometer. They all offer the capability to perform the determination of trace metals at very low levels using micro-samples. They also allow the direct analyses of biological fluids, oils and organic solvents. Minuteness of sample size is of no advantage in applications such as effluent analysis where the available sample volume is not a limiting factor. In fact, care must be exercised to ensure that the micro-sample applied to the flameless device is representative of the bulk sample.

Interference effects are not well documented and these can be severe in even relatively simple matrices. It is important to appreciate that interference effects can change as the resistive element ages.

Another problem associated with electrothermal atomization is acquiring the ability and skill to introduce small sample volumes reproducibly into or on to the device. MATOUSEK[172] has reported a technique where accurate pipetting is not required. The samples are nebulized for a fixed period and the resulting sample mist is directly deposited on to the inner surface of a hot (110°C) Mini Massman graphite tube. Very good reproducibilities were obtained.

The number of operating parameters is greater than that used in flame atomization. For example, the solvent evaporation, the dry-ashing and atomization stage temperatures and time settings, the rate of heating during the atomization stage, the nature and the flow rate of the purge gases, and the sample dissolution procedure must all be carefully selected.

Careful study of operating parameters is essential before a procedure for the estimation of trace metals in a *practical* analytical sample, such as a rock, effluent or a biological fluid, can be accepted as reliable for routine operation.[119]

The field of flameless atomic absorption is rapidly expanding and the interested reader can keep abreast of the latest developments of this and all other aspects of atomic spectroscopy by consulting the Annual Reports on Analytical Atomic Spectroscopy (A.R.A.A.S.) which is published every July by the Chemical Society, Burlington House, London W1V 0BN.

This publication gives a critical report of all published work and major lecture meetings in the preceding year on analytical atomic spectroscopy. It also contains a Table updated each year giving details of commercially available electrothermal atomizers.

6. The Direct Analysis of Solid Samples

Electrothermal atomization would appear to be ideally suited to the direct analysis of solids. A small amount of the solid sample (0·5–5 mg) is weighed out and placed into or on to the device, the sample is then dry-ashed and then atomized, and the absorption signal monitored. However, various problems have been experienced when attempting this apparently simple technique. It is difficult to obtain a homogeneous, uncontaminated and representative sample in a mass of 0·5–5 mg, calibration can cause problems unless suitable solid standard reference standards are available, biological or organic based samples tend to swell up during the dry-ashing stage and physically obstruct part of the light beam. Background absorption can present a serious problem with some samples and the time spent for repeated sample weighings may be greater than that required for a conventional sample dissolution. Despite these problems some successful attempts at the direct analysis of solid samples have been reported.[173–177] ROSSI and PICKFORD[176] have directly determined trace elements in a bovine liver sample using a graphite tube furnace. Five milligrams of the dried powdered liver was introduced through one end of the tube using a specially designed tantalum sampling device. The furnace temperature was then increased to 280°C with the ends of the furnace left open to the atmosphere. This resulted in the formation of a chip of carbonaceous material. The end windows were then replaced and the graphite tube was then taken through the appropriate dry-ashing and atomization stages. For calibration, the procedure was repeated with another 5 mg liver sample except that after the dry-ashing stage the tube was allowed to cool and a 5 μl aliquot of a standard solution was then pipetted on to the porous carbonaceous mass remaining. The liquid addition was then evaporated to dryness, dry-ashed and finally atomized. Good agreement with the certificate values for Ag, Cu, Mn and Pb was obtained. If calibration was attempted by adding pure aqueous solutions into an empty graphite tube errors of up to 100% were encountered.

FULLER and THOMPSON[178] have overcome the limitation of repeated sample weighings for the analysis of soils, rocks, minerals, cement and other similar materials by preparing a stable gel from the finely powdered sample. They ground the sample to pass a 325 mesh sieve (<44 μm particle size) and wetted the sample with a solution of sodium hexametaphosphate. A thixotropic thickening agent (T.T.A.) was then added in order to form a gel and the resulting suspension diluted to a predetermined volume with water. By making the final suspensions 2% v/v with respect to the T.T.A. and 1% m/v with respect to the sample, it was found that the suspensions were stable over a three-day period. Calibration was achieved using aqueous standards containing 2% v/v T.T.A. The disadvantage is that considerable dilution of the sample occurs. It was also possible to nebulize the suspensions directly into a flame providing the pulse nebulization technique (see p. 201) was used. However, it was then essential to grind the samples to a particle size of 10 μm or less in order to achieve efficient atomization.

WILLIS[181] has made a comprehensive investigation of the nebulization of aqueous suspensions of a wide range of powdered rock samples directly into a flame. It was found that only particles below 12 μm diameter contributed significantly to the observed absorption and that the atomization efficiency increased rapidly with decreasing particle size. With suspensions of samples ground to 325 mesh sieve (< 44 μm particle size), the atomization efficiency of a given metal only varied by a factor of about two between very different types of rock. The ground sample (1g) was placed in a beaker and 20 ml water was added, followed by a stirring bar. The beaker was then placed on a magnetic stirrer and directly nebulized into the air-acetylene or nitrous oxide-acetylene flames. A standard atomic absorption unit was used. Very little non-specific background absorption was detected even when nebulizing 5% m/v siliceous rock samples.

Finally mention must be made of sputtering techniques where the sample is atomized by cathodic sputtering in a low pressure inert gas discharge.[183–185] The sputtered atoms can either be detected by atomic absorption or atomic fluorescence techniques. If the sample is metallic it can be directly used with no preparation other than surface cleaning; non-metallic samples are powdered and mixed with a suitable conducting powder (e.g. Ag, Cu, Fe, Ni, etc.). Calibration can cause problems but the use of the internal standard technique would appear to overcome this limitation.[185]

7. Modified Atomic Absorption Systems

THE RESONANCE DETECTOR[186,187]

In a conventional atomic absorption spectrophotometer the required spectral line is isolated by means of a monochromator. WALSH and his team at the C.S.I.R.O. in Melbourne have developed a technique in which the atomic resonance line(s) is isolated by exploiting the phenomenon of resonance. The principle is illustrated in Fig. 7.12.

Modulated radiation from a standard hollow-cathode passes through the flame to a 'resonance lamp', in which the atomic vapour of the same element is produced by cathodic sputtering in a non-standard unmodulated hollow-cathode discharge lamp. The atomic vapour absorbs the resonance lines from

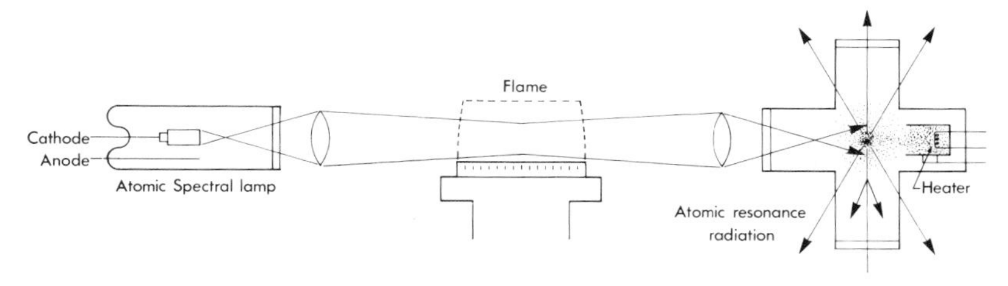

Fig. 7.12 The resonance detector

the source, and some of this absorbed energy is re-emitted in all directions as resonance radiation. A portion, emitted at right-angles to the direction of the light beam from the source, is collected through a side arm of the resonance lamp, and focused upon the photodetector. Since the radiation from the source is modulated, and that from the resonance lamp is unmodulated, only the re-emitted resonance radiation from the latter produces an output signal.

The resonance detector is thus not a true monochromator, since if the element being determined possesses more than one resonance line in its spectrum, energy from all these lines will be transmitted to the photodetector, and the sensitivity will be the average of the absorption sensitivities for all the lines. This will obviously be lower than that for the most absorbing line alone. The linearity of the calibration graph will also suffer. The fact that the analytical sensitivity depends upon the spectral complexity of a particular element constitutes the major disadvantage of this technique. Other disadvantages are that resonance monochromators are difficult to stabilize, expensive to construct and of limited life expectancy.

Some advantages that might lead to the manufacture of commercial resonance instruments and may well encourage their development for multi-channel on-line equipment are listed below:

(a) Resonance detectors cannot be put out of adjustment, in the same way that an optical monochromator can, by temperature changes and mechanical vibration. They could, therefore, be suitable for use under very rigorous conditions.

(b) No adjustment is needed to tune to a given line, so that their operation is even simpler than that of a standard atomic absorption spectrophotometer. They could be ideal for example, where very large numbers of samples are being analysed for a single element, especially if an automatic system were being employed.

SELECTIVE MODULATION[188]

This is a second technique that has been explored in some detail by WALSH and his group. The principle is illustrated in Fig. 7.13. Unmodulated light from a hollow-cathode lamp is passed along the axis of a discharge tube, the open-ended cathode of which is made from the same element as the source lamp. The resonance radiation from the source lamp is thus absorbed by the concentration of atomic vapour within the open-ended cathode, whereas all other radiation passes through undiminished in intensity.

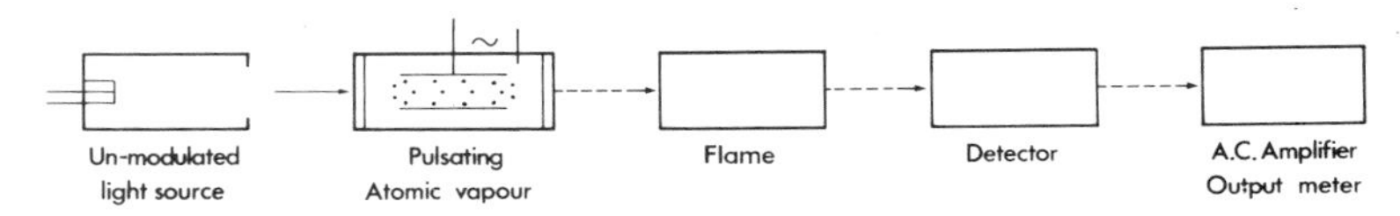

Fig. 7.13 Selective modulation of resonance lines

By modulating the power supply to the open-ended cathode the atomic vapour concentration is made to pulsate, so that the resonance radiation from the source is *selectively modulated*, whereas all other radiation passes through the system unmodulated. In theory an instrument employing this technique does not require a monochromator, but in practice extraneous unmodulated radiation is usually of sufficiently high intensity to cause unacceptable background noise, so that a low-resolution monochromator or a filter has to be incorporated.

WALSH has shown that selective modulation lamps can be built as one unit, with the hollow-cathode source and a modulating loop-electrode inside the same envelope. A gating circuit in the detection system removes any unwanted signals due to the modulation pulse.

The most important advantage of selective modulation derives from the fact that the technique effectively discriminates between the resonance lines and all other radiation.

PULSED CURRENT OPERATION OF HOLLOW-CATHODE LAMPS

DAWSON and ELLIS[(189)] increased the emission intensity from a conventional hollow-cathode lamp, by operating the device at a high current, passed through it in pulses of short duration. The lamp was run at a low d.c. current between pulses and the authors emphasized that this must be at a very low level. They claimed that there was no increase in resonance line width, and that self-absorption did not occur. Also because the signal-to-noise ratio is drastically improved better limits of detection are obtainable.

MALMSTADT and co-workers[(190,191)] have developed an intermittent high-current power supply for hollow-cathode lamps that has been incorporated into an atomic fluorescence spectrophotometer. The lamp is switched on for about 10 ms at a current of about 200 mA and is then switched off for about 80 ms, and this is repeated 10–20 times and the resultant signal averaged. Then there is a pause or wait time of 15–20 s before the lamp is operated again. The advantages of this mode of operation are very high intensities (typically

100–200 times that of conventionally operated hollow-cathode lamps); no warm-up time; a long lamp lifetime and excellent stability of output (typically 0·08% over 10 min and 0·2% over several hours). These sources would appear to be ideal for atomic fluorescence measurements.

Background Absorption Correction Systems

Systems based on Hydrogen or Deuterium Lamps

When performing trace element determinations upon solutions containing appreciable amounts of foreign ions, erroneously high results may be obtained due to non-specific background absorption. This background absorption is almost certainly due to molecular absorption by matrix components that are incompletely atomized in the flame. Background absorption has previously been ascribed to scattering of source radiation by matrix particles in the flame, but is now thought to be almost exclusively due to molecular absorption (see Refs 5 and 14, Chapter 2). Background absorption is more likely to be observed in flameless electrothermal atomization and in this case both scattering of source radiation as well as molecular absorption by matrix components can occur.[97–99]

The analyte atoms absorb radiation over a very small wavelength region of about 0·01 nm (see Chapter 9), thus, with typical spectral bandpass settings of 0·5–1·0 nm trace amounts of analyte will not appreciably absorb the continuous radiation emitted by a deuterium or hydrogen lamp at the analyte absorption line wavelength. The absorbance signal obtained using the analyte lamp is dependent upon both the concentration of the analyte and the concentration of the matrix responsible for the background absorption, whilst the absorbance signal obtained using the deuterium or hydrogen lamp is only dependent upon the concentration of the matrix responsible for the background absorption. Thus, subtraction of the latter signal from the former gives an absorbance signal that is only dependent upon the concentration of the analyte present.

The use of a simple or an automatic background correction system can usually eliminate errors from non-specific background absorption.

In the automatic system (which is now incorporated into many commercially available atomic absorption spectrophotometers), the light beams from the hollow-cathode lamp and a deuterium arc or deuterium hollow-cathode lamp are alternately passed through the flame. The instrumental arrangement is diagrammatically shown in Fig. 7.14. The background absorption is sensibly identical for both beams, whereas the element being determined absorbs only the resonance radiation from the element hollow-cathode lamp. By automatically subtracting the absorbance obtained with the deuterium lamp from that obtained with the element lamp the effect of the background absorption is cancelled out.

Adequate background compensation can be attained with a simple sequen-

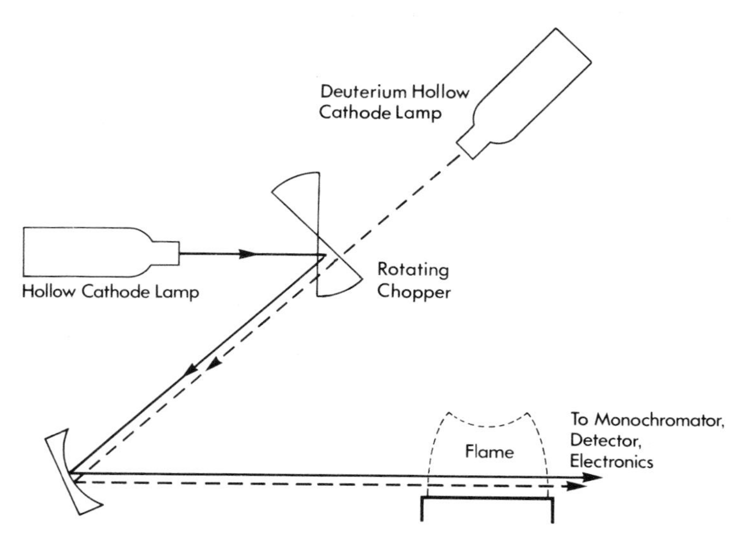

Fig. 7.14 Diagrammatic representation of an automatic background correction system Note. Instead of a rotating chopper the lamps can be pulse operated sequentially.

tial background correction system using an instrument with a multi-turret lamp. This is achieved by measuring the absorbance of a series of solutions, firstly with the relevant hollow-cathode lamp, and then with a deuterium hollow-cathode lamp, and subtracting the latter reading from the former. This procedure is less costly, less instrumentally complex and often gives rise to more precise results when using flames.[192] One reason for this is that the light beams from the two lamps in the non-automatic system should always transverse the same path through the atom reservoir. Another reason is that the lamp currents and monochromator spectral bandpass settings can be optimized independently. With the automatic system compromise settings often have to be used in order to 'balance' the two beams. For transient signals (e.g., flameless electrothermal atomization) the automatic system is more dependable as the background correction signal can vary from one firing to the next.

The magnitude of the background absorption is dependent upon the concentration of the matrix and is most noticeable for determinations carried out below 400 nm. In the past it has been stated that background absorption is only significant at wavelengths below 300 nm. This is *not* true; background absorption can occur at all wavelengths. It is *essential* to check for background absorption whenever a new method is developed. This will only take a few minutes and is time well spent. Some background absorption is usually observed when determining trace amounts of an element in an appreciable concentration of foreign ions especially when using an air–acetylene flame. Background absorption effects are far less severe in the hotter nitrous oxide–acetylene flame.

Table 7.7 Suitable lines above 300 nm for non-automatic background correction measurements

Analyte element	Resonance line wavelength, nm	Element lamp for background correction	Recommended background correction line wavelength, nm
Al	309·3	Fe	310·3
Al	309·3	Pd	311·4
Al	396·2	Pd	395·9
Ba	553·5	Yb	555·6
Ba	553·5	Mo	553·3
Ca	422·7	Pd	421·3
Ca	422·7	Fe	420·2
Cr	357·9	Pd	357·2
Cr	359·4	Pd	360·9
Cu	324·7	Pd	324·3
Cu	324·7	Sn	326·2
Cu	324·7	Ni	324·3
Mo	313·3	Ni	313·4
Mo	313·3	Pd	311·4
Sr	460·7	Cr	461·6
Ti	364·3	Pb	363·9
V	318·4	Fe	320·5
V	318·4	Sn	317·5

Notes: (1) For flameless electrothermal atomization the lamp current should be set to 75–100% of maximum current rating in order to minimize the effect of black body emission from the atomizer.
(2) If the sample matrix contains appreciable quantities of the element used to obtain the background correction line, tests should be made using pure solutions of that element at the maximum concentration likely to be encountered in the sample solutions. If observable atomic absorption occurs the line should not be used.

For wavelengths above 350 nm in the flame, and 300 nm using flameless electrothermal atomization, the light intensity from a deuterium lamp is too low for precise background correction. In these cases a suitable non-resonance line from another hollow-cathode lamp that is within ±2 nm of the resonance line can be used. Table 7.7 lists some suitable lines.

Problems can be experienced if the background absorption spectrum exhibits complex fine structure over the monochromator spectral bandpass wavelength range used for a particular determination. Fortunately this is not a very common occurrence. It has been observed in some rare earth element analyses, some steel analyses (see page 203) and can be observed over certain wavelength regions in flameless electrothermal atomization in the presence of certain molecular species (e.g., C_2, CN, SO_2).[193,194] The author (K.C.T.) has found that nitric acid solutions exhibit a fairly complex background absorption spectrum over the wavelength range 190–240 nm in the air–acetylene flame. For example, using a 3 M solution of nitric acid, the background absorption, as measured with a deuterium hollow-cathode lamp, was equivalent to 0·12 μg/ml lead at 217·0 nm, 0·45 μg/ml lead at 215·0 nm and 0·20 μg/ml lead at 210 nm. Sulphuric acid behaves similarly. If solutions containing appreciable amounts of nitric or sulphuric acid are nebulized, and very scale expanded measurements are made at wavelengths below 240 nm, it is sometimes possible to observe background absorption signals using a hydrogen or a deuterium lamp that are significantly greater than those observed with the analyte element lamp. This effect is almost certainly caused by fine structure in the background absorption spectrum of the matrix over the monochromator spectral bandpass being used.

Hydrochloric acid even at the 5 M level gives negligible background absorption over the wavelength range 200–250 nm.

Systems based on the Zeeman Effect

Finally, a brief mention should be made of background correction using the Zeeman effect. This is normally achieved by placing the cathode of hollow-cathode lamp in a strong magnetic field.[195] For most elements this results in splitting of the spectral lines to produce additional non-absorbable lines at wavelengths just outside the absorption profile of the atomic vapour. It is possible to separate the non-absorbable σ components from the absorbable π components by the use of polarizing filters. Thus, with modulation of the polarizing filters and consequent measurement of the differences of the absorbance signals of the σ and π components it is possible, with only one lamp, conventional single-beam optics and phase sensitive electronics, to obtain a system that is effectively a double-beam instrument that automatically corrects for both changes in the source intensity and for background absorption. Instead of applying the magnetic field to the lamp DAWSON and CO-WORKERS[196,197] have applied the magnetic field (2–9 kG) to the atomic vapour generated by a specially constructed graphite rod atomizer. Eleven elements were investigated. Background absorbances of 2·0 could be compensated to better than 0·005. This was four times better than the correction obtained with a conventional automatic background correction system using a deuterium lamp. A similar technique but using a graphite tube furnace has also been reported.[198,199] A magnetic field strength of 11 kG was used and detection limits for 26 elements given. Applying the magnetic field to the atomizer rather than the hollow-cathode lamp was found to increase considerably the linear region of the calibration graph.

Commercial instrumentation utilizing this technique is available.[200]

MICROPROCESSORS

These semiconductor devices are best described as micro-computers and are now being fitted as standard to the more elaborate atomic absorption spectrophotometers. Basically these devices simply process the data from the signal output of a standard atomic absorption instrument upon prior instructions from a keyboard fitted on to the instrument. For instance, the degree of scale expansion can be keyed in and the instrument will then automatically read out at this desired scale expansion setting. Alternatively, for use in the concentration mode, the blank and standards are nebulized (normally 3–5 standards are nebulized) and the actual concentration level keyed in for each standard and the blank solution. The unit will then select the correct scale expansion and curve correction control settings so that when samples are nebulized the readout is in the desired concentration units. The calibration curve can be updated by simply nebulizing one standard and depressing the appropriate buttons. If the unit is asked to do the impossible (e.g., if one of the standards were nebulized instead of the blank, or a sample were to give an overrange output, an error signal would be displayed). The unit can equally well be used in the emission and fluorescence modes. Another useful facility is the measurement of transient signals (e.g., the hydride generation or the flameless electrothermal atomization techniques). The result can be displayed in any chosen units as a peak height or as a peak area, or in some devices can be sequentially displayed in both modes. In some devices it is also possible to perform various statistical calculations on the results obtained.

Even with this labour saving development it is still essential to correctly align the burner, select the correct burner height and gas-flow settings, regularly clean out the burner head and spray chamber, check for a partial blockage of the nebulizer, etc. It is often assumed, quite erroneously, that once a system is automated that these time consuming procedures are no longer so important. Although the microprocessor is capable of linearizing a grossly non-linear calibration curve caused by poor burner alignment or incorrect setting of the spectral bandpass, the ultimate precision of the results will suffer unless the instrument is carefully optimized prior to each new determination.

The Dual-Channel Double-Beam System

This feature, the object of which is to provide the capability of simultaneous two-element analysis or alternatively internal standardization, is offered with a few commercial atomic absorption units. It comprises a second hollow-cathode lamp, two separate monochromators and special electronic circuitry in addition to the components of a standard atomic absorption spectrophotometer. As in the standard double-beam system the light from the hollow-cathode lamp(s) is split so that half passes through the flame and half by-passes the flame, with subsequent alternate passage to the monochromator (see Fig. 7.15).

Provided the concentrations in the test solution of the two elements both lie in the correct range for estimation, simultaneous two-element analysis is possible.

The other major advantage claimed for the 'dual' double-beam system is for

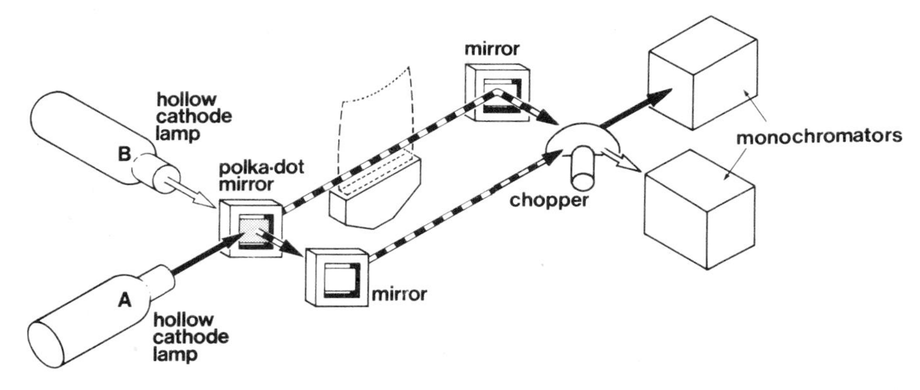

Fig. 7.15 Diagrammatic representation of dual-channel double-beam system

single-element analysis with internal standardization. For this procedure a constant addition of another element, that behaves similarly to the analyte, is made to all solutions. The ratio of the absorbance due to the analyte and that due to the internal standard is taken. This method minimizes the noise and drift due to the flame. In practice it is difficult to find suitable elements to act as reliable internal standards. It is obviously important that the samples should not contain the chosen internal standard element.

The analytical precision demonstrated is not significantly superior to that of a well-constructed single-beam instrument possessing signal integration, and hardly seems to justify the additional complexity and expense of such a unit, particularly for routine analysis.

Chap. 7 References and further reading

NEBULIZATION

1. LANG, R., Ultrasonic atomization of liquids. *J. Acoust. Soc. Amer.* 34 (1962) 6.
2. WENDT, R. H., and FASSEL, V. A., Induction-coupled plasma spectrometric excitation source. *Anal. Chem.* 37 (1965) 920.
3. WEST, C. D., and HUME, D., Radio-frequency plasma emission spectrophotometer. *Anal. Chem.* 36 (1964) 412.
4. WEST, C. D., Ultrasonic sprayer for atomic emission and absorption spectrochemistry. *Anal. Chem.* 40 (1968) 253.
5. HOARE, H. C., MOSTYN, R. A., and NEWLAND, B. T. N., An ultrasonic atomizer applied to AAS. *Anal. Chim. Acta* 40 (1968) 181.
6. VAN RENSBERG, H. C., and ZEEMAN, P. B., The determination of Au, Pt, Pd, Rh by AAS with an ultrasonic nebulizer and a multi-element high-intensity hollow-cathode lamp with selective modulation. *Anal. Chim. Acta* 43 (1968) 173.
7. KORTE, N. E., MOYERS, J. L., and DENTON, M. B., Investigations into the Use of a Pulse Ultrasonic Nebuliser–Burner System for A.A.S. *Anal. Chem.* 45 (1973) 530.
8. DENTON, M. B., and SWARTZ, D. B., An Improved Ultrasonic Nebuliser System for the Generation of High Density Aerosol Dispersions. *Rev. Scient. Instrum.* 45 (1974) 81.
9. KIRSTEN, W. J., and BERTILSSON, G. O. B., Direct continuous quantitative ultrasonic nebulizer for flame photometry and flame absorption spectrophotometry. *Anal. Chem.* 38 (1966) 648.

10. Isaaq, H. J. and Morgenthaler, L. P. Utilization of ultrasonic nebulization in A.A.S., a study of parameters. *Anal. Chem.* 47 (1975) 1661, 1668 and 1748.
11. Sebastiani, E., Ohls, K., and Riemer, G., Results of Atomization of Measured Volumes of Liquid Samples by A.A.S. *Z. Anal. Chem.* 264 (1973) 105.
12. Thompson, K. C., and Godden, R. G., A High Solids Nitrous Oxide–Acetylene Burner with Pulse Nebulisation. *Analyst* 101 (1976) 96.
13. Manning, D. C., Aspirating small volume samples in flame atomic absorption spectroscopy. *Atomic Abs. Newsletter* 14 (1975) 99.
14. Goulden, P. D. Controlling sample flow in flame A.A.S. *Atomic Abs. Newsletter* 16 (1977) 121.
15. Berndt, H. and Jackwerth, E. Automatic injection method for dispensing small volume samples in flame A.A. *Atomic Abs. Newsletter* 15 (1976) 109.
16. Rubeska, I., Miksovsky, M., and Huka, M., A branched capillary for buffering in flame spectrometry. *Atomic Abs. Newsletter* 14 (1975) 28.
17. Ward, D. A., and Biechler, D. G., Rapid, direct determination of calcium in natural waters. *Atomic Abs. Newsletter* 14 (1975) 29.
18. Fuller, C. W., Applications of a branched T-shaped capillary in flame spectrometry. *Atomic Abs. Newsletter* 15 (1976) 73.
19. Miller, T. H., and Edwards, W. H., Direct calcium and magnesium determination using a modified AA aspiration system. *Atomic Abs. Newsletter* 15 (1976) 75.
20. Babington, R. S., U.S. Patents 3421692, 3421699, 3425058, 3425059, 35048599.
21. Fry, R. C., and Denton, M. B., High solids sample introduction in flame A.A. analysis. *Anal. Chem.* 49 (1977) 1413.

Flame Systems

22. Gatehouse, B. M., and Willis, J. B., Performance of a simple atomic absorption spectrophotometer. *Spectrochim. Acta* 17 (1961) 710.
23. Amos, M. D., and Willis, J. B., Use of high temperature premixed flames in atomic absorption spectroscopy. *Spectrochim. Acta* 22 (1966) 1325.
24. Foster, W. H., and Hume, D. N., Factors affecting emission in intensities in flame photometry. *Anal. Chem.* 31 (1959) 2028.
25. Robinson, J. W., Flame photometry using the oxy–cyanogen flame. *Anal. Chem.* 33 (1961) 1226.
26. Baker, M. R., and Vallee, B. L., A theory of spectral excitation in flames as a function of sample flow. *Anal. Chem.* 31 (1959) 2036.
27. Fassel, V. A., and Mossotti, V. G., Atomic absorption spectra of vanadium, titanium, niobium, scandium, yttrium, and rhenium. *Anal. Chem.* 35 (1963) 252.
28. Slavin, W., and Manning, D. C., Atomic absorption spectrophotometry in strongly reducing oxy-acetylene flames. *Anal. Chem.* 35 (1963) 253.
29. Skogerboe, R. K., and Woodriff, R. A., Atomic absorption spectra of europium, thulium and ytterbium using a flame as line sources. *Anal. Chem.* 35 (1963) 1977.
30. Amos, M. D., and Thomas, P. E., The determination of aluminium in aqueous solution by atomic absorption spectroscopy. *Anal. Chim. Acta* 32 (1965) 139.
31. Willis, J. B., AAS with high temperature flames. *Applied Optics* 7 (1968) 1295.
32. Chester, J. E., Dagnall, R. M., and Taylor, M. R. G., Theoretical assessment of the nitrous oxide–hydrogen flame in spectroscopic analysis. *Analyst* 95 (1970) 702.
33. Price, W. J., and Whiteside, P. J., An investigation of the determination of boron by atomic absorption and flame emission spectrophotometry. XVIII Colloquium Spectroscopic Internationale, Grenoble 1975.
34. Slavin, W., Venghiattis, A., and Manning, D. C., Some recent experience with the nitrous oxide–acetylene flame. *Atomic. Abs. Newsletter* 5 (1966) 84.
35. Smithells, A., and Ingle, H., The structure and chemistry of flames. *J. Chem. Soc. (transactions)* 61 (1892) 204.
36. Kirkbright, G. F., Semb, A., and West, T. S., Spectroscopy in separated flames—I The use of the separated air–acetylene flame in thermal emission spectroscopy. *Talanta* 14 (1967) 1011.
37. Kirkbright, G. F., Semb, A., and West, T. S., The separated nitrous oxide–acetylene flame as an atom reservoir in thermal emission spectroscopy. *Spectroscopy Letters* 1 (1968) 7.

38. HINGLE, D., KIRKBRIGHT, G. F., and WEST, T. S., Spectroscopy in separated flames—II The use of the separated air–acetylene flame in long path absorption spectroscopy. *Talanta* 15 (1968) 199.
39. KIRKBRIGHT, G. F., SEMB, A., and WEST, T. S., Spectroscopy in separated flames—III Use of the separated nitrous oxide–acetylene flame in thermal emission spectroscopy. *Talanta* 15 (1968) 441.
40. HOBBS, R. S., KIRKBRIGHT, G. F., SARGENT, M., and WEST, T. S., Spectroscopy in separated flames—IV Application of the nitrogen-separated air–acetylene flame in flame-emission and atomic-fluorescence spectroscopy. *Talanta* 15 (1968) 997.
41. HINGLE, D. N., KIRKBRIGHT, G. F., and WEST, T. S., The determination of beryllium by thermal-emission and atomic fluorescence spectroscopy in a separated nitrous oxide–acetylene flame. *Analyst* 93 (1968) 522.
42. KIRKBRIGHT, G. F., SARGENT, M., and WEST, T. S., Spectroscopy in Separated Flames—VI The Argon or Nitrogen-Sheathed Nitrous Oxide–Acetylene flame in Atomic Absorption Spectroscopy. *Talanta* 16 (1969) 1467.
43. DAGNALL, R. M., KIRKBRIGHT, G. F., WEST, T. S. and WOOD, R., A.F.S. of Al, Mo, Ti, V, and Zr in inert gas separated nitrous oxide-acetylene flame. *Anal. Chem.* 42 (1970) 1029.
44. CRESSER, M. S., and KELIHER, P. N., Analytical Spectroscopy in separated flames. *Amer. Lab.* 2 (1970) 21 (Nov).
45. COKER, D. T., and OTTAWAY, J. A., Formation of free atoms in air–acetylene flames used in atomic absorption spectrometry. *Nature* 230 (1971) 156.
46. COKER, D. T., OTTAWAY, J. M., and PRADHAN, N. K., Metal atom formation processes in flames. *Nature* 233 (1971) 69.
47. ALDOUS, K. M., BAILEY, B. W., and RANKIN, J. M., Burning velocity of the premixed nitrous oxide–acetylene flame and its influence on burner design. *Anal. Chem.* 44 (1972) 191.

IODINE, PHOSPHOTUS AND SULPHUR DETERMINATION

48. KIRKBRIGHT, G. F., WEST, T. S., and WILSON, P. J., The direct determination of iodine by A.A.S. in a nitrogen separated nitrous oxide–acetylene flame. *Atomic Abs. Newsletter* 11 (1972) 53.
49. KIRKBRIGHT, G. F., WEST, T. S., and WILSON, P. J., The application of the Leipert amplification to increase sensitivity in the direct determination of iodine by A.A.S. *Atomic Abs. Newsletter* 11 (1972) 113.
50. KIRKBRIGHT, G. F., and MARSHALL, M., Direct determination of phosphorus by A.A. spectrometry. *Anal. Chem.* 45 (1973) 1610.
51. KIRKBRIGHT, G. F., and MARSHALL, M., The direct determination of sulphur by atomic absorption spectroscopy in a nitrogen separated nitrous oxide–acetylene flame. *Anal. Chem.* 44 (1972) 1288.
52. KIRKBRIGHT, G. F., MARSHALL, M., and WEST, T. S., Direct determination of sulphur in oils by atomic absorption spectrometry using an inert gas shielded nitrous oxide–acetylene flame. *Anal. Chem.* 44 (1972) 2379.
53. KIRKBRIGHT, G. F., and WILSON, P. J., The application of a demountable hollow-cathode lamp as a source for the direct determination of S, I, As, Se and Hg by A.A. flame spectrometry. *Anal. Chem.* 46 (1974) 1414.
54. ADAMS, M. J., and KIRKBRIGHT, G. F., The direct determination of sulphur by A.A.S. using a graphite furnace electrothermal atomizer. *Can. J. Spectrosc.* 21 (1976) 127.

MODIFIED BURNER AND NEBULIZER SYSTEMS

55. FUWA, K., and VALLEE, B. L., The physical basis of analytical atomic absorption spectrometry. *Anal. Chem.* 35 (1963) 942.
56. RUBEŠKA, I., and STUPAR, J., The use of absorption tubes for the determination of noble metals in atomic absorption spectroscopy. *Atomic Abs. Newsletter* 5 (1966) 69.
57. RUBEŠKA, I., and MOLDAN, B., Investigations on long-path absorption tubes in atomic absorption spectroscopy. *Analyst* 93 (1968) 148.
58. SPRAGUE, S., and SLAVIN, W., Performance of a three-slot Boling burner. *Atomic Abs. Newsletter* 4 (1965) 293.
59. BOLING, E. A., A multiple slit burner for atomic absorption spectroscopy. *Spectrochim. Acta* 22 (1966) 425.

60. AGEMAIN, H., ASPILA, K. I. and CHAU, A. S. Y. Comparison of the performance of the single and triple slot air-acetylene burner for A.A.S. *Anal. Chem.* 47 (1975) 1038.
61. ROBINSON, J. W., and HARRIS, B. M., Mechanical feed burner with total consumption for flame photometry and atomic absorption spectroscopy. *Anal. Chim. Acta* 26 (1962) 439.
62. RAWSON, R. A. G., Improvement in performance of a simple atomic absorptionmeter by using pre-heated air and town gas. *Analyst* 91 (1966) 630.
63. RAWSON, R. A. G. A total consumption laminar flow nebulizer and burner system for flame spectroscopy. 3rd International Congress of A.A.S. and A.F.S., Paris 1971.
64. VENGHIATTIS, A. A., A heated chamber burner for A.A.S. *Applied Optics* 7 (1968) 1313.
65. HELL, A., ULRICH, W. F., SHIFRIN, N. and RAMIREZ-MUNOZ, J., Laminar flow burner with improved heated spray chamber and condenser. *Applied Optics* 7 (1968) 1317.
66. UNY, G., N'GUEY LOTTIN, J., TARDIF, J. P., and SPITZ, J. A study of the performance of a preheated burner in A.A.S. *Spectrochim. Acta* 26B (1971) 151.

FLAMELESS ELECTROTHERMAL A.A.S. (see also 201–204)
67. L'VOV, B. V., The Analytical use of atomic absorption spectra. *Spectrochim. Acta* 17 (1961) 761.
68. L'VOV, B. V., The Potentialities of the graphite crucible in AAS. *Spectrochim. Acta* 24B (1969) 53.
69. L'VOV, B. V., Atomic Absorption Spectrochemical Analysis. Elsevier Publishing Co., New York 1970.
70. WEST, T. S., and WILLIAMS, X. K., AAS and AFS with a carbon filament atom reservoir. *Anal. Chim. Acta* 45 (1969) 27.
71. MASSMANN, H., The determination of arsenic by A.A.S. *Z. Analyt. Chem.* 225 (1967) 203.
72. EDIGER, R. D., PETERSON, G. E., and KERBER, J. D., Application of the graphite furnace to saline water analysis. *Atomic Abs. Newsletter* 13 (1974) 61.
73. OTTAWAY, J. M., and SHAW, F., Carbon furnace atomic emission spectrometry—A preliminary appraisal. *Analyst* 100 (1975) 438.
74. OTTAWAY, J. M., and HUTTON, R. C., Determination of volatile elements by carbon furnace atomic emission spectrometry. *Analyst* 101 (1976) 683.
75. HUTTON, R. C., OTTAWAY, J. M., RAINS, T. C. and EPSTEIN, M. S., Determination of barium in calcium carbonate rocks by carbon furnace atomic emission spectroscopy. *Analyst* 102 (1977) 429.
76. OTTAWAY, J. M., and LITTLEJOHN, D. Background emission in carbon furnace atomic emission spectrometry. *Analyst* 102 (1977) 553.
77. OTTAWAY, J. M. and HUTTON, R. C. Use of hydrogen to reduce molecular emission interference in carbon furnace atomic emission spectrometry. *Analyst* 102 (1977) 785.
78. MATOUSEK, J. P., and STEVENS, B. J., Biological applications of the carbon rod atomiser in atomic absorption spectroscopy. Preliminary studies on magnesium, iron, copper, lead and zinc in blood and plasma. *Clin. Chem.* 17 (1971) 363.
79. AMOS, M. D., BENNETT, P. A., BRODIE, K. G., and LUNG, P. W., Carbon rod atomizer in atomic absorption and fluorescence spectrometry and its clinical application. *Anal. Chem.* 43 (1971) 211.
80. ALGER, D., ANDERSON, R. G., MAINES, I. S., and WEST, T. S., Atomic absorption and fluoresence spectroscopy with a carbon filament atom reservoir. Part VI A Study of some matrix effects. *Anal. Chim. Acta* 57 (1971) 271.
81. ANDERSON, R. G., JOHNSON, H. N., and WEST, T. S., Part VII. Atomic absorption under limited field viewing conditions. *Anal. Chim. Acta* 57 (1971) 281.
82. JOHNSON, D. J., WEST, T. S., and DAGNALL, R. M., A Comparative study of the determination of zinc and molybdenum by AAS with a carbon filament atom reservoir. *Anal. Chim. Acta* 66 (1973) 171.
83. ALDER, J. F., and WEST, T. S., Part 12. The determination of nickel in crude and residual fuel oils by A.A.S. *Anal. Chim. Acta* 61 (1972) 132.
84. EVERETT, G. L., WEST, T. S., and WILLIAMS, R. W., Part 14. Determination of vanadium in fuel oils. *Anal. Chim. Acta* 66 (1973) 301.
85. JACKSON, K. W., WEST, T. S., and BALCHIN, L., Determination of vanadium in titanium dioxide by ultra micro AAS on a carbon filament atom reservoir. *Anal. Chem.* 45 (1973) 249.
86. EVERETT, G. L., WEST, T. S., and WILLIAMS, R. W., The determination of tin by carbon

filament AAS. *Anal. Chim. Acta* 70 (1974) 291.
87. THOMPSON, K. C., and THOMERSON, D. R., Recent developments in atomic absorption spectroscopy. *Amer. Lab.* 6 (1974) 53.
88. THOMPSON, K. C., GODDEN, R. G., and THOMERSON, D. R., A method for the formation of pyrolytic graphite coatings and enhancement by calcium addition techniques for graphite rod flameless atomic absorption spectrometry. *Anal. Chim. Acta* 74 (1975) 289.
89. MORROW, R. W., and MCELHANEY, R. J., Determination of chromium in industrial effluent water by flameless atomic absorption spectroscopy. *Atomic Abs. Newsletter* 13 (1974) 45.
90. MANNING, D. C., and EDIGER, R. D., Pyrolysis graphite surface treatment for HGA 2100 sample tubes. *Atomic Abs. Newsletter* 15 (1976) 42.
91. STURGEON, R. E., and CHAKRABARTI, C. I., Evaluation of pyrolytic graphite coated tubes for graphite furnace A.A.S. *Anal. Chem.* 49 (1977) 90.
92. HWANG, J. Y., ULLUCCI, P. A., and MOKELER, C. J., Direct flameless atomic absorption determination of lead in blood. *Anal. Chem.* 45 (1973) 795.
93. HWANG, J. Y., MOKELER, C. J., and ULLUCCI, P. A., Maximization of sensitivities in tantalum ribbon flameless AAS. *Anal. Chem.* 44 (1972) 2018.
94. SCHRENK, W. G., and EVERSON, R. T., Atomic absorption interferences using a tantalum boat atomising system. *Appl. Spectroscopy* 29 (1975) 41.
95. AGGETT, J., and SPROTT, A. J., Non-flame atomisation in AAS. *Anal. Chim. Acta* 72 (1974) 49.
96. MCINTYRE, N. S., COOK, M. G., and BOASE, D. G., Flameless atomic absorption determination of cobalt, nickel and copper: a comparison of tantalum and molybdenum surfaces. *Anal. Chem.* 46 (1974) 1983.
97. ADAMS, M. J., KIRKBRIGHT, G. F., and RIENVATANA, P., Molecular absorption spectra of some simple inorganic salts in the heated graphite atomiser. *Atomic Abs. Newsletter* 14 (1975) 105.
98. CULVER, B. R., and SURLES, T., Interferences of molecular spectra due to alkali halides in non-flame AAS. *Anal. Chem.* 47 (1975) 920.
99. PRITCHARD, M. W., and REEVES, R. D., Non-atomic absorption from matrix salts volatilised from graphite atomisers in AAS. *Anal. Chim. Acta* 82 (1976) 103.
100. THOMPSON, K. C., and GODDEN, R. G., A method for the determination of low levels of barium in calcium carbonate. *Analyst* 100 (1975) 198.
101. CLARK, D., DAGNALL, R. M., and WEST, T. S., The atomic absorption determination of zinc with a graphite furnace. *Anal. Chim. Acta* 63 (1973) 11.
102. FULLER, C. W., Private communication.
103. FULLER, C. W., Electrothermal atomisation, a routine or specialist analytical technique. *European Spectroscopy News* Aug. (1976) 18.
104. FULLER, C. W., Electrothermal atomisation for AAS. Chemical Society, London, 1977.
105. FULLER, C. W., The effects of acids on the determination of thallium by AAS with a graphite furnace. *Anal. Chim. Acta* 81 (1976) 199.
106. EDIGER, R. D., Atomic absorption analysis with the graphite furnace using matrix modification. *Atomic Abs. Newsletter* 14 (1975) 127.
107. HODGES, D. J., Observations on the direct determination of lead in complex matrices by carbon furnace A.A.S. *Analyst* 102 (1977) 66.
108. DOLINSEK, F., and STUPAR, J., Application of the carbon cup atomization technique in water analysis by A.A.S. *Analyst* 98 (1973) 841.
109. REGAN, J. G. T., and WARREN, J., A novel approach to the elimination of matrix interferences in flameless AAS using a graphite furnace. *Analyst* 101 (1976) 220.
110. THOMPSON, K. C., WAGSTAFF, K., and WHEATSTONE, K., A method for the minimisation of matrix interferences in the determination of cadmium and lead in non-saline waters using electrothermal atomisation. *Analyst* 102 (1977) 310.
111. KERBER, J. D., RUSSO, A. J., PETERSON, G. E., and EDIGER, R. D., Performance improvements with the graphite furnace. *Atomic Abs. Newsletter* 12 (1973) 106.
112. KREMLING, K., and PETERSEN, H., APDC-MIBK Extraction system for the determination of copper and iron in 1 cm^3 of sea water by AAS. *Anal. Chim. Acta* 70 (1974) 35.
113. SEGAR, D. A., and GONZALEZ, J. G., Evaluation of atomic absorption with a heated graphite atomizer for the direct determination of trace transition metals in sea water. *Anal. Chim. Acta* 58 (1972) 7.

114. DONELLY, T. H., FERGUSON, J., and ECCLESTON, A. J., Direct determination of trace metals in seawater using the Varian Techtron carbon rod atomizer model 63. *Appl. Spectroscopy* 29 (1975) 158.
115. CAMBELL, W. C., and OTTAWAY, J. M., Direct determination of cadmium and zinc in seawater by carbon furnace A.A.S. *Analyst* 102 (1977) 495.
116. THOMPSON, K. C., and GODDEN, R. G., The determination of lead in cast iron using the mark 2 longitudinal graphite rod head. Shandon Southern Application Report No. 661.
117. SHAW, F., and OTTAWAY, J. M., Determination of trace amounts of lead in steel and cast iron by atomic absorption spectrometry with the use of carbon furnace atomisation. *Analyst* 99 (1974) 184.
118. FRECH, W., Rapid determination of lead in steel by flameless AAS. *Anal. Chim. Acta* 77 (1975) 43.
119. CRUZ, R. B., and VAN LOON, J. C., A critical study of the application of graphite furnace non-flame AAS to the determination of trace base metals in complex heavy-matrix sample solutions. *Anal. Chim. Acta* 72 (1974) 231.
120. IHNAT, M., Atomic absorption determination of selenium with carbon furnace atomisation. *Anal. Chim. Acta* 82 (1976) 293.
121. GLENN, M., SAVORY, J., HART, L., GLENN, T., and WINEFORDNER, J. D., Determination of copper in serum with a graphite rod atomizer for atomic absorption spectrophotometry. *Anal. Chim. Acta* 57 (1971) 263.
122. SCHRAMEL, P., Determination of eight metals in the international biological standard by flameless A.A.S. *Anal. Chim. Acta* 67 (1973) 69.
123. GROSS, S. B., and PARKINSON, E. S., Analyses of metals in human tissues using base (TMAH) digests and graphite furnace AAS. *Atomic Abs. Newsletter* 13 (1974) 107.
124. KUBASIK, N. P., VOLOSIN, M. T., and MURRAY, M. H., Carbon rod atomizer applied to measurement of lead in whole blood by AAS. *Clin. Chem.* 18 (1972) 410.
125. NORVAL, E., and BUTLER, L. R. P., The determination of lead in blood by AA with the high temperature graphite tube. *Anal. Chim. Acta* 58 (1972) 47.
126. KUBASIK, N. P., and VOLOSIN, M. T., A simplified determination of urinary Cd, Pb and Tl with use of carbon rod atomization and AAS. *Clin. Chem.* 19 (1973) 954.
127. MAESSEN, F. J. M. J., POSMA, F. D., and BALKE, J., Direct determination of Au, Co and Li in blood plasma using the mini-massmann carbon rod atomiser. *Anal. Chem.* 46 (1974) 1445.
128. POSMA, F. D., BALKE, J., HERBER, R. F. M., and STUIK, E. J., Microdetermination of cadmium and lead in whole blood by flameless AAS using carbon tube and carbon cup as sample cell and comparison with flame studies. *Anal. Chem.* 47 (1975) 834.
129. DOLINSEK, F., STUPAR, J., and SPENKO, M., Determination of aluminium in dental enamel by the carbon cup atomic absorption method. *Analyst* 100 (1975) 884.
130. KAMEL, H., BROWN, D. H., OTTAWAY, J. M., and SMITH, W. E., Determination of gold in blood fractions by AAS using carbon rod and carbon furnace atomisation. *Analyst* 101 (1976) 790.
131. FULLER, C. W., A kinetic theory of atomisation for AAS with a graphite furnace. Part IV—Assessment of interference effects. *Analyst* 101 (1976) 798.
132. BRATZEL, M. P., CHAKRABARTI, C. L., STURGEON, R. E., MCINTYRE, M. W., and AGEMIAN, H., Determination of gold and silver in parts per billion or lower levels in geological and metallurgical samples by atomic absorption spectrometry with a carbon rod atomiser. *Anal. Chem.* 44 (1972) 372.
133. WELCHER, G. G., KRIEGE, O. H., and MARKS, J. Y., Direct determination of trace quantities of Pb, Bi, Se, Te and Tl in high temperature alloys by non-flame AAS. *Anal. Chem.* 46 (1974) 1227.
134. GUERIN, B. D., The determination of the noble metals by AAS with a carbon rod furnace. *J. South African Chemical Institute* 25 (1972) 230.
135. ADRIAENSSENS, E., and KNOOP, P., A study of the optimal conditions for flameless AAS of iridium platinum and rhodium. *Anal. Chim. Acta* 68 (1974) 37.
136. RADCLIFFE, D. B., BYFORD, C. S., and OSMAN, P. B., The determination of arsenic antimony and tin in steels by flameless AAS. *Anal. Chim. Acta* 75 (1975) 457.
137. SHAW, F., and OTTAWAY, J. M., The determination of trace amounts of aluminium and other elements in iron and steel by AAS with carbon furnace atomisation. *Analyst* 100 (1975) 217.

138. EVERETT, G. L., The determination of the precious metals by flameless AAS. *Analyst* 101 (1976) 348.
139. PERSSON, J., FRECH, W., and CEDERGREN, A. Determination of aluminium in low alloy steels and stainless steels by flameless AAS. *Anal. Chim. Acta* 89 (1977) 119.
140. PRICE, W. J. and WHITESIDE, P. J. Determination of phosphorus in steel by AAS with electrothermal atomization. *Analyst* 102 (1977) 618.
141. SIGHINOLFI, G. P., Determination of beryllium in standard rock samples by flameless atomic absorption spectroscopy. *Atomic Abs. Newsletter* 11 (1972) 96.
142. HEINRICHS, H., and LANGE, J., Trace element analysis and micro analysis of silicate carbonate rocks by flameless AAS. *Z. Anal. Chem.* 265 (1973) 256.
143. BRATZEL, M. P., and CHAKRABARTI, C. L., Determination of lead in petroleum and petroleum products by AAS with a carbon rod atomizer. *Anal. Chim. Acta* 61 (1972) 25.
144. BRODIE, K. G., and MATOUSEK, J. P., Application of the carbon rod atomizer to atomic absorption spectrometry of petroleum products. *Anal. Chem.* 43 (1971) 1557.
145. ROBBINS, W. K., Determination of lead in gasoline by heated vaporization AAS. *Anal. Chim. Acta* 65 (1973) 285.
146. FULLER, C. W., The application of the heated graphite furnace to the determination of trace metal impurities in some refractory materials. *Atomic Abs. Newsletter* 12 (1973) 40.
147. FULLER, C. W., and WHITEHEAD, J., The determination of trace metals in high-purity sodium calcium silicate glass, sodium borosilicate glass, sodium and calcium carbonate by flameless AAS. *Anal. Chim. Acta* 68 (1974) 407.
148. FULLER, C. W., The determination of trace metals in high purity lead silicate glasses by flameless AAS. *Atomic Abs. Newsletter* 14 (1975) 73.
149. WOODRIFF, R., Atomization chambers for atomic absorption spectrochemical analysis: A Review. *Appl. Spectroscopy* 28 (1974) 413.
150. KIRKBRIGHT, G. F., Application of non-flame atom cells in atomic absorption and atomic fluorescence spectroscopy. A Review. *Analyst* 96 (1971) 609.
151. THOMPSON, K. C., and THOMERSON, D. R., Sample atomization in A.A.S. *Chemistry in Britain* 11 (1975) 316.
152. SYTY, A., Developments in methods of sample injection and atomisation in atomic spectrometry. *CRC Critical Reviews in Analytical Chemistry* 4 (1974) 155.
153. OSBOURNE, A. C., and WEST, T. S., Determination of trace amounts of metals in soils by ultra-micro A.A.S. *Proc. Soc. Analyt. Chem.* 9 (1972) 198.
154. TRACHMAN, H. L., TYBERG, A. J. and BRANIGAN, P. D. A.A.S. determination of sub-ppm quantities of tin in extracts and biological materials with a graphite furnace. *Anal. Chem.* 49 (1977) 1090.
155. HOCQUELLET, P. and LABEYRIE, N. Determination of tin in foods by flameless A.A. *Atomic Abs. Newsletter* 16 (1977) 124.
156. HENNING, S., and JACKSON, T. L., Determination of molybdenum in plant tissue using flameless AA. *Atomic Abs. Newsletter* 12 (1973) 100.
157. DUDAS, M. J., The quantitative determination of cadmium in soils by solvent extraction and flameless A.A.S. *Atomic Abs. Newsletter* 13 (1974) 109.
158. BRADY, D. V., MONTALVO, J. G., JUNG, J., and CURRAN, R. A., Direct determination of lead in plant leaves via graphite furnace A.A. *Atomic Abs. Newsletter* 13 (1974) 118.
159. SIMMONS, W. J., and LONERAGAN, J. F., Determination of copper in small amounts of plant material by AAS using a heated graphite atomiser. *Anal. Chem.* 47 (1975) 566.
160. SIMMONS, W. J., Determination of low concentrations of cobalt in small samples of plant material by flameless AAS. *Anal. Chem.* 47 (1975) 2015.
161. SMEYERS-VERBEKE, J., MICHOTTE, Y., VAN DEN WINKEL, P., and MASSART, D. L., Matrix effects in the determination of copper and manganese in biological materials using carbon furnace A.A.S. *Anal. Chem.* 48 (1976) 125.
162. FERNANDEZ, F. J., and MANNING, D. C., Atomic absorption analysis of metal pollutants in water using a heated graphite atomiser. *Atomic Abs. Newsletter* 10 (1971) 65.
163. PICKFORD, C. J., and ROSSI, G., Analysis of high purity water by flameless AAS. *Analyst* 97 (1972) 647.
164. HENN, E. L., Determination of selenium in water and industrial effluents by flameless AA. *Anal. Chem.* 47 (1975) 428.

165. CERNIK, A. A. Determination of blood lead using a 4 mm paper punched disc carbon sampling cup technique. *Brit. J. Industr. Med.* 31 (1974) 239.
166. CERNIK, A. A., and SAYERS, M. H. P., Application of blood cadmium determination to industry using a punched disc technique. *British J. Industr. Med.* 32 (1975) 155.
167. CERNIK, A. A., Lead in blood, the analyst's problem. *Chemistry in Britain* 10 (1974) 58.
168. POSMA, F. D., SMIT, H. C., and ROOZE, A. F., Optimisation of instrumental parameters in flameless AAS. *Anal. Chem.* 47 (1975) 2087.
169. MAESSEN, F. J. M. J., and POSMA, F. D., Fundamental aspects of flameless A.A. using the Mini-Massmann carbon rod atomiser. *Anal. Chem.* 46 (1974) 1439.
170. ISSAQ, H. J., and ZIELINSKI, W. L., Modification of a graphite tube atomizer for flameless AAS. *Anal. Chem.* 47 (1975) 2281.
171. MATOUSEK, J. P. and CZOBIK, E. J. Effect of anions on atomization temperatures in furnace A.A. *Talanta* 24 (1977) 573.
172. MATOUSEK, J. P. Aerosol deposition in furnace atomization. *Talanta* 24 (1977) 315.

THE DIRECT ANALYSIS OF SOLID SAMPLES

173. KERBER, J. D., KOCH, A., and PETERSON, G. E., The direct analysis of solid samples by A.A. using a graphite furnace. *Atomic Abs. Newsletter* 12 (1973) 104.
174. HENN, E. L., Determination of trace metals in polymers by flameless A.A. with a solid sampling technique. *Anal. Chim. Acta* 73 (1974) 273.
175. GRIES, W. H., and NORVAL, E., New solid standards for the determination of trace impurities in metals by flameless A.A.S. *Anal. Chim. Acta* 75 (1975) 289.
176. PICKFORD, C. J., and ROSSI, G., Determination of some trace elements in N.B.S. (SRM-1577) bovine liver using flameless atomic absorption and solid sampling. *Atomic Abs. Newsletter* 14 (1975) 78.
177. LANGMYHR, F. J. and RASMUSSEN, S. A.A.S. determination of gallium and indium in inorganic materials by direct atomization from the solid state in a graphite furnace. *Anal. Chim. Acta*, 72 (1974) 79.
178. FULLER, C. W., and THOMPSON, I., Novel sampling system for the direct analysis of powders by A.A.S. *Analyst* 102 (1977) 141.
179. GONG, H. and SUHR, N. H. The determination of cadmium in geological materials by flameless A.A.S. *Anal. Chim. Acta.* 81 (1976) 297.
180. FULLER, C. W. Determination of trace elements in titanium (IV) oxide pigments by A.A.S. using an aqueous slurry technique. *Analyst* 101 (1976) 961.
181. WILLIS, J. B. A.A.S. analysis by direct introduction of powders into the flame. *Anal. Chem.* 47 (1975) 1752.
182. L'VOV, B. V., Trace characterisation of powders by A.A.S. The state of the art. *Talanta* 23 (1976) 109.
183. GOUGH, D. S., HANNAFORD, P., and WALSH, A., The application of cathodic sputtering to the production of atomic vapours in A.F.S. *Spectrochim. Acta* 28B (1973) 197.
184. GOUGH, D. S. Direct analysis of metals and alloys by A.A.S. *Anal. Chem.* 48 (1976) 1926.
185. MCDONALD, D. C. Use of internal standards for A.A.S. analyses of samples atomized by sputtering. *Anal. Chem.* 49 (1977) 1336.

MODIFIED ATOMIC ABSORPTION SYSTEMS

186. SULLIVAN, J. V., and WALSH, A., Resonance radiation from atomic vapours. *Spectrochim. Acta* 21 (1965) 727.
187. SULLIVAN, J. V., and WALSH, A., The application of resonance lamps as monochromators in atomic absorption spectroscopy. *Spectrochim. Acta* 22 (1966) 1843.
188. BOWMAN, JUDITH A., SULLIVAN, J. V., and WALSH, A., Isolation of atomic resonance lines by selective modulation. *Spectrochim. Acta* 22 (1961) 205.
189. DAWSON, J. B., and ELLIS, D. J., Pulsed current operation of hollow cathode lamps to increase the intensity of resonance lines for atomic absorption spectroscopy. *Spectrochim. Acta* 23A (1967) 565.
190. MALMSTADT, H. V., and CORDOS, E., Characteristics of hollow-cathode lamps operated in an intermittent high current mode. *Anal. Chem.* 45 (1973) 27.
191. MALMSTADT, H. V., WOODRUFF, T. A., and DEFREESE, J. D., New type of programmable current-regulated power supply for operation of hollow cathode lamps in a high intensity programmed mode. *Anal. Chem.* 46 (1974) 1471.

192. ZANDER, A. T., O'HAVER, T. C., and KELIHER, P. N., Continuum source atomic absorption spectrometry with high resolution and wavelength modulation. *Anal. Chem.* 48 (1976) 1166 (see Table 7).
193. MASSMANN, H., and GÜCER, S., Physical and chemical processes in atomic absorption analyses with a graphite tube furnace. *Spectrochim. Acta* 29B (1974) 283.
194. MASSMANN, H., EL GOHARY, Z., and GÜCER, S., Measurement of spectral background in A.A.S. *Spectrochim. Acta*, 31B (1976) 399.
195. STEPHENS, R., and RYAN, D. E., An application of the Zeeman effect to analytical atomic spectroscopy (Parts I and II). *Talanta* 22 (1975) 655 and 659.
196. DAWSON, J. B., GRASSAM, E., ELLIS, D. J., and KEIR, M. J., Background correction in electrothermal atomic absorption spectroscopy using the Zeeman effect in the atomized sample. *Analyst* 101 (1976) 315.
197. GRASSAM, E., DAWSON, J. B., and ELLIS, D. J., Application of the inverse Zeeman effect in electrothermal A.A. analysis. *Analyst* 102 (1977) 804.
198. KOIZUMI, H., and YASUDA, K., Determination of lead, cadmium and zinc using the Zeeman Effect in A.A.S. *Anal. Chem.* 48 (1976) 1178.
199. KOIZUMI, H., YASUDA, K., and KATAYAMA, M., A.A.S. based on the polarization characteristics of the Zeeman Effect. *Anal. Chem.* 49 (1977) 1106.
200. BROWN, S. D., Zeeman effect-based background correction in A.A.S. *Anal. Chem.* 49 (1977) 1269A. (A useful review with 40 refs).

ADDITIONAL FLAMELESS ELECTROTHERMAL A.A.S. REFERENCES
201. PRÉVÔT, A., and GENTE-JAUNIAUX, M., Rapid determination of phosphorus in oils by flameless A.A. *Atomic Abs. Newsletter* 17 (1978) 1.
202. EDIGER, R. D., KNOTT, A. R., PETERSON, G. E., and BEATY, R. D., The determination of phosphorus by A.A. using the graphite furnace. *Atomic Abs. Newsletter* 17 (1978) 28.
203. CHAO, T. T., SANZOLONE, R. F., and HUBERT, A. E., Flame and flameless AA determination of tellurium in geological materials. *Anal. Chim. Acta*, 96 (1978) 251.
204. REGAN, J. G. T., and WARREN, J., The influence of ascorbic acid on the matrix interferences observed during the flameless AA determination of lead in some drinking waters. *Analyst* 103 (1978) 447.

8 *Flame Emission and Atomic Fluorescence Spectroscopy*

8.1 Flame Emission Spectroscopy

8.1.1 Introduction

The beginnings of flame emission spectroscopy (F.E.S.) extend back to the middle of the 18th century (e.g., GEOFFRAY, 1732; MELVILL, 1752; and MARGGRAF, 1758). The colour imparted to candle and alcohol flames by various metallic salts was observed and used as a crude identification method. TALBOT (1826) showed remarkable insight for his time and extended this work. The main limitation at that time was that only the relatively luminous diffusion-type flames were known (e.g., candle and alcohol). KIRCHHOFF and BUNSEN (1860) developed the Bunsen burner and spectroscope, and were able to demonstrate that the visible lines observed in the flame were due to elements and not compounds. Their observations laid the foundations to the flame emission technique. The first quantitative flame emission measurement was performed in 1873 by CHAMPION, PELLETT, and GRENIER. They determined sodium using two flames. One flame was saturated with sodium chloride from a wick whilst the sample was drawn through the other flame on a platinum wire. A blue wedge was gradually brought in front of the first flame until they both appeared to be of equal brightness.

The Swedish agronomist LUNEGARDH (1928) demonstrated the versatility of F.E.S. by using a pneumatic nebulizer and an air–acetylene flame. The first commercial flame photometers appeared in 1937 and more sophisticated instruments from the early 1950s. The advent of atomic absorption pushed flame emission techniques (except for sodium and potassium) into the background, but recently there has been renewed interest.

8.1.2. Theoretical Considerations

When an atomic vapour is formed in a typical premixed flame, a small number of atoms are transferred to an excited state by vibrationally excited flame gas molecules.

e.g.

$$F^* + A \rightarrow F + A^*$$
$$A^* \rightarrow A + h\nu$$

A = Atom (produced from the nebulized solution)
F = Flame gas molecule

A^* = Electronically excited atom
F^* = Vibrationally excited flame gas molecule.

The number of atoms (N_1) that are in an excited state compared to the number (N_0) in the ground state (for a system in effective thermal equilibrium) is given by the Boltzmann equation:

$$N_1 = N_0 \frac{g_1}{g_0} \exp\left(\frac{-E}{kT}\right)$$

where E is the energy of the excited state above the ground state (J)

g_1, g_0 are the statistical weights of the excited and ground states, respectively
k is the Boltzmann constant (JK^{-1})
T is the absolute temperature

The calculated values of N_1/N_0 for some resonance lines at different temperatures are given in Table 8.1. It can be seen that this ratio (N_1/N_0) rapidly decreases with decreasing wavelength (increasing E). Except for elements with long-wavelength resonance lines (K, Cs) at extremely high temperatures, the number of atoms in the lowest excited state is negligible in comparison to the number in the ground state.

Table 8.1 Ratio of the number of atoms in the excited state to the number in the ground state—

Resonance line nm	g_1/g_0	Excitation energy (ev)	N_1/N_0 T = 2000	2500	3000 K
Cs 852·1	2	1·46	$4{\cdot}44 \times 10^{-4}$	$2{\cdot}42 \times 10^{-3}$	$7{\cdot}24 \times 10^{-3}$
Na 589·0	2	2·10	$0{\cdot}99 \times 10^{-5}$	$1{\cdot}14 \times 10^{-4}$	$5{\cdot}83 \times 10^{-4}$
Ca 422·7	3	2·93	$1{\cdot}22 \times 10^{-7}$	$3{\cdot}67 \times 10^{-6}$	$3{\cdot}55 \times 10^{-5}$
Cu 324·8	2	3·81	$4{\cdot}82 \times 10^{-10}$	$4{\cdot}04 \times 10^{-8}$	$6{\cdot}65 \times 10^{-7}$
Mg 285·2	3	4·35	$3{\cdot}35 \times 10^{-11}$	$5{\cdot}20 \times 10^{-9}$	$1{\cdot}50 \times 10^{-7}$
Zn 213·9	3	5·80	$7{\cdot}45 \times 10^{-15}$	$6{\cdot}22 \times 10^{-12}$	$5{\cdot}0 \times 10^{-10}$

The ratio N_1/N_0 increases exponentially with temperature and wavelength. Thus, it can be seen that at flame temperatures of 2–3000 K effectively 100 per cent of the atoms are in the ground state.

8.1.3. Equipment Considerations

Most atomic absorption spectrophotometers can be used for flame emission studies. The various essential components are discussed below:
Excitation Source.
The most common source is a premixed laminar flame.

Flame	Maximum flame temperature	Uses
Air–propane	1925°C	Na, K, Rb, Cs
Air–acetylene	2300°C	Na, K, Li, Cs, Ca, Sr, Mn, Ag, Cu
Nitrous oxide–acetylene	2750°C	Applicable to over 40 elements
Nitrogen–hydrogen-entrained air (diffusion flame)	ca. 1400°C	S, P

The premixed oxygen–acetylene flame has a higher temperature than the nitrous oxide–acetylene flame. Unfortunately, the risk of explosive flashback with oxygen–acetylene mixtures is so great that no manufacturer has developed a safe burner system that will support this flame.

The ideal flame shape is circular. This minimizes cooling and self-absorption effects. The standard 120 mm slot air–acetylene atomic absorption flame is not suitable because of cooling of the large flame surface area (emission intensity exponentially dependent on temperature) and self-absorption considerations. The nitrous oxide–acetylene flame, supported on a 50 mm atomic absorption slot burner, is a useful flame emission source.[1,2] The main advantage of this type of burner is that carbonization of the burner slot, of modern atomic absorption nitrous oxide–acetylene burners, is far less serious than with circular emission burners. The primary reaction zone of the flame, which is situated just above the top of the burner, should not be viewed by the detector. This zone emits intense flame background radiation and is not in thermal equilibrium. Flame emission measurements are normally made from the interconal zone, situated above the primary reaction zone.

Flame emission limits of detection are normally limited by the stability of the flame background emission, with a small flame emission signal being measured above a large flame background signal. Thus, careful optimization of flame conditions, and also very precise regulation of the flame gas flow rates, are essential. For most elements, when using the nitrous oxide–acetylene flame, the acetylene flow rate should be adjusted while nebulizing distilled water to give the minimum flame background radiation intensity at the wavelength of interest. This results in a slightly fuel-rich flame. In most cases the flame background intensity is far more dependent on the acetylene flow rate than the element emission intensity. The flame background signal increases when no solution is nebulized. This is because the fine water droplets from the nebulized solution cool the flame. Hence, it is very important that the matrix and viscosity of standards, samples and blank are carefully matched.

Monochromator

Although filters have been used to isolate the emissions from Ca, Li, K and Na, in general a monochromator is required. Ideally, the monochromator should have a high resolution (0·2 nm or better), low stray light figures, and a low *f* number (see page 193).

Radiation Detector

Photomultipliers are used, the requirement being adequate response over spectral range 260–780 nm (852 nm if caesium is to be determined) and a low stable dark current.

Amplifier and Readout

In most cases the emission signal is directly proportional to the concentration of the emitting species. The output from the photomultiplier is normally electronically modulated, amplified, and then usually read out on a recorder. The use of wavelength scanning (see 8.1.10), and integration of the output signal for a fixed period, are also useful readout techniques.

8.1.4. SAMPLE REQUIREMENTS

The essential requirements are similar to those for atomic absorption. A representative sample is dissolved to form a homogeneous solution. (Water, acids, alkali, or organic solvents can be used.) In general, the maximum concentration of total solids in solution that can be nebulized is 5–10% m/v. For routine analysis the total solid content should not exceed 2–3% m/v to avoid burner encrustation during long periods of operation.

8.1.5. SHAPE OF CALIBRATION CURVES

The emission intensity is directly proportional to the concentration of nebulized element over a wide concentration range, normally to about 200–2000 times the detection limit. At higher concentrations self-absorption causes the signal to become proportional to the square root of the concentration of the nebulized element, and the calibration curve then bends towards the concentration axis.

8.1.6. STANDARDIZATION AND INTERFERENCE EFFECTS

It is essential to realize that flame emission spectroscopy is a comparative technique and, if meaningful results are to be obtained, extreme care must be taken to ensure that the standards match the behaviour of the samples to be analysed.

In some cases simple aqueous standards of the element to be determined are adequate, but it should not be assumed that this is always the case. It is always advisable to test the suitability of proposed standards against known sample solutions. The flame background signal is very dependent on the solution uptake rate of the nebulizer. The greater the solution uptake rate the lower the flame background signal. This effect is caused by cooling of the flame by the nebulized solvent. Thus, careful matching of the standards and samples is essential.

Spectral interference is the most serious form of interference, but any extraneous substance that causes a variation in the number of ground state atoms in the flame will interfere.

Some causes of interference and methods of overcoming them are given in Table 8.2.

Table 8.2 Causes of interference in flame emission spectroscopy

Cause of interference	Method of overcoming interference
Change of aspiration rate (e.g., wine and spirit analyses, variable matrix analyses)	Match viscosity of samples and standards. Use method of standard additions combined with wavelength scanning.
Compound formation (e.g., determination of calcium in varying amounts of phosphate or sulphate)	Use releasing agent, e.g. La, E.D.T.A. Use a nitrous oxide–acetylene flame.
Ionization (e.g., determination of rubidium)	Add ionization suppressant (e.g., potassium chloride) to standards and samples.
Spectral interference	Reduce spectral bandpass (e.g., Mn and K lines at 403·3 and 404·4 nm, Ag and Cu lines at 328·1 and 327·4 nm). Use an alternative emission line.
	Use method of standard additions combined with wavelength scanning (e.g., low levels of strontium in variable levels of calcium. The broad calcium oxide band spectrum overlaps the strontium emission line at 460·7 nm).

8.1.7. Range of Elements

In theory all elements that can be determined by atomic absorption techniques will give flame atomic emission spectra, but only 40–45 elements are amenable to direct determination by flame atomic emission techniques. Elements with their main resonance lines below 270 nm (As, B, Be, Cd, Hg, Sb, Se, Si, Te, Zn, etc.) cannot be determined satisfactorily as the available flame temperatures are insufficient to excite sufficient atoms to the first excited state (see Table 8.1). Similarly, certain elements are not appreciably broken down to atoms even by a fuel-rich nitrous oxide–acetylene flame (Ce, La, Th, U, etc.). For certain elements, e.g. B, Be, La, P, S, it is possible to utilize molecular band emission, but with considerably reduced selectivity.

The inductively coupled R.F. excited argon plasma, which has an effective temperature of 6–10 000 K, is capable of exciting almost all elements. The source consists of an argon plasma maintained by a radio frequency electromagnetic field. The sample is nebulized directly into the plasma. Few interelement effects have been observed.[3]

8.1.8. Limit of Detection

The definition of the limit of detection is now generally agreed as 'The minimum amount of an element that can be detected with 95% certainty.' This is that quantity of the element that gives a reading equal to twice the standard deviation, of a series of at least ten determinations, at or near the blank level. The actual detection limit will depend on the stability of the nebulizing system, the stability of the flame, the flow control of flame gases, the electronics, the photomultiplier, and the monochromator. The limits of detection in flame emission are usually far more dependent upon the instrumentation used than in atomic absorption.

In general the use of the separated flame technique (see p. 208) does not result in any marked improvement in detection limits. The exceptions are the few elements that have resonance lines coincident with very intense flame background bands (e.g. Bi at 306·8 nm). Although flame separation techniques considerably reduce the flame background signal, the flame temperature is lowered resulting in a considerable decrease in the atomic emission signal for many elements. Thus, in many cases no overall improvement in the detection limit is observed. Useful wavelengths and some detection limits obtained in the author's laboratory (K.C.T.) for some elements that can be determined by flame emission in the nitrous oxide–acetylene flame are given in Table 8.3.

8.1.9. Precision

The precision obtainable depends on many factors (e.g. concentration, wavelength, flame stability). In general, at concentration levels above 100–200 times the detection limit in simple matrices, the precision should be 0·3–1% R.S.D. (Relative Standard Deviation) if signal integration techniques are used. With care wavelength scanning techniques can give precisions of 1–5%.

8.1.10 Wavelength Scanning Technique

This is a useful rapid qualitative (or even semi-quantitative) technique that can be used for multielement analysis. Alternatively it can be used for most single element quantitative analysis. Most emission lines lie in the wavelength range 320–450 nm. The maximum scan speed, time constant of the readout system, and spectral bandpass are linked by the equation.

$$\text{Maximum scan speed (nm/s)} = \frac{\text{spectral bandpass (nm)}}{4 \times \text{time constant (s)}}$$

Table 8.3 Some nitrous oxide–acetylene flame-emission detection limits (5 cm slot burner). 0·2 nm spectral bandpass.

Element	Wavelength (nm)	2σ Limit of detection (μg/ml)	Other useful but less sensitive lines (nm)
Ag	328·1	0·1	338·3
Al*	396·2	0·01	394·4
Ba*	553·5	0·01	
Ca*	422·7	<0·001	
Cr*	425·4	0·02	427·5, 429·0
Co	345·4	0·1	340·5, 341·2
Cs*	852·1	—	455·5
Cu	327·4	0·1	324·7
Fe	372·0	0·05	373·7, 374·7, 386·0
Ga*	417·2	0·05	403·3
In*	451·1	0·02	410·2
K	766·5	—	404·4
Li*	670·8	0·0001	
Mg*	285·2	0·1	
Mn	403·3	0·01	
Mo	390·3	0·5	379·8, 386·4
Na*	589·0	<0·0002	589·6, 330·3
Nb*	405·9	—	
Ni	341·5	—	352·5, 362·0
Os	426·1	—	
Pb	405·7	1	
Pd	363·5	—	
Rb*	780·0	—	420·2
Re	346·1	—	
Sr*	460·7	0·001	
Ti*	399·8	1	
Tl*	377·6	0·03	535·0
V	437·9	0·05	
W	400·9	—	
Zr*	360·1	—	

* Excess potassium chloride to suppress ionization.

Note: Li, Na, K, Rb, Cs can be determined satisfactorily in the air–acetylene (or even air–propane) flames. Common elements that cannot be satisfactorily determined by atomic flame emission are: As, B, Be, Bi, Cd, Hg, Pt, Sb, Se, Si, Te, Zn.

When using the wavelength scanning technique for a single element in a variable matrix (e.g. Al in river water and effluents, Sr in bone) a slow scan speed with a correspondingly long time constant should be used. Figure 8.1 illustrates the potential of the technique.

8.1.11. Comparison of Atomic Absorption and Flame Emission Spectroscopy

F.E.S.——Advantages in relation to A.A.S.

(a) A hollow-cathode lamp is required for every element to be determined by atomic absorption.

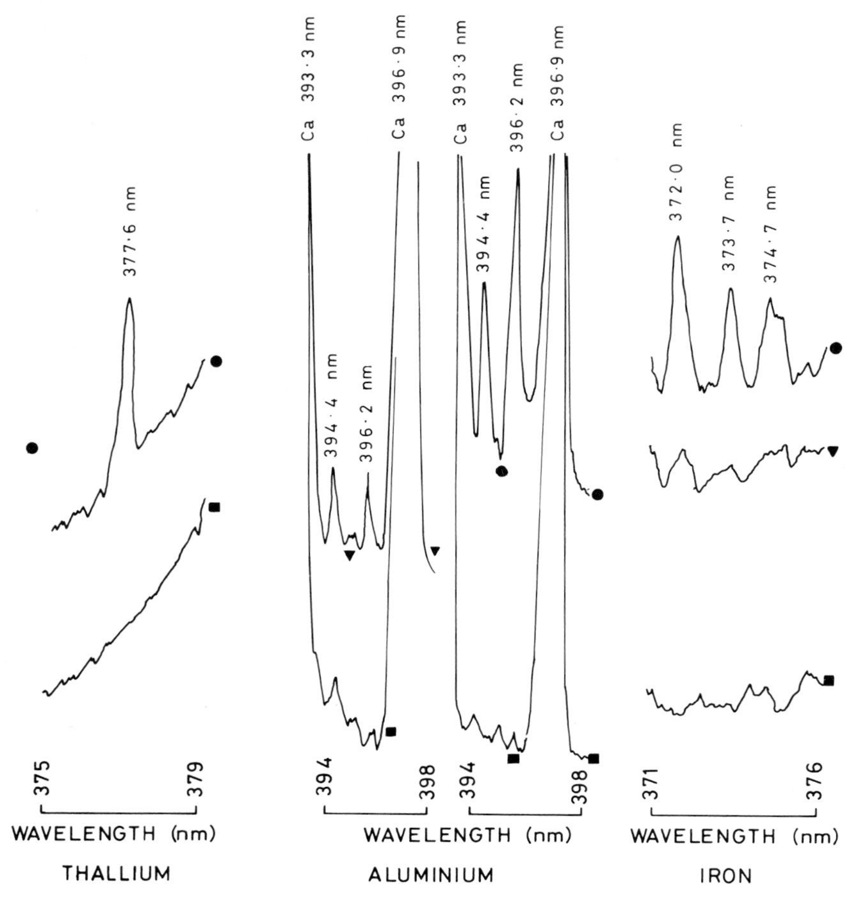

Fig. 8.1 Some scanning flame emission results
Nitrous oxide–acetylene flame. Tap water matrix, 0·2 nm spectral bandpass
Distance from top of burner grid to bottom of monochromator entry slit—10 mm
Time constant 0·3s, scan speed 10 nm/min

Thallium
● Tap water + 1 μg/ml Tl
■ Tap water blank
Aluminium
● Tap water + 1 μg/ml Al
▼ Tap water + 0·2 μg/ml Al
■ Tap water blank
Iron
● Tap water 1 μg/ml Fe
▼ Tap water + 0·2 μg/ml Fe
■ Tap water blank

Note. For rapid survey work, a spectral bandpass of 0·2 nm, a scan speed of 10 nm/min and a time constant of 0·3 s are suitable. For single element determinations, a spectral bandpass of 0·1 nm, a scan speed of 1 nm/min and a time constant of 1·5 s are suitable.

(b) It is possible to perform rapid qualitative flame emission analysis by the wavelength scanning technique.
(c) With simple instrumentation it is possible to obtain very good precision when using signal integration with the flame emission technique.
(d) Calibration curves can often be used over wider concentration ranges in flame emission than in atomic absorption.

F.E.S.—Disadvantages in Relation to A.A.S.

(a) The number of excited atoms varies exponentially with temperature (see Table 8.1) whilst the number of ground state atoms remains essentially constant. Thus, any slight temperature fluctuation in an atomic vapour will have a more pronounced effect on the excited state atom population utilized in flame emission, than on the ground state atom population utilized in atomic absorption. Hence, careful regulation of the flame gas flow rates is very critical.
(b) Spectral interference is often observed in flame emission owing to the relatively large number of atomic lines, and broad molecular bands (e.g. CaOH, MgO, CuOH), that can be observed in emission. In atomic absorption studies, the primary radiation source is modulated either electronically or mechanically, and an a.c. amplifier tuned or phase-locked to the modulation frequency is connected to the detector. Thus, any atomic or molecular emission from the flame is not directly detected by the measuring system. However, intense emission from the flame at the absorption line wavelength can cause an increase in the noise level of atomic absorption determinations.

Most elements have simple atomic absorption spectra consisting of relatively few absorption lines, and spectral interference between absorption lines of different elements is seldom observed. Non-specific background absorption in hot flames is usually weak, and can generally be ignored unless appreciable concentrations of substances (>200 μg/ml) are being nebulized.

Spectral interference in flame emission can be reduced by improving the monochromator resolution and minimizing the stray light figures. Hence, the monochromator requirements are more stringent for flame emission than atomic absorption.

It is often assumed that flame emission techniques are more prone to interelement effects than atomic absorption techniques. Although this is true of spectral interference, it must be realized that any factor that affects the ground-state atom population, e.g., ionization, compound formation in the flame, will affect absorption and emission techniques equally.
(c) From Table 8.1 it can be seen that flame emission techniques are not very sensitive for elements whose main resonance lines lie below 270 nm (As, Bi, Cd, Hg, Sb, Se, Te, Zn, etc.). The sensitivity of the A.A. technique does not depend upon the wavelength of the resonance line. The techniques of flame emission and atomic fluorescence are complementary. Often elements that cannot be determined by flame emission (As, Bi, Cd, Hg, Sb, Se, Te, Zn, etc.)

can be determined with high sensitivity by atomic fluorescence.
(d) More operator skill is required in the interpretation of flame emission results.

8.1.12. Conclusions

Flame emission should be regarded as a complementary technique to atomic absorption. Although it cannot be applied to as many elements as can atomic absorption techniques, it can prove very useful for detecting very low levels of certain elements (alkali, and alkaline earth metals). Additionally, the technique of wavelength scanning allows a rapid qualitative or semi-quantitative analysis of a completely unknown solution to be made. Nearly all commercially available atomic absorption instruments have facilities for flame emission measurements. It is to be hoped that more users will avail themselves of this technique.

8.2. Atomic Fluorescence Spectroscopy (A.F.S.)

8.2.1. Introduction

Atomic fluorescence spectroscopy may be defined as the measurement of radiation from discrete atoms that are being excited by the absorption of radiation from a given source which is not seen by the detector. Atomic fluorescence may be considered to be the analogue of molecular spectrofluorimetry and was first reported by Wood[(4)] in 1904, when he succeeded in exciting fluorescence of the D lines of sodium vapour. This was achieved by illuminating sodium vapour contained in an evacuated test tube with light from a gas flame containing sodium chloride, and visually observing the yellow D lines.

Wood called this fluorescence 'resonance radiation' because it was predicted by the classical theory of a light wave vibrating with the same frequency as the dipole oscillations of the medium. Soon after this initial discovery, resonance radiation was observed for mercury, cadmium, zinc and many other elements. This early work has been summarized by Mitchell and Zemansky[(5)] in their treatise *Resonance Radiation and Excited Atoms*.

The fluorescence of atoms in flames was first reported by Nichols and Howes[(6)] in 1923 who obtained weak fluorescence from Ba, Ca, Li, Na, Sr, and Tl atoms present in high concentration in a hydrogen–air flame irradiated by light containing the resonance line of the appropriate metal.

Little work was reported on atomic fluorescence spectroscopy after the 1930s until quite recently, when it has attracted new interest for two reasons. Firstly, as a method of investigating the physical and chemical processes that occur in flames. Secondly, as a basis of a new analytical technique theoretically possessing some advantages over both atomic absorption and flame emission methods for the detection and estimation of trace metallic elements.

In 1961 Robinson[(7)] observed weak fluorescence of the 285·2 nm Mg line in an oxygen–hydrogen flame using a magnesium hollow-cathode lamp. The following year ALKEMADE[8] used the atomic fluorescence of sodium to study mechanisms of excitation and deactivation of atoms in flames and to measure the quantum efficiency for the 589·0 nm sodium D line. He was the first to point out the possible analytical value of the technique.

The first analytical method was developed by WINEFORDNER and his co-workers and was described in a series of four papers published in 1964–1965.[9–12] A year later WEST and his co-workers[13–14] at Imperial College used commercially available equipment for measuring atomic fluorescence and since then both research teams have greatly extended the application of this technique.

In atomic absorption spectrophotometry, the subsequent history of the energy absorbed by the atoms is of little concern. Much of the energy is lost by collisional deactivation within the flame gases. However, some of the energy imparted to the atoms is re-emitted in all directions and this phenomenon is the basis of atomic fluorescence.

8.2.2. THEORETICAL CONSIDERATIONS

The basic atomic fluorescence arrangement, as shown in Fig. 8.2, consists of an intense source focused on to an atomic population in a flame or atom reservoir. Fluorescent radiation, which is emitted in all directions, then passes to a detector in the same plane usually positioned at right angles to the incident light.

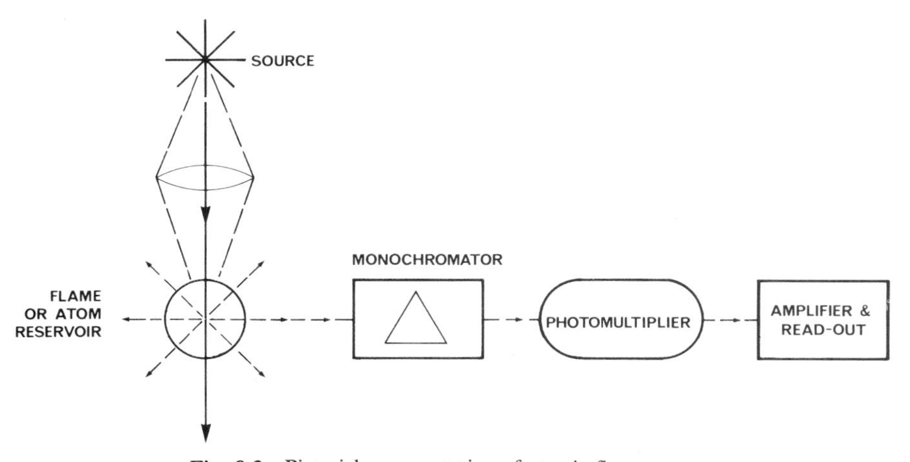

Fig. 8.2 Pictorial representation of atomic fluorescence

The source may be either an atomic line source or a continuum and serves to excite atoms by the absorption of radiation of the proper wavelength. The atoms are then deactivated partly by collisional quenching with flame gas molecules and partly by emission of fluorescent radiation in all directions.

The wavelength of the fluorescent radiation is generally the same or longer than the incident radiation. The wavelength of the emitted radiation is characteristic of the absorbing atoms and the intensity of the emission may be used as a measure of their concentration.

At low concentrations, this intensity is governed by the following relationship:

$$I_f = K\text{Ø}IoC$$

where

I_f is the intensity of fluorescent radiation
C is the concentration of metal ion in solution
K is a constant
Io is the intensity of the source at the absorption line wavelength
Ø is the quantum efficiency for the fluorescent process
Ø may be defined as the ratio of the number of atoms which fluoresce from the excited state to the number of atoms which undergo excitation to the same excited state from the ground state in unit time.

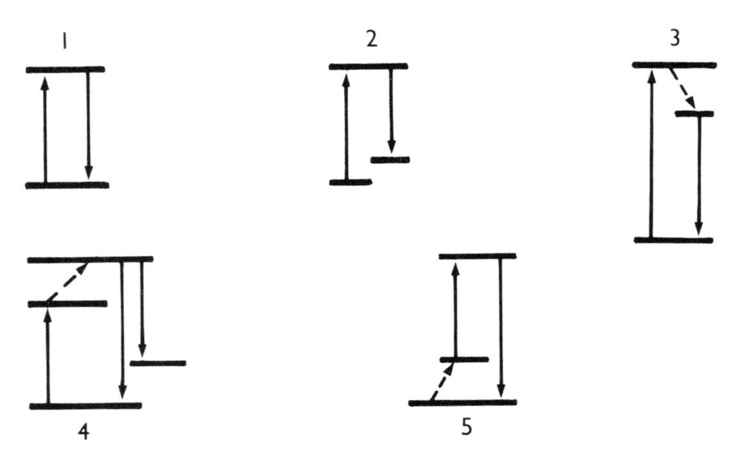

Fig. 8.3 Diagrammatic representation of the five basic types of fluorescence

There are five basic types of fluorescence which occur in flame measurement: (1) resonance fluorescence, (2) direct-line fluorescence, (3) stepwise-line fluorescence, (4) thermally assisted direct-line fluorescence, (5) thermally assisted anti-Stokes fluorescence.

Resonance Fluorescence. Resonance fluorescence occurs when an atom emits a spectral line of the same wavelength as used for excitation of the atom.

Direct-line Fluorescence. This occurs when an atom emits a spectral line of longer wavelength than the spectral line used for excitation (both the exciting and emitted lines have the same excited state). Thus, after the fluorescence emission occurs the atom is in a metastable state above the ground state, e.g., emission of the 238·6 nm tellurium line after excitation by the 214·3 nm tellurium line.

Stepwise-Line Fluorescence. This occurs when an atom emits radiation after excitation of the atom to a higher energy state than the first resonance state followed by a radiationless deactivation to the first excited state, e.g., emission of the 306·8 nm bismuth line after excitation by the 206·2 nm line.[15]

Thermally Assisted Direct-line Fluorescence. This type of fluorescence was first observed in a study of bismuth[15] atomic fluorescence characteristics and occurs when an atom is raised to an excited state by absorption of a resonance line and then is further excited to a slightly higher level by energy from the flame (i.e., collisions with flame gas molecules). Fluorescence emission then occurs from this higher energy excited state either to the ground state, or to a metastable state above the ground state, e.g., emission of the 293·8 nm bismuth line after excitation by the 206·2 nm bismuth line.[15]

Thermally Assisted Anti-Stokes Fluorescence. This type of process was reported for indium[16] and occurs when an atom is raised to a metastable state above the ground state by thermal excitation from the flame, then is excited by absorbing radiation to a higher energy excited state. Fluorescent emission may then occur from this excited state to the ground state. The wavelength of the fluorescent emission is then shorter than the wavelength of the exciting radiation, e.g., emission of the 410·1 nm indium line after excitation by the 451·1 nm indium line.[16]

Combinations of the above processes therefore make the final fluorescence spectrum observed when a flame containing an atomic population is excited either by an atomic line source or continuum.

Sensitized fluorescence has been observed in non-flame systems and occurs when an atom emits radiation after collisional activation by a foreign atom, which has been previously excited by the absorption of resonance radiation, e.g., excitation of thallium by a mercury vapour excited by the 253·7 nm mercury resonance line.

The following process takes place with the resulting sensitized fluorescence of the thallium 377·6 and 535·0 nm lines.

$$Hg^* + Tl \rightarrow Tl^* + Hg$$
$$Tl^* \rightarrow Tl + h\nu$$

* Note: An asterisk denotes an electronically excited atom.

This type of fluorescence has been reported in non-flame cells[5] but has never been observed in flames. This is thought to be due to the relatively low pressures of the elements in a flame even when nebulizing concentrated solutions.

8.2.3. Equipment Considerations

Fig. 8.2 depicts the layout of an atomic fluorescence spectrophotometer. Some commercial atomic absorption units have atomic fluorescence facilities fitted as standard, most others can be easily adapted for fluorescence measurements.

Excitation Sources

Microwave lamps (E.D.L.s) are the most commonly used source.[17,18] Ideally the light output should be modulated and a phase-sensitive detection system used so that the flame background and any thermal emission signal are not recorded at the output.

Passing heated air over microwave lamps whilst operating the lamp in conventional cavities has considerably simplified lamp operation and considerably improved the light output stability (see p. 190). In general, microwave lamps are between 200 and 2000 times more intense than corresponding hollow-cathode lamps.

Pulsing hollow-cathode lamps[19,20] with regular intermittent large current pulses gives a large (approximately 100–300 times) increase in peak intensity. The overall power dissipated is similar to that of a conventionally operated lamp. This type of system gives some useful detection limits when coupled with a phase-sensitive detection system (see p. 238).

Winefordner *et al.*[21,22] have used a dye laser for exciting atomic fluorescence of a large number of elements. The output of the laser could be easily varied over the wavelength range 360–650 nm. Malmstadt and Co-workers[23] have described a microprocessor controlled scanning dye laser that functioned from 360–650 nm. Multi-element scans of Ca, In, Mn, Na and Sr were demonstrated. An advantage of this type of source (for a two-level system) is that if the light output is sufficiently intense over the absorption linewidth to cause saturation fluorescence then the fluorescence intensity for a given concentration of atoms is independent of the light intensity, the quantum efficiency or any self-absorption effects. The main disadvantage is that the method is restricted to elements with resonance lines above 360 nm unless a frequency doubling (second harmonic generation) technique is used. A detection limit of 0·00004 μg ml^{-1} lead has been obtained[24] by combining a frequency doubled dye laser and a graphite rod atomizer (see p. 220). The excitation wavelength was 283·3 nm with the direct-line fluorescence measured at 405·8 nm.

The xenon arc lamp has been used as a multielement source. The main disadvantage is the relatively low integrated intensity in the u.v. region over absorption line half widths. This results in relatively poor detection limits and consequently a limited range of calibration curve linearity. When a continuum source is used interferences due to light scattering by refractory oxide particles in the flame increase in proportion to the square of the spectral bandpass of the monochromator. However, the fluorescence signal (for a single fluorescence line) only increases in direct proportion to the spectral bandpass. Thus, increasing the spectral bandpass to increase the sensitivity will reduce the selectivity. Spectral interferences are also more likely to occur when continuum source excitation is used; all the elements in the matrix that are atomized in the flame (or atom reservoir) will fluoresce.

Vapour discharge lamp sources can be used, but these are only available for a relatively limited number of elements (Cd, Ga, Hg, In, Tl and Zn).

Table 8.4 Comparison of atomic fluorescence sources

Source	Advantages	Disadvantages
Microwave lamp (Thermally stabilized by hot air).	Intense. Stable. Long life. Sources can be made for most elements that are favourably determined by A.F.S.	Relatively expensive. Microwave generator and cavity required.
Pulsed hollow-cathode lamps.	Intense. Stable. Very short warm-up time. Relatively inexpensive power supply.	In most cases not as intense as microwave lamps.
Dye laser	One source will excite many elements. With saturation, fluorescence, signal is independent of source intensity, quantum efficiency and self absorption.	Very expensive. Dangers of laser radiation. At present restricted to wavelengths above 360 nm unless frequency doubling is used.
Xenon arc lamp	One source will excite all elements. Multielement analyses using wavelength scanning techniques possible.	Poor detection limits and consequently limited range of linearity of calibration curves. Spectral and source scattering interferences more likely to occur.
Vapour discharge lamps.	Intense. Relatively stable. Inexpensive power supply.	Limited range of elements

OMENETTO[25] has made a review of suitable sources for A.F.S. Table 8.4 gives a comparison of atomic fluorescence sources.

Atom Reservoir

Most studies, to date, have been carried out using flames. Ideally, a flame with a high temperature, low radiative background and high fluorescence quantum efficiency is required. Unfortunately a high temperature flame (e.g., nitrous oxide–acetylene) usually has a high radiative background and a low fluorescence quantum efficiency whilst a low radiative background flame with a high fluorescence quantum efficiency (e.g., argon–hydrogen-entrained air) has a very low temperature. Some impressive detection limits have been obtained with pure solutions in the argon–hydrogen-entrained air flame,[15,26–28] but interelement effects are very severe. The use of the air or nitrous oxide–acetylene premixed flames minimizes interelement effects but considerably degrades the detection limit often to a level above that obtained by conventional atomic absorption techniques. The optimum flame source would appear to be the argon separated air–acetylene flame pioneered by KIRKBRIGHT and co-workers[29,30] (see p. 208). This flame exhibits a high-temperature relatively low radiative background with moderate quantum efficiency.

An alternative method of atomization uses a West-type graphite rod atomizer (see p. 220) with the source and detector optics focused just above the rod surface. The fluorescence is measured in an argon atmosphere just above the rod in a region of very low radiative background and high fluorescence quantum efficiency. Some impressive detection limits in pure solutions have been obtained using this technique[31,32] (e.g., Cd $1{\cdot}5 \times 10^{-15}$ g).

Monochromator

It has been found that the atomic fluorescence detection limit is often limited by the noise component in the flame background emission especially when a hot flame such as air or nitrous oxide–acetylene is employed. The flame background signal is proportional to the square of the spectral bandpass employed whilst the fluorescence signal (for a single line) is directly proportional to the spectral bandpass. Ideally, the monochromator should have a high resolution, large aperture and low stray light figures. The optics should be arranged so that the excited region of the flame (or atom reservoir) completely fills the monochromator aperture. Specular reflection direct from the source to the monochromator should be carefully minimized.

It is possible to dispense with the monochromator and use a solar blind photomultiplier which only responds to radiation with wavelengths less than 320 nm. This system uses an atomic line source to excite the fluorescence which is focused directly onto the photocathode of the solar blind photomultiplier. It is essential that the source does not contain any other element or this will excite fluorescence of this other element and considerably reduce the selectivity of the technique. A good example of this is a cadmium lamp containing a trace of zinc. The zinc resonance line emitted by the 'impure' cadmium lamp will cause any zinc present in the sample to fluoresce and consequently be detected together with any cadmium by the non-specific solar blind detector.

LARKINS and WILLIS[33,34] have made a comprehensive study of this technique and obtained good detection limits but relatively poor selectivity.

Radiation Detector

A photomultiplier with a low dark current and high sensitivity over the wavelength range 190–500 nm is required.

Amplifier and Readout

The fluorescence signal is directly proportional to the concentration of the fluorescing species over a wide concentration range. The modulated output from the photomultiplier is amplified, usually phase sensitively demodulated, and then displayed on a meter or recorder. HUBBARD and MICHEL[35] have listed various inexpensive instrumental refinements for improving atomic fluorescence detection limits.

8.2.4. SAMPLE REQUIREMENTS

The essential requirements are similar to those for atomic absorption and flame emission (see p. 256).

8.2.5. SHAPE OF CALIBRATION CURVES

At low concentrations the fluorescence signal is directly proportional to concentration. In general for resonance line fluorescence, curvature of the calibration graph becomes pronounced at a similar concentration to that observed for the corresponding line in atomic absorption. Hence, if the fluorescence detection limit is better than the corresponding absorption limit a longer linear fluorescence calibration range is observed.

8.2.6. STANDARDIZATION AND INTERFERENCE EFFECTS

Atomic fluorescence is subject to the same interference effects as atomic absorption (see p. 5). Molecular absorption by matrix components is far more serious in atomic absorption than in atomic fluorescence.[36] The reverse applies to scattering of the source radiation by matrix components (i.e., unvaporized refractory oxide clotlets in the flame). A good example of the former is the determination of low levels of selenium in copper. Strong molecular absorption at 196·0 nm by large amounts of copper species in the flame completely swamps the weak selenium atomic absorption signal. Solvent extraction of the selenium prior to atomic absorption determination is essential. Using the atomic fluorescence technique it was possible to determine 0·05 μg/ml selenium in the presence of 20 000 μg/ml copper. However, the presence of elements forming unvolatilized refractory oxide clotlets in the flame (e.g., Al) causes severe source scattering in atomic fluorescence (when using the air–acetylene flame) and much less relative interference in atomic absorption. 0·05% scattered light would cause a large apparent signal in atomic fluorescence, but would hardly be detectable in atomic absorption.

An automatic correction system for light scatter in A.F.S. has been described[37] where the light from a microwave lamp atomic line source and a xenon arc lamp with an adjustable iris diaphragm were alternately passed through the flame. The resulting signal from the photomultiplier was fed to a phase-sensitive amplifier which corrected for the contribution of the light scattering to the fluorescence signal. The system was used to determine cadmium in orchard leaves and liver. The two exciting sources were initially balanced by adjusting the xenon lamp iris diaphragm whilst nebulizing a cadmium-free 1% m/v lanthanum solution.

The possibility of source scattering interference should never be ignored when developing a new atomic fluorescence method. It is difficult to estimate the contribution from scattered radiation to the analytical signal unless typical samples known to contain negligible amounts of the analyte element are available. Otherwise a relatively complex correction technique, such as a pulsed xenon arc lamp, a dual monochromator system,[38] a scanning dye laser,[23] or a source that can easily be self-reversed without significant loss of intensity,[39] must be used. The typical magnitude of the scatter contribution in the air–acetylene flame when nebulizing sea-water samples has been reported to be equivalent to +0·006 μg/ml cadmium or zinc.[39] In most surface water samples the corresponding figure was +0·001–0·002 μg/ml. The scatter signal could be significantly reduced by addition of perchloric acid to the samples.

Stray light reaching the photomultiplier can sometimes result in unpredictable interference effects when working in matrices containing elements that emit strongly in flames (e.g., calcium and sodium). Although the phase-sensitive detection system should not respond to this extraneous radiation, if it is excessive a significant interference effect as well as an increased noise level can be observed. This effect can be minimized by using a monochromator with a low stray light specification or by fitting a solar blind photomultiplier that does not respond to radiation above 320 nm. Obviously in this case the fluorescence line must be below 320 nm.

8.2.7. Range of Elements

In theory, all elements that can be determined by atomic absorption spectroscopy can be detected using atomic fluorescence techniques. In practice atomic fluorescence analysis of real samples using hot flame atom reservoirs and non-laser excitation sources is only practicable for elements with resonance lines below 400 nm. At wavelengths greater than 400 nm, the combined flame background and thermal emission signal are usually much more intense than the fluorescence signal. Although these 'background' signals are rejected by a phase-sensitive detection system, they cause a large increase in the noise level of the determination, thus resulting in poor detection limits.

Atomic fluorescence flame techniques show greatest promise for elements that have their main resonance lines below 270 nm and that can be appreciably atomized in the separated air–acetylene flame. At wavelengths below 270 nm the flame background signal is not very intense and thermal emission is negligible. As, Bi Cd, Hg, Sb, Se, Te and Zn can be determined by atomic fluorescence using microwave sources and a separated air–acetylene flame. For these elements the reported detection limits are 2–20× better than those that can be obtained using flame atomic absorption techniques.

It should be pointed out that As, Sb, Bi, Se, and Te (as well as Ge and Sn) can now be determined by atomic absorption using hydride generation techniques (see p. 79) which improves the conventional flame absorption detection limits 50–500 times.

Replacing the flame with an argon-sheathed graphite filament atom reservoir[31] extends the range of elements that can be usefully determined by atomic fluorescence as there is then negligible background radiation or thermal emission, and minimal quenching. The fluorescence signal is normally viewed from the non-luminous region just above the rod. The intense black body continuum emission from the hot rod is shielded from the detector. However source scattering interference can be severe.

8.2.8. Detection Limits

Atomic fluorescence detection limits are much more dependent on the instrumentation used than corresponding atomic absorption detection limits. Increasing the source intensity and reducing the intensity and associated noise of the flame background radiation results in improvements to atomic fluorescence detection limits. Much work has been performed to extend the range of microwave excited electrodeless lamps (E.D.L.'s) and increase their intensity. The intensity of tuneable dye lasers can be sufficient to excite saturated fluorescence;[42] the fluorescence signal is then independent of the source intensity and then the detection limit mainly depends on the flame background and scattered radiation (when measuring resonance fluorescence).[43] The absence of self-absorption results in a very wide linear range of the calibration curves. The main disadvantage of this type of source is that at present measurements have been mainly restricted to wavelengths above 360 nm. Frequency doubling should improve this situation.[24]

Many impressive fluorescence detection limits have been obtained using low temperature low background flames[26–28] such as air–propane argon–hydrogen-entrained air, argon–oxygen–hydrogen. These flames exhibit low background emission, but exhibit quite pronounced interelement effects. For routine analysis an air–acetylene or separated air–acetylene flame should be used. Although it is possible to use an argon separated nitrous oxide–acetylene flame to atomize refractory oxide forming elements (Al, Be, Mo, Ti, V)[40]the quoted detection limits are similar or slightly worse than those quoted for atomic absorption (see p. 33). The flame background even of the separated nitrous oxide–acetylene flame is relatively intense compared to that of the separated air–acetylene flame.

Table 8.5 gives a list of detection limits for some elements that can be

Table 8.5 Typical atomic fluorescence detection limits for air–acetylene flame

Element	Line (nm)	2σ Limit of detection (μg/ml)	
		Normal flame	Argon-separated flame
As	193·7	0·25	0·07
Bi	223·1	0·07	0·025
Cd	228·8	0·0004	0·00015
Sb	217·6	0·06	—
Se	196·1	0·02	0·008
Te	214·3	0·03	0·013
Zn	213·9	0·0003	0·0001

favourably determined by atomic fluorescence. The results were obtained in the authors' (K.C.T.) Laboratory using a Shandon Southern (Baird Atomic) A3400 Spectrophotometer with microwave excited light sources electronically modulated at 325 Hz.

8.2.9. Comparison of Atomic Absorption and Atomic Fluorescence Spectroscopy

Atomic fluorescence is essentially an emission technique, i.e., an absolute intensity is measured, whilst atomic absorption is a comparative technique, i.e., the ratio of two intensities are measured. The respective merits and demerits of the two techniques are summarized below.

A.F.S.—Advantages in relation to A.A.S.

(a) A.F.S. is essentially an emission technique and the sensitivity can be increased by either increasing the source intensity or the photometric sensitivity of the instrument. In A.A.S. the absorbance is a ratio ($\log_{10} I_0/I_T$), hence any increase in the source intensity I_0 is accompanied by a corresponding increase in I_T. Hence, the ratio $\log_{10} I_0/I_T$ remains unaltered.
(b) A.F.S. requires somewhat less sophisticated apparatus and burners than A.A.S. Optical alignment of the various components is usually simpler. Multielement instrument design is easier.
(c) A continuous source can be used in A.F.S. with a normal monochromator (the detector does not 'see' the source), whilst the use of a continuous source in A.A.S. requires the use of a high resolution monochromator.
(d) In general, for most elements, there are more fluorescence lines than absorption lines. This gives a greater choice in the wavelength of the measurement, e.g., the main arsenic resonance lines are at 189·0, 193·7, and 197·2 nm where absorption by optics and flame gases can be quite serious, and also the photomultiplier response is often poor at these wavelengths, especially for the most sensitive 189·0 nm line. Although these resonance lines must be

used in absorption measurements (often the 189·0 nm line cannot even be detected), the fluorescence resulting from the absorption of the 189·0 nm line can be measured at 228·8 or 235·0 nm.[27]

(e) Instruments that use a solar blind photomultiplier (a photomultiplier that does not respond above 320 nm) in place of a monochromator have been described.[33,34] The detector does not 'view' the source and can be positioned close to the flame thus increasing the sensitivity. The atomic line exciting source then effectively acts as a monochromator by only exciting the element that is contained in the source.

(f) If the wavelength range of tuneable lasers can be extended down to 190 nm atomic fluorescence could have many advantages, e.g., single source for all elements, wavelength scanning techniques for correction of source scattering interference and multi-element analysis would be possible, and saturation fluorescence would give high precision and a long linear calibration curve range.[41,42]

A.F.S.—Disadvantages in relation to A.A.S.

(a) Atomic absorption measurements are independent of the quantum efficiency but fluorescence measurements are directly proportional to the quantum efficiency (unless laser-induced saturation fluorescence is excited). As long as the flame temperature and composition are kept constant and the partial pressures of the nebulized species are low, which is generally the case, inter-element effects on the quantum efficiency should be negligible.

(b) The range of elements that can be determined by atomic fluorescence is somewhat limited compared to atomic absorption, e.g., the determination of the alkali or alkaline earth metals is not very feasible. The determination of refractory metals by A.F.S. using conventional or even argon separated nitrous oxide–acetylene flames is not very satisfactory. This is because A.F.S. is essentially an emission technique and larger slit widths and higher gains than are commonly employed in A.A.S. are necessary. Although the light from the excitation source can be modulated so that the high intensity background radiation from the flame is not detected. This background radiation can nevertheless seriously overload the photomultiplier and markedly increase the noise level.

(c) Light scattering by particles, e.g., refractory oxides, in the flame is seldom observed in A.A.S. In fact, using conventional A.A. instruments it is difficult to detect 0·05 per cent scatter of the incident radiation.

Light scatter is a greater potential source of interference in A.F.S. than in A.A.S. This is because the integrated intensity in all directions of a given fluorescence line at a given wavelength is seldom greater than 0·2 per cent of the integrated intensity of the incident line from the source reaching the flame at the same wavelength.[44] At low concentrations the figure is much lower. Thus, 0·05 per cent scatter of the incident radiation represents a far greater source of interference in A.F.S. than in A.A.S. It is difficult to compensate for source scattering when using an atomic line excitation source. Source scattering is far more serious when a continuous source is used in A.F.S. because the integrated intensity of the scatter signal reaching the detector is then dependent on the square of the monochromator band-pass.

However, with a well-designed nebulizer–burner unit and using line sources, source scattering can be minimized for many A.F.S. applications.

(d) The sensitivity of atomic absorption is almost independent of the source intensity. If only a low-intensity source is available atomic fluorescence is not very sensitive, because the sensitivity of this technique is proportional to the source intensity.

8.2.10. Conclusions

The technique of atomic fluorescence is unlikely to replace atomic absorption in the near future, but for certain elements with resonance lines below 270 nm it can be used to advantage. Most commercial atomic absorption instruments can be easily modified to make fluorescence measurements if a microwave generator and suitable burner are available. The most serious drawback, at present, is the increased operator skill required to set up the source and interpret the results obtained for real samples. It must be admitted that at present there is no instrument designed solely for atomic fluorescence on the market, but this is mainly because of the limited applicability of the technique (e.g., the A.F.S. detection limits of Ca, Ba, Sr, Li, Na, K, are unlikely to rival those obtained by A.A.S. or F.E.S. in the near future). Another factor is that alternative techniques, viz., atomic absorption flameless electrothermal atomization and hydride generation techniques, give adequate or better detection limits at a relatively modest cost for these elements favourably determined by A.F.S. An obvious exception to this is the cold vapour mercury fluorescence detector.[45–47]

The main hope for atomic fluorescence is the development of inexpensive dye lasers with outputs tuneable over the wavelength range 190–770 nm.

Chap. 8 References and further reading

Flame Emission References

1. Pickett, E. E., and Koirtyohann, S. R., The Nitrous Oxide–Acetylene Flame in Emission Analysis Parts 1 and 2. *Spectrochim. Acta* 23B (1968) 235 and 673.
2. Christian, G. D., and Feldman, F. J., Optimum Parameters for Flame Emission Spectrometry with the Nitrous Oxide–Acetylene Flame. *Anal. Chem.* 43 (1971) 611.
2. (a) Ramuson, J. O., Fassel, V. A., and Kniseley, R. N., An Experimental and Theoretical Evaluation of the Nitrous Oxide Acetylene Flame as an Atomisation Cell for Flame Spectroscopy. Spectrochim. Acta 28B (1973) 365.
3. Boumans, P. W. J. M., and De Boer, F. J., Studies of Flame and Plasma Torch Emission for Simultaneous Multielement Analysis. Parts 1, 2 and 3. *Spectrochim. Acta* 27B (1972) 391; 30B (1975) 309 and 31B (1976) 355.

BOOKS

DEAN, J. A., and RAINS, T. C., Flame Emission and Atomic Absorption Spectrometry, Volume 2. Components and Techniques, Marcel Dekker, New York, 1971.

MAVRODINEAU, R., Analytical Flame Spectroscopy, Macmillan & Co., London, 1970.

ALKEMADE, C. T. J., and HERRMANN, R., Flame Spectroscopy. Adam Hilger Ltd., (The Institute of Physics) Bristol, 1978.

ATOMIC FLUORESCENCE REFERENCES

4. WOOD, R. W., The fluorescence of sodium vapour and the resonance radiation of electrons. *Phil. Mag.* 10 (1905) 513.
5. MITCHELL, A. C. G., and ZEMANSKY, M. W., Resonance Radiation and Excited Atoms. University Press, Cambridge, 1961.
6. NICHOLS, E. L., and HOWES, H. L., The photo-luminescence of flames. *Phys. Rev.* 22 (1923) 425; 23 (1924) 472.
7. ROBINSON, J. W., Mechanism of elemental spectral excitation in flame photometry. *Anal. Chim. Acta* 24 (1961) 254.
8. ALKEMADE, C. TH. J., Proc. Xth Colloquium Spectroscopium Internationale 1962, p. 143. Spartan Books, Washington, D.C., 1963.
9. WINEFORDNER, J. D., and VICKERS, T. S., Atomic fluorescence spectrometry as a means of chemical analysis. *Anal. Chem.* 36 (1964) 161.
10. WINEFORDNER, J. D., STAAB, R. A., Determination of zinc, cadmium and mercury by atomic fluorescence flame spectrometry. *Anal. Chem.* 36 (1964) 165.
11. WINEFORDNER, J. D., and STAAB, R. A., Study of experimental parameters in atomic fluorescence flame spectrometry. *Anal. Chem.* 36 (1964) 1367.
12. WINEFORDNER, J. D., MANSFIELD, J. M., and VEILLON, C., High sensitivity determination of zinc, cadmium, mercury, thallium, gallium and indium by atomic fluorescence flame spectrometry. *Anal. Chem.* 37 (1965) 1049.
13. DAGNALL, R. M., WEST, T. S., and YOUNG, P., Determination of cadmium by atomic fluorescence spectroscopy and atomic absorption spectroscopy. *Talanta* 13 (1966) 803.
14. DAGNALL, R. M., THOMPSON, K. C., and WEST, T. S., An investigation of some experimental parameters in atomic fluorescence spectroscopy. *Anal. Chim. Acta* 36 (1966) 269.
15. DAGNALL, R. M., THOMPSON, K. C., and WEST, T. S., Studies in atomic fluorescence spectroscopy—VI. The fluorescence characteristics and analytical determination of bismuth with an iodine electrodeless discharge tube as source. *Talanta* 14 (1967) 1151.
16. OMENETTO, N., and ROSSI, G., Some observations on direct line fluorescence of thallium, indium, and gallium. *Spectrochim. Acta* 24B (1969) 95.
17. DAGNALL, R. M., THOMPSON, K. C., and WEST, T. S., Microwave-excited electrodeless discharge tubes as spectral sources for atomic fluorescence and atomic-absorption spectroscopy. *Talanta* 14 (1967) 551.
18. HAARSMA, J. P. S., DE JONG, G. J., and AGTERDENBOS, J., The Preparation and Operation of Electrodeless Discharge Lamps—A Critical Review. *Spectrochim. Acta* 29B (1974) 1.
19. CORDOS, E., and MALMSTADT, H. V., Characteristics of Hollow Cathode Lamps Operated in an intermittent high current mode. *Anal. Chem.* 45 (1973) 27.
20. MITCHELL, D. G., and JOHANSSON, A., Simultaneous Multielement analysis using sequentially excited atomic fluorescence radiation—II. An improved instrument for simultaneous analysis. *Spectrochim. Acta* 26B (1971) 677.
21. WINEFORDNER, J. D., and FRASER, L. M., Laser-excited atomic fluorescence flame spectrometry. *Anal. Chem.* 43 (1971) 1693.
22. WINEFORDNER, J. D., and FRASER, L. M., Laser-excited atomic fluorescence flame spectrometry as an analytical method. *Anal. Chem.* 44 (1972) 1444.
23. PERRY, J. A., BRYANT, M. F. and MALMSTADT, H. V. Microprocessor controlled scanning dye laser for spectrometric analytical systems. *Anal. Chem.* 49 (1977) 1702.
24. NEUMANN, S., and KRIESE, M., Sub-Picogram Detection of Lead by Non-Flame. AFS with Dye Laser Excitation. *Spectrochim. Acta* 29B (1974) 127.
25. OMENETTO, N., Pulsed sources for A.F.S. *Anal. Chem.* 48 (1976) 75A.
26. DAGNALL, R. M., THOMPSON, K. C., and WEST, T. S., Studies in Atomic fluorescence spectroscopy—IV. The atomic fluorescence spectroscopic determination of selenium and tellurium. *Talanta* 14 (1967) 557.

27. DAGNALL, R. M., THOMPSON, K. C., and WEST, T. S., Studies in atomic fluorescence spectroscopy—VII. Fluorescence and analytical characteristics of arsenic with microwave EDT's as source. *Talanta* 15 (1968) 677.
28. BROWNER, R. F., DAGNALL, R. M., and WEST, T. S., The determination of tin by atomic fluorescence spectroscopy, with an electronically modulated EDT as source. *Anal. Chim. Acta* 46 (1969) 207.
29. HOBBS, R. S., KIRKBRIGHT, G. F., SARGENT, M., and WEST, T. S., Application of the nitrogen separated air–acetylene flame in flame emission and atomic fluorescence spectroscopy. *Talanta* 15 (1968) 997.
30. HINGLE, D. N., KIRKBRIGHT, G. F., SARGENT, M., and WEST, T. S., Shielded flame emission burner assembly for use with atomic absorption spectrophotometers. *Lab. Practice* 18 (1969) 1069.
31. ALDER, J. F., and WEST, T. S., A study of the determination of cadmium by atomic fluorescence spectroscopy with an unenclosed atom reservoir. *Anal. Chim. Acta* 51 (1970) 365.
32. KIRKBRIGHT, G. F., and WEST, T. S., Atomic fluorescence for chemical analysis. *Chemistry in Britain* 8 (1972) 428.
33. LARKINS, P. L., Non-dispersive systems in atomic fluorescence spectroscopy—I. A single channel system employing a solar-blind detector and an air–acetylene flame. *Spectrochim. Acta* 26B (1971) 477.
34. LARKINS, P. L., and WILLIS, J. B., Non-dispersive systems in atomic fluorescence spectroscopy—II. Comparison of nitrous oxide supported and air supported flames. *Spectrochim. Acta* 26B (1971) 491.
35. HUBBARD, D. P., and MICHEL, R. G., Studies in A.F.s. Part 1. Inexpensive Methods of Improving Signal to Noise Ratios. *Anal. Chim. Acta* 67 (1973) 55.
36. WEST, C. D., Relative Effect of Molecular Absorption on Atomic Absorption and Atomic Fluorescence. *Anal. Chem.* 46 (1974) 797.
37. RAINS, T. C., EPSTEIN, M. S., and MENIS, O., Automatic Correction System for Light Scatter in AFS. *Anal. Chem.* 46 (1974) 207.
38. DOOLAN, K. J., and SMYTHE, L. E., On-line dual wavelength instrumentation for automatic correction of light scattering interference in dispersive A.F.S. *Spectrochim. Acta*, 32B (1977) 115.
39. HAARSMA, J. P. S., VLOGTMAN, J., and AGTERDENBOS, J., Investigation on light sources and on scattering in analytical A.F.S. *Spectrochim. Acta* 31B (1976) 129.
40. DAGNALL, R. M., KIRKBRIGHT, G. F., WEST, T. S., and WOOD, R., Atomic fluorescence spectrometry of Al, Mo, Ti, V and Zr in inert gas separated nitrous oxide–acetylene flames. *Anal. Chem.* 42 (1970) 1029.
41. OMENETTO, N., BENNETTI, P., HART, L. P., WINEFORDNER, J. D., and ALKEMADE, C. TH. J., Non-linear optical behaviour in atomic fluorescence flame spectrometry. *Spectrochim. Acta* 28B (1973) 289.
42. OMENETTO, N., BENETTI, P., HART, L. P., and WINEFORDNER, J. D., On the Shape of A.F. Analytical Curves with a Laser Excitation Source. *Spectrochim. Acta* 28B (1973) 301.
43. OMENETTO, N., BOUTILIER, G. D., WEEKS, S. J., SMITH, B. W. and WINEFORDNER, J. D. Pulsed vs. continuous wave A.F.S. *Anal. Chem.* 49 (1977) 1076.
44. THOMPSON, K. C., and WILDY, P. C., The use of electronically modulated microwave-excited discharge tubes in atomic spectroscopy. *Analyst* 95 (1970) 776.
45. THOMPSON, K. C. and GODDEN, R. G., Improvements in the A.F. determination of mercury. *Analyst* 100 (1975) 544.
46. THOMPSON, K. C., and REYNOLDS, G. D., The atomic fluorescence determination of mercury by the cold vapour technique. *Analyst* 96 (1971) 771.
47. THOMPSON, K. C., A cold vapour mercury atomic fluorescence detector. *Lab. Practice* 21 (1972) 645.

ATOMIC FLUORESCENCE REVIEW PAPERS

WEST, T. S., and CRESSER, M. S., Atomic Fluorescence Spectrometry. *Applied Spectroscopy Reviews* 7 (1973) 79.

BROWNER, R. F., Atomic Fluorescence Spectrometry as an Analytical Technique. *Analyst* 99 (1974) 617.

See also Ref. 32.
BOOKS
SYCHRA, V., SVOBODA, V., and RUBESKA, I., Atomic Fluorescence Spectroscopy. Van Nostrand Reinhold, London 1975.
KIRKBRIGHT, G. F. and SARGENT, M., Atomic Absorption and Fluorescence Spectroscopy. Academic Press, London 1974.

9 *Theory*

History

The phenomenon of atomic absorption was first noticed by WOLLASTON[1] in 1802, when he observed a few dark lines in the solar spectrum. Then in 1814, FRAUNHOFER[2] observed the numerous lines in the solar spectrum that now bear his name. Although FRAUNHOFER could not explain their origin he made a map of about 700 lines and assigned the letters A to H to the eight most prominent ones (e.g., the sodium D lines).

The basic principles underlying atomic absorption were established by KIRCHHOFF[3] in 1860, when he put forward the general law relating to the absorption and emission of light for a given system. KIRCHHOFF showed that a flame containing sodium chloride would not only emit the yellow sodium D lines, but also absorb the same yellow light from a continuous source placed behind the flame, no other wavelengths being absorbed. Thus, the FRAUNHOFER lines were attributed to the absorption by certain elements, in the outer cooler solar atmosphere, of the continuous spectrum emitted by the hot interior of the sun.

KIRCHHOFF was the first person to emphasize the great significance of the characteristic spectra of the different elements. This important observation was the foundation of analytical spectroscopy.

The theory of atomic spectroscopy was evolved in the early twentieth century by physicists and astrophysicists. Much of the work of these early researchers (summarized in a treatise by MITCHELL and ZEMANSKY[4]) was performed at low pressures in enclosed vessels, and was not directed towards analytical purposes, except for some astrophysical studies on the determination of the compositions of the solar and stellar atmospheres. A special case was the estimation of the contamination of air by mercury, which has an appreciable vapour pressure at room temperature.

Although emission methods of analysis (arc, spark, and flame) became firmly established, it was not until 1953 that WALSH[5] realized the analytical potential of atomic absorption and demonstrated the superiority of the technique over flame emission spectroscopy. The first commercial instruments appeared about 1960, and since that time there has been an exponential increase in the number of atomic absorption papers published.

Absorption and Emission Line Profiles

A typical absorption line profile of an atomic vapour in a conventional flame

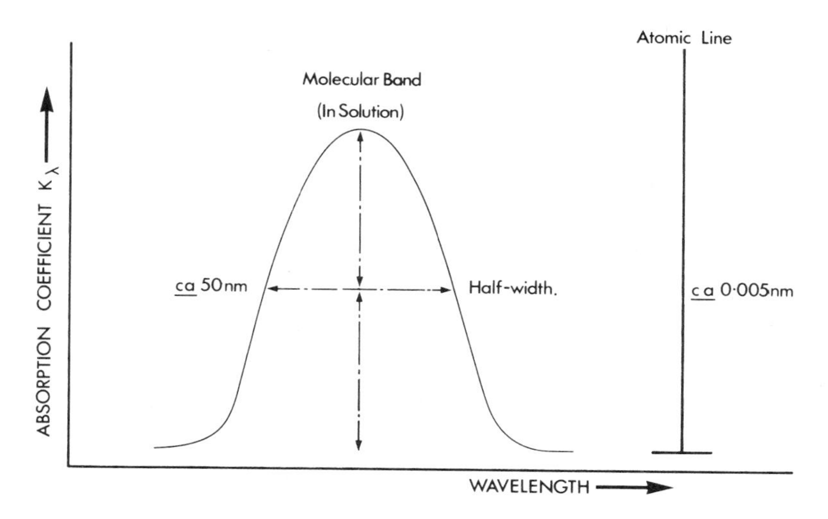

Fig. 9.1 Typical molecular absorption band (in solution) and atomic absorption line (gas phase)

and a typical profile of a molecular absorption band in solution are shown in Fig. 9.1.

The ordinate represents the absorption coefficient (K_λ) and the half width is the width across the profile where the absorption coefficient is half its maximum value. The absorption coefficient for a path length L of uniform atomic vapour (Fig. 9.2) is defined by the equation:

$$I_{T\lambda} = I_{0\lambda} \exp(-K_\lambda L)$$

where

$I_{0\lambda}$ = the intensity of the incident radiation at wavelength λ per unit wavelength with no atomic vapour in the beam.
$I_{T\lambda}$ = the intensity of the transmitted radiation at wavelength λ per unit wavelength with an atomic vapour in the beam.
L = the path length of the uniform vapour (cm).
K_λ = the absorption coefficient (cm^{-1}).

The large half-widths of molecular electronic absorption bands allow the absorption to be measured using a continuous source and an inexpensive low-resolution monochromator. The small half-width of atomic absorption line profiles means that if a continuous source is used to measure the absorption, an expensive very high-resolution monochromator must be used in order to obtain any appreciable absorption. This is possibly one reason why the analytical uses of atomic absorption were not exploited earlier.

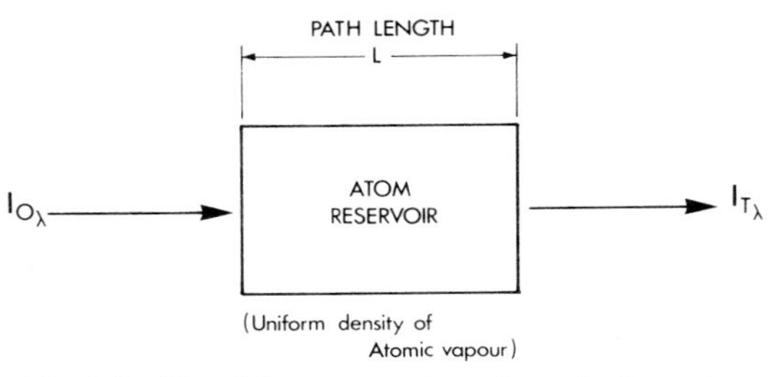

Fig. 9.2 Pictorial representation of atomic absorption

The large half-widths of molecular electronic absorption bands in solution (10–50 nm) can be ascribed to vibrational and rotational fine structure of the electronic energy levels and also to solvent-solute interaction in the condensed phase. This is one reason why spectral interference due to overlapping of absorption line profiles is common in solution spectrophotometry.

The half-widths of absorption and emission lines is of considerable importance in atomic absorption spectroscopy and a summary of the broadening processes involved is given below.

Broadening Processes of Atomic Spectral Lines

For a given system in thermal equilibrium the profile of a resonance emission line is the same as the profile of the same resonance line in absorption (i.e., they have the same half-width). If a solution containing sodium atoms is nebulized into a flame the half-width of the D_1 emission line will be the same as the half-width of the D_1 absorption line. The main types of broadening processes of atomic lines are:

Natural Broadening

Natural broadening is due to the finite lifetime of the atom in the excited state (i.e., Heisenberg's uncertainty principle) and is independent of the environment of the atom. For most resonance lines the natural width is of the order of 10^{-5} nm which is negligible compared to that due to other causes in analytically used flames or sources.

Doppler Broadening

Doppler broadening is caused by the absorbing or emitting atoms having different component velocities along the line of observation. The broadening is symmetrical about the mean wavelength of the line. The Doppler half-width is proportional to the square root of the absolute temperature, independent of the pressure, and is given by the equation:

$$\Delta\lambda_D = \frac{\lambda_0}{c}\sqrt{\frac{8(\log_e 2)RT}{M}} \tag{1}$$

where

$\Delta\lambda_D$ = the Doppler half-width (nm)
λ_0 = the wavelength of the centre of the line (nm)
c = the velocity of light (cm s^{-1})
R = the gas constant (ergs K^{-1} $mole^{-1}$)
M = the atomic weight of the absorber (g $mole^{-1}$)
T = the temperature of the atoms (K).

For an element of atomic weight 75 in a flame at 2500 K the Doppler half-width varies from 0·0008 nm at 200 nm to 0·0032 nm at 800 nm. Therefore lines in the visible region have a greater Doppler half-width than lines in the ultra-violet. Also at a given wavelength the smaller the atomic weight of the element, the greater the Doppler half-width.

COLLISIONAL BROADENING (ALSO KNOWN AS PRESSURE OR LORENTZ BROADENING)

Collisional broadening is due to the perturbation of the absorbing or emitting atoms by foreign gas atoms. The effect of this type of broadening varies for different foreign gases and for different atomic states. Assuming the broadening is symmetrical, the half-width is given by the equation:[6]

$$\Delta\lambda_c = \frac{2\lambda_0^2\sigma_c^2 P_f}{\pi c k T}\left[2\pi R T\left(\frac{1}{M_a} + \frac{1}{M_f}\right)\right]^{\frac{1}{2}} \qquad (2)$$

where:

$\Delta\lambda_c$ = the collisional half-width (cm)
λ_0 = the wavelength of the centre of the line (cm)
σ_c = the effective cross-section for collisional broadening for a given line under given conditions (cm)
P_f = the pressure of the foreign gases (dynes cm^{-2})
M_a = the atomic weight of the absorbing or the emitting species (g $mole^{-1}$)
M_f = the effective molecular weight of the foreign gas molecules (g $mole^{-1}$)
R = the gas constant (erg K^{-1} $mole^{-1}$)
T = the temperature of the foreign gases (K)
c = the velocity of light (cm s^{-1})
k = the Boltzmann constant (erg K^{-1}).

Thus

$$\Delta\lambda_c\ (\text{nm}) = 10^7 \times \Delta\lambda_c\ (\text{cm}).$$

It can be seen, therefore, that the collisional half-width is proportional to the pressure, and inversely proportional to the square root of the temperature of the emitting or absorbing atoms. The half-width is also proportional to the square of the wavelength of the line. Owing to an uncertainty in the values of the collisional cross-sections, the direct calculation of collisional half-widths is difficult and the results are not always consistent.[6,7,8]

It has been stated that collisional broadening can cause a shift of the wavelength of the centre of the absorption line profile and can also cause asymmetry of this profile.[4,9–11a] Thus, the centre of a given absorption line in a high-pressure system (e.g., conventional flame) can be at a slightly different wavelength relative to the same line emitted by a low-pressure source (e.g., hollow-cathode lamp) (Fig. 9.3). In conventional flames, according to

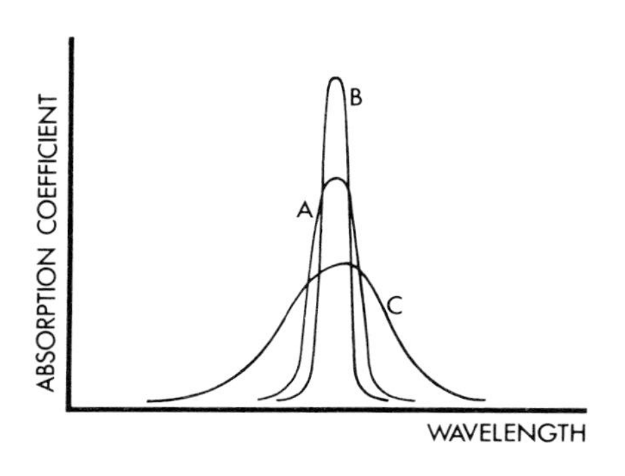

Fig. 9.3 Typical profiles of atomic absorption lines

A. Typical absorption line profile in a flame. (Doppler and collisional broadening with slight wavelength shift)
B. Doppler broadened line. (Negligible collisional broadening)
C. Doppler collision broadened absorption line at very high pressure showing exaggerated asymmetry and wavelength shift

Lindholm's impact theory, the red shift of the centre of the absorbing line should be given by 0·36 $\Delta\lambda_c$.[10] HOLLANDER *et al.*[9] obtained values of this red shift of 0·2–0·3 $\Delta\lambda_c$ for calcium, strontium and barium in the air–acetylene flame. WAGENAAR and L. DE GALAN[11a] reported a general red shift in flames of approximately 0·2 $\Delta\lambda_c$.

Fortunately these wavelength shifts are relatively small when compared to the total width of absorption line profiles in flames. The asymmetry of most resonance lines in conventional flames is thought to be small.

RESONANCE BROADENING

Resonance broadening is due to the perturbation of the absorbing or emitting atom by atoms of the same type. In conventional flames the partial pressure of metal atoms, even at very high concentrations, seldom exceeds

10^{-1} Torr, thus this type of broadening process can be considered negligible in flames at 760 Torr.[12]

Stark and Zeeman Broadening

Stark and Zeeman broadening are caused by electric and magnetic fields respectively. Although Stark broadening can cause lines to become diffuse in the arc and spark, both these effects are negligible in flames. It is just feasible that the Stark effect could very slightly influence the lines from a hollow-cathode lamp, but at the voltages and currents employed in most lamps this effect is thought to be negligible.

Hyperfine Structure

Hyperfine structure, which is not an actual broadening process, can be attributed to a non-zero value of the nuclear spin and/or the presence of several isotopes.[4,13,14] Thus each line consists of a number of separate hyperfine components each acting as an independent line. For all isotopes with an even atomic number and mass the nuclear spin is zero. Thus for an element with an even atomic number and just one even atomic mass isotope (e.g., naturally occurring thorium exists solely as Th^{232}), there is no hyperfine structure to the spectrum. This is why lamps containing an element with an even atomic number and a single even atomic mass isotope are used as wavelength standards (Kr^{86}, Hg^{198}, Th^{232}).[15,16,17]

The separation of the hyperfine components due to nuclear spin, and different isotopes, is small and difficult to measure, and it is difficult to evaluate the total hyperfine separation compared to the total absorption line half-width in conventional flames. Although, in the past, it has been stated that hyperfine splitting is negligible compared to absorption line half-widths, this cannot be considered to be true for all elements because isotopic analysis of lithium 6 and 7 in flames[18] by atomic absorption has been reported, and Mitchell and Zemansky[4] state that the 253·7 nm mercury line, from naturally occurring mercury, is split into 9 components, of which 5 are resolvable using an interferometer.

It has been stated[11a] that except for a few elements (e.g., calcium), hyperfine structure is the most important factor in determining the profile of atomic transitions. Willis[10] quotes the extreme hyperfine splitting of copper 324·7 nm, the silver 328·1 nm and gold 267·6 nm lines as 0·0043 nm, 0·00081 nm and 0·00152 nm respectively. The values for the copper and gold lines are comparable to the absorption line half-widths in conventional flames. Other values are quoted in Tables 9.1 and 9.2.

Each hyperfine component will be broadened by the processes described above.

Total line half-widths. In conventional flames the two most important broadening processes (hyperfine structure not being an actual broadening

Table 9.1 Hyperfine splitting due to nuclear spin*

Element	Naturally occurring isotopes	Natural % abundance	Spin of nucleus	Line (nm)
Hg	198	10·0	0	any
	199	16·8	1/2	253·7
	200	23·1	0	any
	201	13·2	3/2	253·7
	202	29·8	0	any
	204	6·8	0	any
Bi	209	100	9/2	306·8
				222·8
				206·2
Th	232	100	0	any
Tl	203	29·5	1/2	377·6
	205	70·5	1/2	
Li	6	7·4	1	670·8
	7	92·6	3/2	
Na	23	100	3/2	589·0
K	39	93·1	3/2	766·5
	41	6·9	3/2	
Rb	85	72·2	5/2	780·0
	87	27·8	3/2	780·0
Cs	133	100	7/2	852·1
Au	197	100	3/2	242·8
U	235	0·72	7/2	415·3
				502·7
	238	99·3	0	any
Pb	204	1·4	0	any
	206	25·1	0	any
	207	21·2	1/2	283·3
	208	52·3	0	any
Mn	55	100	5/2	403·3

(a) Extreme hyperfine splitting is the sum of the splitting of the upper state and the splitting of the lower state.

* A useful bibliography on hyperfine structure is given in the following reference. HEILIG, K., Bibliography on experimental optical isotope shifts 1918–October 1976. *Spectrochim Acta* 32B (1977) 1.

Extreme[a] hyperfine splitting (nm)	Number of components		References	Element
	Upper energy level	Lower energy level		
0	1	1	4, 19	Hg
0·0047	2	1		
0	1	1		
0·0047	3	1		
0	1	1		
0	1	1		
>0·0078[b]	2	4	13	Bi
>0·0070[b]	4	4	13	
	6	4		
0	1	1		Th
0·016	2	2	13	Tl
0·0012[c]	Splitting of the upper $^2P_{1\frac{1}{2}}$ state, at least an order of magnitude smaller than that of ground state $^2S_{\frac{1}{2}}$.	2	14	Li
0·0020[c]		2	4, 14	Na
0·0009[c]		2	14	K
		2	14	
0·0061[c]		2	14, 20	Rb
0·014[c]		2	20	
0·022[c]		2	14, 21	Cs
0·0012[c]		2	14	Au
smaller than isotopic shift	–	–	22	U
0	1	1		
0	1	1		Pb
0	1	1	58	
0·0036	2	1		
0	1	1		
0·0042	6	6	31	Mn

[b] Represents splitting of the upper level only, no results for the splitting of the ground $^4S_{1\frac{1}{2}}$ state could be found.

[c] Represents splitting of lower $^2S_{\frac{1}{2}}$ level only. The splitting of the upper ^2P levels is at least one order of magnitude smaller.[14]

process) are Doppler and Collisional broadening. The total line half-width of each hyperfine component is given to a good approximation by:

$$\Delta\lambda_T = [(\Delta\lambda_D)^2 + (\Delta\lambda_c)^2]^{1/2} \tag{3}$$

where

$\Delta\lambda_T$ = the total line half-width of each hyperfine component (nm)
$\Delta\lambda_D$ = the half-width due to Doppler broadening alone (nm)
$\Delta\lambda_c$ = the half-width due to collisional broadening alone (nm).

In order to calculate the effective total line half-width, knowledge of the hyperfine structure of the line is required.

Another important parameter is known as the damping constant (a) which is defined by the equation:

$$a = \frac{\Delta\lambda_c}{\Delta\lambda_D}\left(\log_e 2\right)^{\frac{1}{2}} \tag{3a}$$

For most absorption lines in conventional flames the damping constant is thought to have values between 0·5 and 1·5. Thus, $\Delta\lambda_c$ and $\Delta\lambda_D$ are of the same order of magnitude.

The measurement of the total line half-widths of absorption line profiles in flames can be made by various methods:

(1) Plotting the logarithm of total line emission intensity versus the logarithm of concentration to determine the damping constant (a).[8,9,26] Measurement of the flame temperature then gives the Doppler half-width, hence the total half-width can then be calculated using equations (3) and (3a). This method is limited to elements that emit relatively strongly in flames, and neglects hyperfine structure.

N.B. For a given line, under conditions of thermal equilibrium, the emission and absorption line profiles are the same, i.e., they have the same half-width.

(2) Zeeman scanning, whereby the σ component of a ‘narrow’ low-pressure atomic emission line is scanned, using a variable magnetic field, over the wider absorption line profile. Using this method HOLLANDER *et al.*[9] found the total half-widths of the calcium 422·7 nm strontium 460·7 nm and barium 553·5 nm lines in the air–acetylene flame to be 0·0029 nm, 0·0035 nm and 0·0041 nm respectively. This method is only really suitable for elements that exhibit small or negligible degrees of hyperfine splitting and show a simple Zeeman spectrum.

(3) Direct scanning of the absorption[27] or emission[28] line in the flame using a piezo-electric Fabry–Perot interferometer. Using this technique KIRKBRIGHT *et al.*[27,28] found the half-width of the calcium 422·7 nm line to be 0·0040–0·0042 nm in the nitrous oxide–acetylene flame and 0·0033–0·0040 nm in the air–acetylene flame.

Table 9.2 Hyperfine splitting due to naturally occurring isotopes*

Element	Naturally occurring isotopes	Natural % abundance	Spin of nucleus	Isotope shift (nm)	Isotopes shift measured between	Line (nm)	References
Hg	198	10·0	0	0·0010	198–200	253·7	4, 19
	199	16·8	1/2	0·0012	200–202		
	200	23·1	0	0·0010	202–204		
	201	13·2	3/2	0·0032	198–204		
	202	29·8	0				
	204	6·8	0				
Bi	209	100	9/2	0		any	
Th	232	100	0	0		any	17
Tl	203	29·5	1/2	0·0016	203–205	535·0	13
	205	70·5	1/2				
Li	6	7·4	1	0·016	6–7	670·8*	23, 24
	7	92·6	3/2				
B	10	18·7	3	0·001	10–11	249·7	25
	11	81·3	3/2			249·8	
Na	23	100	3/2	0		any	
Rb	85	72·2	5/2	Less than hyperfine splitting of 87 Rb		780·0	20
	87	27·8	3/2				
U	235	0·72	7/2	0·007	235–238	415·3	22
	238	99·3	0	0·010	235–238	502·7	22
Cs	133	100	7/2	0		any	
Pb	204	1·4	0				
	206	25·1	0	0·0007	206–208	283·3	58
	207	21·2	$\frac{1}{2}$	0·0021	206–207a**		
	208	52·3	0	0·0008	207b–208**		

* Doublet (see Fig. 9.12).
** See Fig. 9.13.

N.B. The calcium 40 isotope has a natural abundance of 97% so that isotopic fine structure is negligible. There is no hyperfine splitting due to nuclear spin of this isotope.

WAGENAAR and DE GALAN[11a] have made a comprehensive study, using a pressure scanning Fabry–Perot Interferometer, on the atomic line profile of nine elements emitted by a hollow-cathode lamp and a nitrous oxide–acetylene

* See reference on page 284.

flame. In nearly all cases hyperfine structure was found to have a decisive influence upon the observed line profiles.

WINEFORDNER, PARSONS and MCCARTHY[6] have theoretically calculated the maximum and minimum values of the total line half-widths of 60 elements in various types of flame (hyperfine splitting was ignored). The Doppler half-widths were calculated by substitution of the known flame temperatures and the atomic weights of the elements in equation (1). The collisional half-widths were calculated, using the same flame temperatures, an effective molecular weight of the flame gases, and assuming a minimum collisional cross-sectional area of 0·2 nm^2 in conjunction with equation (2). This gave a minimum value of the collisional half-width. By assuming a maximum collisional cross-sectional area of 1 nm^2 a maximum value of the collisional half-width was obtained. Then using equation (3) the maximum and minimum total half-widths were calculated. For most elements the average total line half-width lies in the range $10^{-2} - 10^{-3}$ nm, the total half-width tending to increase with increasing wavelength. This is because the Doppler half-width varies directly with wavelength, and the collisional half-width varies with the square of the wavelength.

The effect of hyperfine structure will be to increase the effective line half-width values for many elements. This is borne out by studies of the overlap of emission line with absorption line profiles. For instance, the 206·163 nm iodine line (emitted by a heavily cooled iodine electrodeless discharge tube) is fairly strongly absorbed by bismuth atoms[29,30] in air–propane and air–acetylene flames. The corresponding bismuth absorption line is at 206·170 nm, a wavelength difference of 0·007 nm. Even assuming the effective Doppler temperature of the electrodeless discharge tube to be about 1000 K, and neglecting collisional broadening, the total line half-width of the 206·163 nm iodine line is only 0·0004 nm. Using WINEFORDNER and PARSONS' method,[6] the maximum bismuth 206·170 nm absorption line half-width in the air–acetylene flame is 0·0007 nm. Thus it would appear that the line profiles must have a somewhat greater half-width than the maximum values calculated ignoring hyperfine structure. This is further substantiated by the fact that the 228·812 nm arsenic line from an electrodeless discharge tube is weakly absorbed by cadmium atoms in the air–acetylene flame.[29,30] The corresponding cadmium absorption line is at 228·802 nm, a wavelength difference of 0·01 nm. (The arsenic lamp did not emit a cadmium spectrum, and this means that the absorption was due to overlap of the arsenic and cadmium line profiles.) Assuming a Doppler temperature of the source of 1000 K and neglecting collisional broadening, the total line half-width of the arsenic 228·812 nm line is 0·0006 nm. Using WINEFORDNER and PARSONS' method, the maximum cadmium 228·802 nm absorption line half-width in the air–acetylene flame is 0·0011 nm. Similarly, ALLAN[31] has stated that hyperfine structure is the major factor involved in the overlap of the manganese 403·307 nm and gallium 403·298 nm lines.

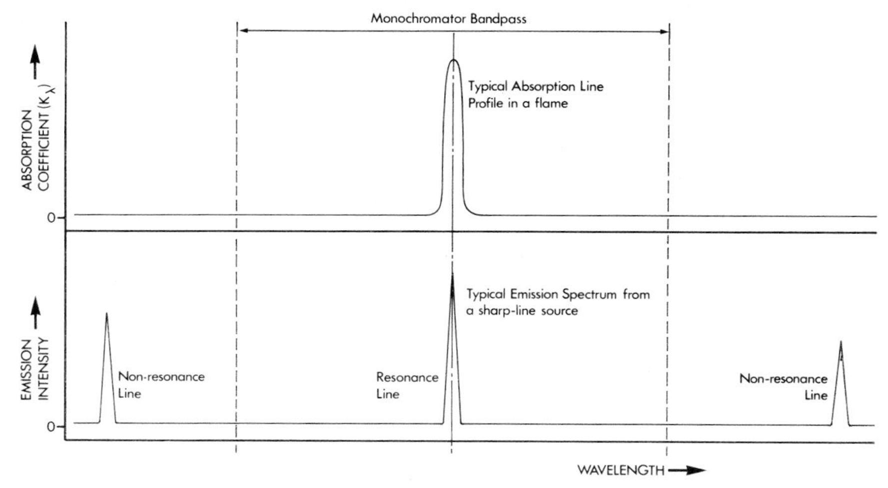

Fig. 9.4 Diagrammatic representation of atomic absorption using a hollow-cathode lamp and a relatively low resolution monochromator

It is important to realize that the total effective width of an absorption line is not necessarily twice the half-width. In fact the wings of a line can extend some way beyond this, but the above overlaps are thought to be more likely due to the hyperfine structure than absorption by the wings of the absorption line. (For bismuth see Table 9.1.)

The narrow half-width of absorption line profiles means that a continuous source cannot be used to measure the absorption unless a high-resolution monochromator is used (<0·02 nm). WALSH[5] overcame this problem by using an atomic line source (hollow-cathode lamp) that emitted the atomic line spectrum of the element to be determined. The low pressure in hollow-cathode lamps means that there is negligible collisional broadening, also the Doppler temperature of the discharge is somewhat less than that of a flame.[11,32] Thus the total half-width of a resonance emission line from a hollow-cathode lamp is expected to be less than the half-width of the absorption profile for the same line in a conventional flame. An inexpensive, low-resolution monochromator can, therefore, be used to isolate the desired resonance line emitted by the hollow-cathode lamp, from any other unwanted lines.

The combination of an atomic line source and a low-resolution monochromator is equivalent to a very high-resolution monochromator in conjunction with a continuum source (see Fig. 9.4).

Spectral Interference in Atomic Absorption

The narrow width of absorption lines results in very little spectral interference in atomic absorption. The average line half-width is about 0·005 nm and the wavelength region over which usable lines occur is about

600 nm (200–800 nm), thus there is only remote possibility of overlap of the limited number of absorption line profiles. In solution spectrophotometry with molecular band half-widths of about 10–50 nm the possibility of overlap is much greater.

Although spectral interference in atomic absorption is rare it has been observed. FASSEL, RAMUSON and COWLEY[33] have observed spectral interference between the following absorption lines: copper 324·754 nm and europium 324·753 nm ($\Delta\lambda$ = 0·001 nm), iron 271·903 nm and platinum 271·904 nm ($\Delta\lambda$ = 0·001 nm), silicon 250·690 nm and vanadium 250·691 nm ($\Delta\lambda$ = 0·001 nm), aluminium 308·216 nm and vanadium 308·211 nm ($\Delta\lambda$ = 0·005 nm). MANNING and FERNANDEZ[34] observed spectral interference between the mercury 253·652 nm and cobalt 253·649 nm lines ($\Delta\lambda$ = 0·003 nm). MANNING[35] has also observed overlap of the chromium 359·349 nm and neon 359·353 nm lines ($\Delta\lambda$ = 0·004 nm). Overlap of the calcium 422·673 nm line and the germanium 422·657 nm line has been reported (see Ref. 23 Chapter 4) ($\Delta\lambda$ = 0·016 nm).

ALLAN[31] has observed interference between the gallium 403·298 nm and the manganese 403·307 nm lines ($\Delta\lambda$ = 0·009 nm). Thus if it were necessary to determine copper in the presence of europium, the 327·4 nm line should be used rather than the 324·8 nm line. Also a correction would be necessary if mercury were to be determined in the presence of large amounts of cobalt.

All other examples quoted were for little-used absorption lines, hence it is true to say that spectral interference is seldom observed under normal working conditions in atomic absorption. NORRIS and WEST[35a] have reviewed the application of spectral overlap in atomic absorption.

The Shape of Calibration Curves in Atomic Absorption

CASE 1

The absorption line half-width is greater than the source line half-width, but is less than the monochromator bandpass.

This is generally the case for hollow-cathode lamps and electrodeless discharge tubes. These sources operate at low pressures and should show very little collisional broadening. The main cause of broadening is due to the Doppler effect. YASUDA[11] has estimated the effective temperature of a calcium hollow-cathode lamp, operated at 50 mA to be about 1400 K. Most lamps are run at considerably lower currents and should have an even lower Doppler temperature. In fact, BRUCE and HANNAFORD[32] have shown that the effective Doppler temperature of the atomic vapour in a calcium hollow-cathode lamp run at 5–15 mA is 350–450 K. They have also shown that the shape of the 422·7 nm resonance line (this line exhibits negligible hyperfine structure) emitted at these low currents is largely due to Doppler broadening and self absorption broadening (see p. 299). For the above range of lamp currents the 422·7 nm resonance line total half-width was found to be

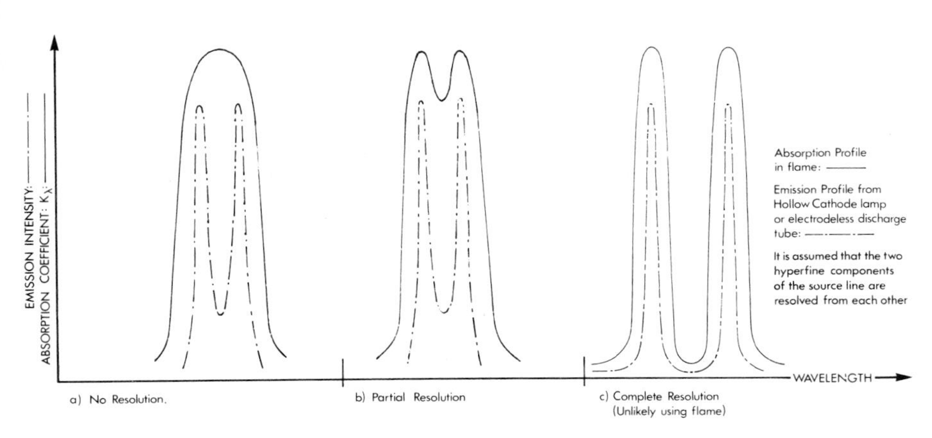

Fig. 9.5 Possible shapes of the absorption profile and the emission profile of a line with two hyperfine components

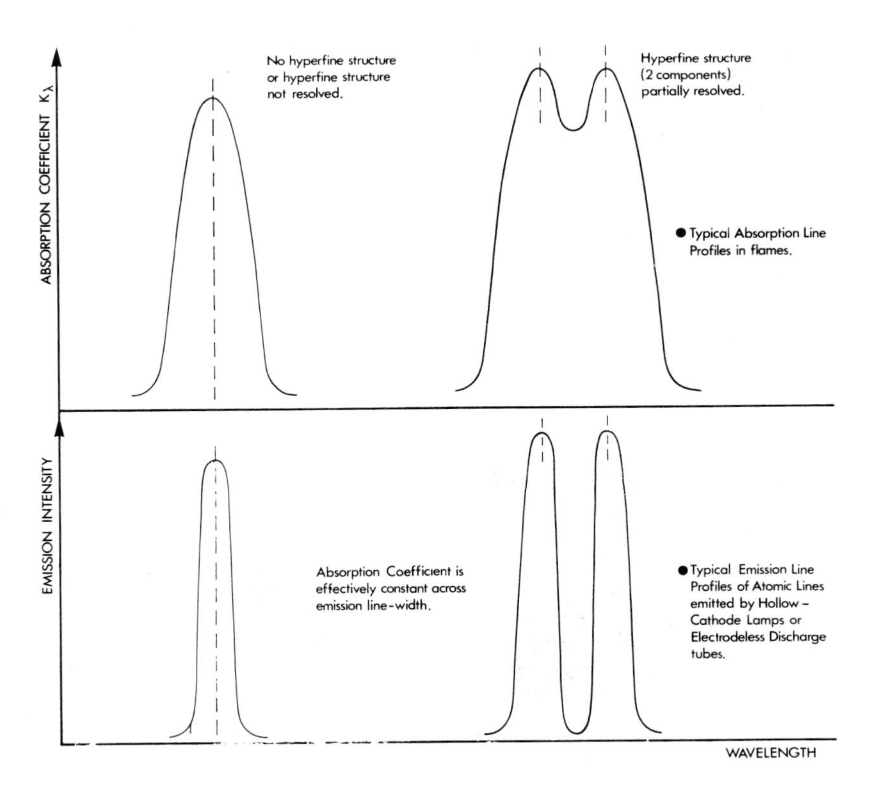

Fig. 9.6 Atomic absorption using a narrow line source

0·00092–0·00154 nm. Comprehensive descriptions of piezo electric[36] and pressure scanning[11a] Fabry–Perot interferometers to the study of the line shapes emitted by various atomic line sources have been made. For most hollow-cathode lamps studied the Doppler temperatures were found to be in the range 400–600 K.[11a]

The actual operating temperature of most electrodeless discharge tubes is relatively low, and these should also exhibit a low Doppler temperature, and negligible collisional broadening. Thus, the half-widths of resonance lines from both hollow-cathode lamps and electrodeless discharge tubes are expected to be less than the half-widths of absorption lines in flames, because there will be appreciable collisional broadening and a relatively high Doppler temperature in most conventional flames.

The effect of the hyperfine splitting of spectral lines, due to nuclear spin and different isotopes, is difficult to assess, as relatively little work has been performed on this aspect of atomic absorption. It is possible that the half-widths of resonance lines emitted by electrodeless discharge tubes (and also by hollow-cathode lamps) are sufficiently small for some of the hyperfine structure to be partially or completely resolved.[11a,20,21,36] There is less chance that some of the hyperfine components of the wider absorption lines in flames are resolvable.

There are three cases to be considered. Firstly, the hyperfine splitting will be small compared to the absorption line half-width, secondly, there might be partial resolution of some of the absorption line components, and lastly the main hyperfine components will be resolved. In Fig. 9.5 it is assumed that the two hyperfine components of the emission line are resolved.

It is thought that the last possibility (complete resolution) is very unlikely in conventional flames, although it has been reported in low-pressure absorption cells[19] (see isotopic analysis, later).

Assuming that the absorption coefficient (K_λ) is effectively constant across the emission line width(s), then the Beer–Lambert law can be applied (see Fig. 9.6).

$$-\int_{I_T}^{I_0} \frac{dI}{I} = \int_L^0 K\,N\,dl \qquad (4)$$

where

I_0 = the integrated intensity of the resonance line from the source passing through the flame and reaching the detector when nebulizing a blank solution (erg s^{-1} cm^{-2})

I_T = the integrated intensity of the source line after absorption by atoms in the flame when a sample solution is nebulized into the flame (erg s^{-1} cm^{-2})

L = the path length through the flame (cm)

N = the concentration of atoms in the flame (cm^{-3})
K = constant ($K\,N = \bar{K}_\lambda$) (cm^2)
$\bar{K}_\lambda$ = effective absorption coefficient across the emission line profile (cm^{-1})

$$\therefore \qquad \log_e \frac{I_0}{I_T} = KNL \qquad (5)$$

$$\therefore \qquad \text{Absorbance} = A = \log_{10} I_0/I_T = 0{\cdot}434KNL.$$

The absorbance ($\log_{10} I_0/I_T$) is proportional to the number of atoms in the flame. In general, for a resonance line from a sharp line source such as a hollow-cathode lamp or an electrodeless discharge tube (assuming that the resonance line is resolved from all other lines), the absorbance is directly proportional to concentration of the sample for absorbances less than about 0·5–0·7, and for absorbances greater than this the absorbance increases less than proportionally with concentration (see Fig. 9.8). This curving off of the calibration curve is probably due to the fact that at high absorbances the absorption coefficient cannot be considered to be effectively constant across the emission line width, or line widths in the case of hyperfine structure (see Fig. 9.5). At high absorbances the shape of the wings of the line become important (Fig. 9.3). The curvature of the calibration curve is unlikely to be due to resonance broadening,[12] as even when high concentrations of the test element are nebulized into the flame, the partial pressure of the test element will seldom exceed 10^{-1} Torr.

CASE 2

The absorption line half-width is less than the source line half-width and is less than the monochromator bandpass.

This is the case for a continuum source, e.g., xenon arc lamp, deuterium lamp, quartz–iodine lamp, and the absorption is now critically dependent on the bandpass of the monochromator (Fig. 9.7). If a continuous source is used it is essential to employ a monochromator with a spectral bandpass $<0{\cdot}02$ nm otherwise very poor limits of detection will be obtained. The spectral bandpass of most monochromators is greater than the total absorption line profile width (Fig. 9.7) and this fact is assumed for the derivation given below.[37]

ASSUMPTIONS

1. Constant temperature throughout the flame.
2. Constant and uniform atomic concentrations throughout the flame.
3. Constant transmission of light within monochromator spectral bandpass and zero transmission at all other wavelengths (Fig. 9.7).

4. The intensity of the continuum source is constant over the wavelength region of the absorption line profile.
5. Flame uniformly illuminated by a parallel beam of light.
6. The spectral bandpass of the monochromator is greater than the total absorption line profile width (Fig. 9.7).

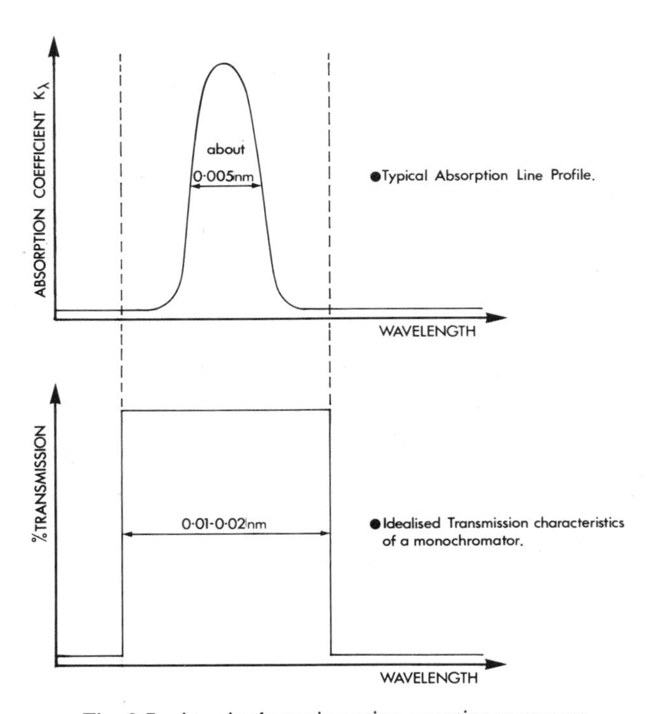

Fig. 9.7 Atomic absorption using a continuous source

Let

λ_0 = the peak wavelength of the absorption line profile (cm)
$I_{0\lambda}$ = the intensity of the continuum source at wavelength λ that is passing through the flame (erg s^{-1} cm^{-2} cm^{-1}), i.e., energy per unit time per unit area per unit wavelength
I_A = the integrated intensity of the absorbed radiation (erg s^{-1} cm^{-2}), i.e., total energy absorbed per unit time per unit area
K_λ = the absorption coefficient at wavelength λ (cm^{-1})
L = the path length of uniform atomic vapour (cm)
S = the spectral bandpass of the monochromator (cm)

$$I_A = \int_0^\infty I_{0\lambda}(1 - e^{-K_\lambda L})\, d\lambda \text{ erg s}^{-1}\text{ cm}^{-2}$$

I_0 is constant over absorption line profile

$$\therefore \qquad I_A = I_{0\lambda} \int_0^\infty (1 - e^{-K_\lambda L})\, d\lambda \tag{7}$$

At low values of $K_\lambda L$ (i.e., weak absorption), the above expression reduces to

$$I_A = I_{0\lambda} \int_0^\infty K_\lambda L \, d\lambda$$

$$= I_{0\lambda} \frac{\pi e^2 N f \lambda_0^2 L}{mc^2} \tag{8}$$

where

e = the electronic charge (c.g.s units)
m = the mass of the electron (g)
N = concentration of ground-state atoms (cm^{-3})
f = the oscillator strength of the absorption transition (no units)
c = the velocity of light ($cm\ s^{-1}$).

Thus if

α = the fraction of the radiation absorbed,

$$\alpha = \frac{I_A}{I_{0\lambda} S} = \frac{\pi e^2 N f \lambda_0^2 L}{Smc^2} \tag{9}$$

Hence the calibration curves are plotted as percentage absorption versus the concentration of atoms supplied to the flame. The relationship shown in equation (9) is independent of the line broadening factors or hyperfine splitting of the absorption line. The 1 per cent absorption figures are very dependent on the monochromator bandpass and are in general somewhat poorer than those obtained using hollow-cathode lamps. It has been claimed that the stability of most continuum sources is better than that of hollow-cathode lamps, thus allowing high degrees of scale expansion to be used. It is, however, thought that modern hollow-cathode lamps are of comparable stability to continuum sources. The main disadvantage of using continuum sources is that the relationship shown in equation (9) only holds at very low percentage absorptions.

When more than 5–10 per cent of the light is absorbed (depending on the resolution of the monochromator), the calibration curve rapidly becomes convex with respect to the concentration axis (see Fig. 9.8). This is the major disadvantage of the technique, since a convex calibration curve reduces the precision of the method. It is important to bear in mind that the majority of atomic absorption analyses are performed some way above the limit of

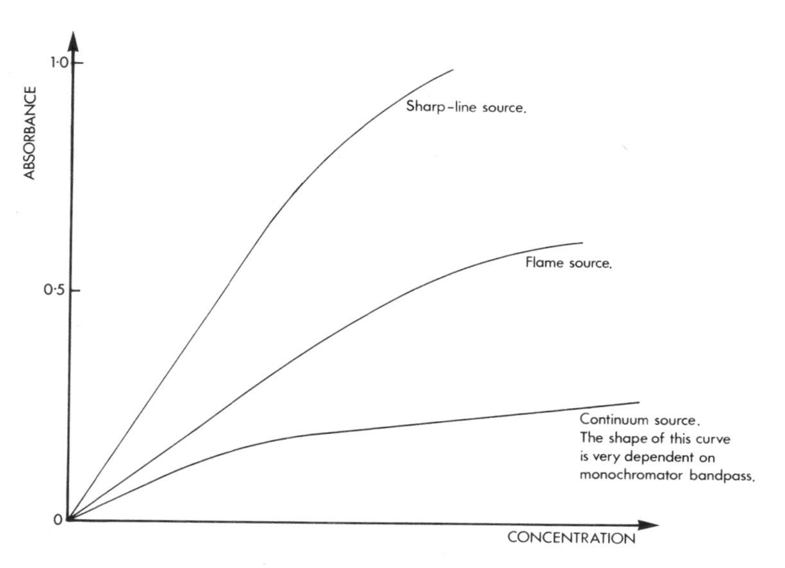

Fig. 9.8 Typical calibration curves for sharp-line, flame, and continuum source

detection and that precision is usually more important than an absolute limit of detection; this is one reason why commercial manufacturers do not produce units incorporating a continuous source. The use of continuum sources has been reported by quite a few workers.[38–41a]

Elser and Winefordner[41] have described a double modulation technique where source and wavelength modulation were used. The instrumental system did not respond to light scattering or molecular absorption. The limits of detection for cadmium, calcium, copper, magnesium and silver (at their main resonance lines) were quoted as 10, 0·1, 0·5, 0·1 and 2 μg/ml respectively. These limits are about two orders of magnitude poorer than those obtained using an average single beam atomic absorption instrument. Keliher *et al*[41a] using a similar technique, but with a very high resolution echelle monochromator have reported some impressive detection limits for a number of elements.

The main disadvantages and advantages of the use of continuous sources are shown in Table 9.3.

Case 3

The absorption line half-width is similar to the source line half-width but less than the monochromator bandpass.

In this case the source is a flame supplied with the metal to be determined, and the emission from this source flame is passed through the absorbing flame

Table 9.3 Comparison of types of source

Source	Advantages	Disadvantages
Hollow-cathode lamp	(1) Only one operating parameter (current) (2) Easy to stabilize (3) Moderate intensity (4) Emits very sharp lines (5) Easy to change (6) Commercially available for all elements that can be determined using atomic absorption	(1) Cannot be easily prepared (2) Relatively expensive
Microwave-excited electrodeless discharge lamp	(1) Easily prepared for some elements (2) Relatively cheap (3) Long shelf life (4) Very intense, about 10–1000 times more intense than most hollow-cathode lamps (5) Emits very sharp lines	(1) Microwave generator required (2) More operating parameters than hollow-cathode lamp (cooling, position of tube within microwave cavity, and operating power) (3) At present, not commercially available for all elements that can be determined by atomic absorption (4) Stability poorer than hollow-cathode lamps
Radiofrequency (R.F.) excited electrodeless discharge lamp	(1) Long shelf life (2) About 5–100 times more intense than most hollow-cathode lamps (3) Only one operating parameter, input power (4) Emits very sharp lines	(1) R.F. power supply required (2) Only a limited number of lamps are commercially available
Continuum source (xenon arc lamp)	(1) Only one source required for all elements (2) Background correction easily applied (3) Multi-element analysis by wavelength scanning techniques is possible	(1) Calibration graphs only linear over a relatively short concentration range. This leads to poor precision unless all solutions are diluted to within linear range of calibration graphs (2) Expensive, high resolution monochromator required. Sensitivity very dependent on monochromator bandpass (3) Sensitivity usually much poorer than those obtained using sharp line sources, especially for elements with resonance lines in the far u.v. ($<$250 nm) (4) High resolution monochromators require precise temperature stabilization.

Table 9.3—*contd.*

Source	Advantages	Disadvantages
Flame source	(1) Simple (2) Inexpensive (3) Useful for rare elements where only a few analyses are required (4) Multi-element sources can easily by made	(1) Sensitivities poorer than those obtained with sharp line sources (2) Stability of the emission from a flame system is not very good, hence very restricted use of scale expansion (3) Only limited number of elements will emit suitable intense resonance lines from flames (4) Non-linear calibration curves
Plasma source	(1) All elements that can be determined by atomic absorption should be excited (2) Multi-element sources can easily be made	(1) Relatively poor sensitivities compared to sharp line sources (2) Expensive high power R.F. generator required (3) Stability of the emission is not very good

which is usually of the same type as the source flame. Thus the atoms in both flames are in identical environments and the emission and absorption line profiles will be identical. The source flame is usually shielded by an outer flame to prevent self-reversal of the resonance lines emitted. The choice of the concentration of the test element that is supplied to the source flame is a compromise between obtaining a suitably intense signal (requiring a high concentration) and minimizing self-absorption and self-reversal of the emitted radiation (requiring a low concentration). When there is a significant concentration of absorbing atoms between the emitting atoms and the detector, additional broadening of the resonance lines occurs. This is due to preferential absorption of the centre of the emission line by the absorbing atoms and is known as self-absorption. If the absorbing atoms are at a lower temperature than the emitting atoms, self-reversal will occur. In this case the emitted line profile will show two peaks (Fig. 9.9) because the cooler absorbing atoms absorb over a smaller wavelength region than the total emitted line profile.

Several workers have reported the use of flame sources[42–44] and RANN[44] has given a theoretical treatment to derive the shape of the calibration curves. The theory is rather complex and makes use of a Voigt profile[4] to describe the emission- and absorption-line profiles. By assuming various values of the damping constant a[$a = \sqrt{\log_e 2}(\Delta\lambda_c/\Delta\lambda_D)$] and the degree of self absorption of the resonance line emitted by the source flame, absorbance versus concentration calibration curves were constructed with the aid of a computer, and the closest fit to an experimental calibration curve corresponded to an *a* value of 0·46 for copper at the 324·8 nm line in the air–acetylene flame.

The shape of the calibration curves are somewhat dependent on the concentration of the element supplied to the source flame (depending on the degree of self-absorption and self-reversal of the resonance line), but are generally intermediate between those of a sharp line source and those of a continuum. RANN[44] obtained sensitivities for cobalt, copper, magnesium and silver that were 30–70 per cent of those obtained using hollow-cathode lamps. The main disadvantage of using flame sources is that only relatively few metals can be readily excited to give intense enough resonance emission. Elements with lines in the far-ultra-violet, such as arsenic, cadmium and zinc give very poor emission, even in hot flames, whilst elements that formed refractory oxides, such as aluminium, beryllium and molybdenum, will only emit from hot high background fuel-rich flames (e.g., nitrous oxide–acetylene) thus giving a relatively high noise-to-signal ratio.

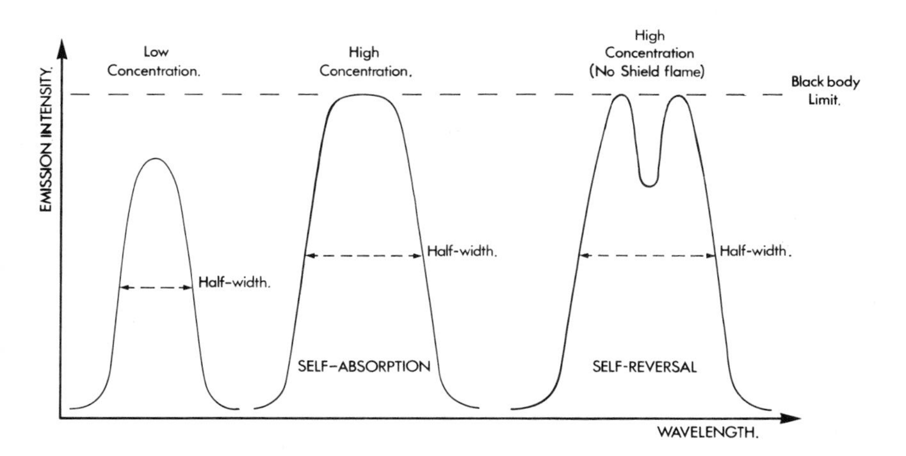

Fig. 9.9 Pictorial representation of self-absorption and self-reversal of a resonance line emitted by a flame (system in thermal equilibrium)

NICKLESS and CHEICK HUSSEIN[45] have reported the use of a plasma source, which can be used to excite nearly all elements, but at the high temperature of a plasma the emitted line half-widths will tend to be wider than the absorption line profiles of atoms in flames, thus giving more convex calibration curves and poorer sensitivities than flame sources. The advantages and disadvantages of the various types of source are given in Table 9.3.

Atomization in Flames

The term atomization refers to the breakdown of a compound to free atoms. When a solution is nebulized into a flame, atomization must occur before absorption of the source radiation by free atoms can occur. Most commercially available pneumatic nebulizing units have solution uptake rates of about 2–4 ml/min of which 5–15 per cent finally reaches the flame. The total flow rate of flame gases (e.g., acetylene and air or nitrous oxide) is about 8–12 litres/min. Hence, it can be seen that the sample is highly diluted in the flame gases.

For example, consider a nebulizer with an uptake rate of 3 ml/min, 10% of which reaches the flame. If a 10 μg/ml silver solution were nebulized the amount of silver reaching the flame would be 3 μg/min (3×10^{-8} moles min^{-1}). Assuming the total flow rate of the flame gases to be 11 litres/min (0·5 mole min^{-1}) and ignoring air entrainment in the flame:

$$\text{The dilution factor} = \frac{0{\cdot}5}{3 \times 10^{-8}} = 1{\cdot}7 \times 10^{7} \text{ times}$$

Assuming a uniform atomic vapour density, the analytical sensitivity (1% absorption figure) is directly proportional to the dilution factor multiplied by the path length over which absorption occurs.

Few elements are completely atomized in the flames, thus a further effective dilution factor is involved. If the solution uptake rate is increased appreciably, the nebulized sample causes appreciable cooling of the flame gases which results in poorer atomization of the sample.

The sensitivity of atomic absorption techniques can be increased considerably by using a graphite tube atomizer, graphite rod atomizer, or similar device, as an atom reservoir in place of the flame.[46] Using these techniques a small volume of sample can be completely atomized in a relatively small volume. This has the advantage that the large dilution factor associated with flames is avoided and very good limits of detection are obtained. The number of recent publications indicate that there is considerable interest in these techniques (see Chapter 7).

Factors affecting the Percentage Atomization

The idealized processes of atomization for an indirect nebulizer, when the nebulized droplets of a solution reach the flame, are depicted in Fig. 9.10. The last two steps, volatilization and atomization, are only essentially complete for relatively few elements, e.g., Ag, Na. It would appear, for the majority of elements, that after evaporation of the water, equilibria are set up between the sample and the flame gases resulting mostly in the formation of oxide species (in a few cases hydroxide species are also formed, e.g., Ca, Cs, K). Although the sample might be nebulized as the chloride or bromide, the halides (other than perhaps certain fluorides in HF media) are usually less stable than the oxides in conventional oxygen-supported flames (in hydrogen–fluorine flames, fluorides are, in general, more stable than oxides). The course of reaction can be depicted:

$$\text{Sample} \xrightarrow[\text{hydrolysis}]{\text{evaporation}} \text{Clotlets} \xrightarrow{\text{volatilization}} \begin{cases} \xrightarrow[\text{flame gases}]{\text{reaction with}} \text{MO(MOH)} \\ \qquad\qquad\quad \downarrow\uparrow \\ \xrightarrow[\text{atomization}]{} \text{M} + (\text{O}) \\ \qquad\qquad\quad \downarrow\uparrow \\ \qquad\qquad\quad \text{M}^{+} + e^{-} \ (\text{ionization}) \end{cases}$$

Some oxide species have such a low vapour pressure that the oxide is present in the flame as unevaporated solid or liquid particles, e.g., Al, Mo, U, etc. This is borne out by the fact that when solutions containing aluminium are nebulized into an air–acetylene flame, irradiated with a very intense mercury lamp, weak scattering of the mercury radiation (at all the intense mercury lines) occurs. No scattering is observed when elements like zinc and cadmium, which form relatively unstable oxides, are nebulized into the same air–acetylene flame.[30]

The fuel-rich air–acetylene flame has been found to give higher absorbance

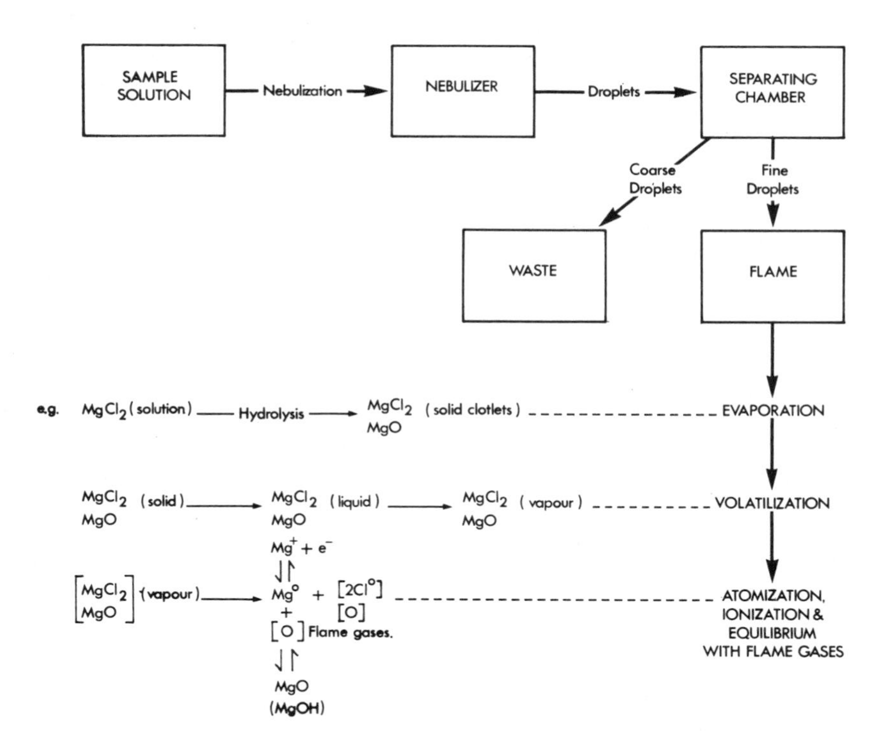

Fig. 9.10 Idealized processes of atomization

values for certain elements (Cr, Mg, Mo, Sn) than the hotter stoichiometric flame. This indicates that a higher degree of atomization occurs in the fuel-rich flame, and this can be attributed to the more favourable chemical environment of the fuel-rich flame although it has a lower temperature than the stoichiometric flame.

Consider the following simple dissociation reaction of a metal oxide occurring within a flame:

$$MO_{vap} \leftrightharpoons M_{vap} + [O]$$

This is not meant to imply that all metals, within an oxygen-supported flame, are in a simple equilibrium with their monoxides, but this simple representation can be used to explain certain observations. Assuming that the above reaction has attained a state of equilibrium, then:

$$K_p = \frac{[M]\,[O]}{[MO]}$$

$$\frac{[M]}{[MO]} = \frac{K_p}{[O]}$$

where K_p is the equilibrium constant of the above reaction. The concentration of oxygen [O] is effectively controlled by the composition and temperature of the flame gases and should be independent of any substance nebulized into the flame. It can be seen that the [M]/[MO] ratio is controlled by the value of the equilibrium constant (K_p) and the oxygen concentration [O]. The value of (K_p) will increase with increasing temperature whilst the oxygen concentration is reduced by making the flame fuel-rich. The fact that certain elements are atomized to a higher degree in a fuel-rich flame compared to a slightly hotter stoichiometric flame indicates that the small decrease in the value of (K_p) of the fuel-rich flame, compared to the hotter stoichiometric flame, is more than compensated for by a reduction in the oxygen concentration in the fuel-rich flame.

COWLEY, FASSEL and KNISELEY[47] have observed a striking enhancement in the degree of atomization of many elements in the pre-mixed fuel-rich oxygen–acetylene flame compared to the hotter stoichiometric flame. The enhancement is attributed to a favourable chemical environment provided by the interconal zone of the fuel-rich flame (for the formation and existence of free atoms).

Similar results have been observed by KIRKBRIGHT, PETERS and WEST[48] with a nitrous oxide–acetylene flame. The reducing zone of the fuel-rich flame was mainly confined to the region of red CN emission. Above the red zone of the fuel-rich flame and also in stoichiometric flame (with no red zone) poor atomization of elements with refractory oxides, e.g., Al, Mo, Ti, etc., was observed.

The necessity of the correct chemical environment is clearly demonstrated by the fact that tin is scarcely atomized in the stoichiometric air–acetylene flame (temperature approximately 2300°C), but appreciably atomized in the very fuel-rich air–hydrogen flame and also in the cool nitrogen–hydrogen diffusion flame.[49] This is presumably due to reduction of the oxide clotlets by hydrogen in a relatively oxygen-free atmosphere.

WILLIS[50] has made a comprehensive study of the mode of operation of a typical nebulizer, spray chamber and burner. The sensitivity of an atomic absorption instrument and the effect of chemical interferences in the flame were found to be critically dependent on the construction of the nebulizer and particularly on the rate of liquid uptake. The drop size distribution of the spray entering the flame indicated the importance of small drop size for the efficient atomization of metals which tend to form refractory compounds in the flame.

The above discussion has assumed that the nebulized substances attain equilibrium with the flame gases. The rate of attaining equilibrium will be dependent on the nebulizer and the flame temperature. A nebulizer that produces very small droplets is better in this respect than one that produces large droplets. In general, equilibria will be established more rapidly in a hot flame than in a cool flame.

For instance, when calcium solutions containing phosphate are nebulized into an air–acetylene flame, the absorbance just above the primary reaction zone is less than that of an equivalent pure calcium solution. This decrease in the calcium-free atom population is attributed to the formation of involatile calcium phosphate clotlets after evaporation of the droplets reaching the flame. These clotlets are then slowly thermally decomposed on passage through the flame thus re-establishing equilibrium between calcium atoms and the flame gases. (The interference of phosphate upon calcium almost disappears when the absorbances are measured some considerable distance above the burner.) This type of interference can be overcome by using a hotter flame in order to decompose the involatile clotlets very low down in the flame, so that equilibrium can rapidly be set up between the metal species and the flame gases. (Phosphate has little effect on the calcium absorbance in the hot nitrous oxide–acetylene flame.)

Calcium phosphate is almost certainly formed in the clotlets, rather than by reaction in the flame gases of the air–acetylene flame, because when an air–acetylene flame is supplied by two independent nebulizers and calcium solution is supplied to one nebulizer and phosphoric acid to the other, no interference is observed. Similarly, the presence of phosphorus compounds in petroleum has no effect on the calcium absorbance in the air–acetylene flame, because calcium phosphate is not readily formed in non-aqueous media, and thus involatile clotlets will not be formed upon evaporation of the non-aqueous droplets.

Most of the interferences observed in the air–acetylene flame that can be attributed to involatile clotlet formation can be overcome by using the hotter nitrous oxide–acetylene flame.

The determination of the percentage of the sample that is atomized in the flame is a complicated procedure.[10,51–56] Initially the total concentration (N_T) of the nebulized sample in the flame is determined from the sample uptake rate, the transfer efficiency of the spray chamber and the flow rate of the flame gases. It is then possible to calculate the degree of atomization by measuring the integrated absorption signal (I_A) given by equation (8). This is determined using a continuum source in conjunction with a monochromator set to a spectral bandpass somewhat greater than four times the total line half-width in the flame. (Under these conditions the integrated absorption signal (I_A) is independent of the line-broadening factors or hyperfine splitting of the absorption line.) Knowledge of the oscillator strength of the absorption line allows the value of N (the concentration of atoms in the flame) to be deduced from equation (8). Hence, knowing the total concentration (N_T) of the sample in the flame, the percentage atomization figure is then given by $N/N_T \times 100$. Lack of reliable values of the oscillator strength for certain elements limits the accuracy obtainable.

Instead of measuring the integrated absorption signal, the absorption signal using a conventional hollow-cathode lamp can be measured.[53,56] In this case knowledge of the emission and absorption line profiles (including any hyperfine structure) is required as well as the oscillator strength of the absorption line. The former method is thought to give more accurate results especially when there is appreciable hyperfine structure.[10] Table 9.4 shows some values that have been obtained by various workers in the air–acetylene and nitrous oxide–acetylene flames.

Table 9.4(a) Percentage atomization in the pre-mixed air–acetylene flame

	Percentage atomization					
	Reference					
Element	10	51	52	53	54	55
Ag		66		100	70	—
Al		$<10^{-3}$			$<5 \times 10^{-3}$	—
Au					40	63
Ba	0·21	0·11	0·84		0·18	0·26
Ca	4·7	14	8·6		7·0	7·0
Cr	13	6·4			7·1	—
Cu	82	98		37**	88	100*
Mg	—	59			106	84
K	43	25			32	28
Na	100*	100		40**	104	53
Sr	11	13	20		6·3	—
Zn	—	45			—	110

* Assumed value.
** These results have been shown to be low owing to neglect of hyperfine structure.
In most cases the flame conditions were optimized for maximum percentage atomization.

Table 9.4(b) Percentage atomization in the pre-mixed nitrous oxide–acetylene flame

Element	Percentage atomization		
	Reference		
	54	55	56
Al	13	28	29
B	0·35	—	—
Ba	17	20	7·4
Be	9·5	—	—
Mg	88	100*	107
Si	5·5	—	—
Ti	11	21	—
V	32	65	—

* Assumed value.
In most cases the flame conditions were optimized for maximum percentage atomization.

The published results are not all in agreement because of the inherent errors of the measurement.

For good atomization of elements that form refractory oxides and involatile clotlet species, the hot reducing nitrous oxide–acetylene flame is to be recommended. Such elements are Al, Ba, B, Be, Ca, Ge, Mo, Si, Sr, Ti, V, W, Zr, etc. For good atomization of elements that do not form refractory oxides or involatile clotlet species, the cooler air–acetylene flame is satisfactory. Such elements are Bi, Cd, Fe, Hg, K, Na, Pb, Sb, Se, Te, Zn, etc. The degree of atomization of these elements is not very dependent on flame conditions or flame temperature, and these elements tend to give better sensitivities in the cooler air–propane or air–coal gas flames than in the hotter air–acetylene flame.

This increase in sensitivity is probably due to the increased density of the atomic vapour at the lower temperature. The drawback of these low-temperature flames is that if the solution being nebulized has a very high solids content of elements that form refractory oxides, large clotlets will be formed. These clotlets will be more slowly decomposed in the cooler flames than in the hotter air–acetylene flame, thus the element will take longer to reach effective equilibrium with the flame gases, i.e., interference is observed.

Ionization in Flames

The population of ground-state atoms can be depleted by ionization.

$$M \rightleftharpoons M^+ + e^-$$

The degree of ionization can be calculated from the Saha equation.

$$\log \frac{x^2}{1 - x^2} = -\log P - \frac{5040\,V}{T} + 2{\cdot}5 \log T + \log U^+ - \log U^\circ - 6{\cdot}18$$

Where

x = fraction of atoms ionized
P = total pressure of the electrons and metal in all forms in the burnt gases (atm)
T = flame temperature (K)
V = ionization potential of the element (e.v)
U^+ = partition function of the ion
U° = partition function of the atom.

WOODWARD[57] has calculated the percentage ionization of 48 elements in the air–propane, air–acetylene and nitrous oxide–acetylene flames using a computer programme. The total pressure (P) was taken as 10^{-6} atm.

In the air–acetylene flame only the alkali metals and alkaline earth metals showed appreciable ionization but in the nitrous oxide–acetylene flame the following elements were shown to be more than 10% ionized: Al, Ba, Ca, Cs, Ga, In, K, La, Li, Na, Rb, Sc, Sr, Ti, Tl, Y.

Isotopic Analysis by Atomic Absorption

It is possible to determine isotopic compositions using atomic emission techniques, but these usually require expensive very high-resolution monochromators (resolution $\leqslant 0{\cdot}001$ nm) in order to resolve the hyperfine structure.

In theory it should be possible to determine the isotopic composition of an element by atomic absorption if certain conditions are met:

(1) Single (or at least highly enriched) isotope sources or isotope absorption tubes must be available for all the major isotopes of the element. The line half-widths from these devices must be somewhat less than the isotopic displacements between neighbouring components of the absorption line.

(2) The absorption line profile half-widths must be somewhat less than the isotopic displacements between neighbouring components of the absorption line.

(3) For a given isotope, the hyperfine components due to nuclear spin must themselves be partially resolved from the other isotopic components of the absorption line.

Principle of Isotopic Analysis

Consider an element with two isotopes A and B. Initially the absorption of a sample, in a suitable atom reservoir, is measured with a lamp containing pure isotope A. Assuming that the emission and absorption line profiles of the isotope A are completely resolved from those of isotope B, then the absorption reading is proportional to the quantity of isotope A in the atom reservoir. (Isotope B in this case will not absorb radiation emitted from isotope A.) Similarly, if this procedure is repeated using a lamp containing pure isotope B, the absorption reading will be proportional to the concentration of B in the atom reservoir. It can be seen that a high-resolution monochromator is not required. Unfortunately this simple case is rarely observed in practice.

There are three main cases to be considered.

CASE 1

The isotopic line absorption and emission profiles do not overlap with any other isotopic line (see Fig. 9.11). The source contains the pure isotope to be determined. This case is very rare owing to the small wavelength shifts involved. It has been observed[19] for mercury 202, which was determined using a mercury 202 cooled low-pressure microwave source and a low-pressure absorption cell, giving negligible collisional broadening (of the emission and

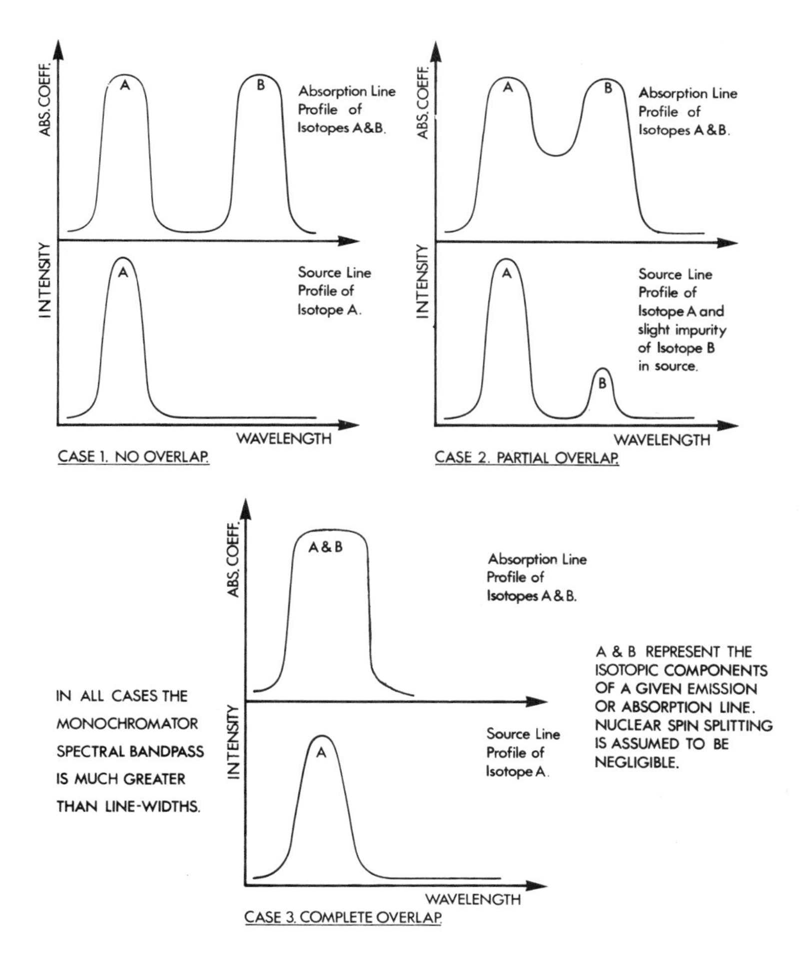

Fig. 9.11 Some possible cases of isotopic overlap of absorption line profiles for an element with two isotopes A and B

absorption lines). The absorption cell was operated at room temperature and the Doppler width of the 253·7 nm mercury absorption line under these conditions was 0·0002 nm. The mercury 202 emission and absorption lines were resolved from all other isotopic lines and thus the measured absorption was directly proportional to the mercury 202 concentration. Direct determination of mercury isotopes other than mercury 200 and 202 was not possible because their absorption lines showed appreciable overlap. It is almost impossible to atomize other elements at such a low temperature, thus this case is very rare.

CASE 2

The isotopic line absorption profiles partially overlap (Fig. 9.11) and/or the source contains small quantities of other isotopes.

In this case part of the absorption observed using a given isotopic source is due to other isotopes. Consider a case with just two isotopes A and B. The absorption is measured using the source containing isotope A (or enriched isotope A) and then measured using the source containing isotope B (or enriched isotope B), a calibration curve then being plotted of isotopic proportion of A in the standards versus (absorbance using source A/absorbance using source B).

This situation has been observed in the case of lithium, the 670·8 nm resonance line is a doublet (0·016 nm separation) each line of which is split into isotopic components also 0·016 nm apart (see Fig. 9.12). (The nuclear spin splitting is about 0·001 nm—see Table 9.1.) The determination of lithium isotopes has been reported using hollow-cathode lamps,[24] flame sources[18] of lithium 6 and lithium 7, and a flame as an absorption cell. (The absorption and

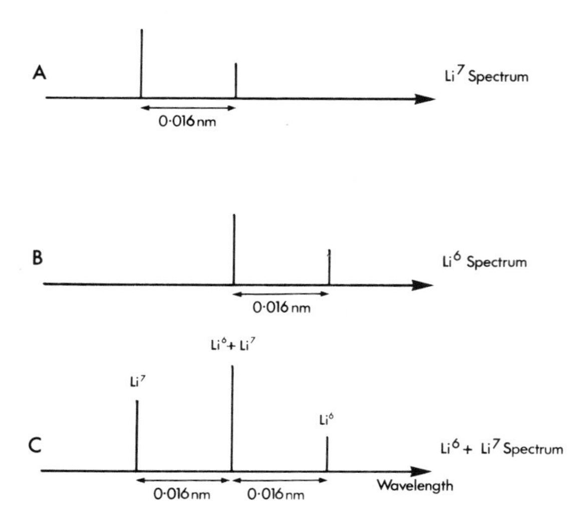

Fig. 9.12 Isotopic splitting of the lithium 670·8 nm doublet

emission line half-width of the 670·8 nm lines in the air–acetylene flame is reported as 0·013 nm.[26]) Using flames there is appreciable overlap of the isotopic components but it was possible to determine lithium isotopes using a flame as the absorption cell.[18,24]

A more accurate measurement was made by GOLEB and YOKOYAMA[23] whose used two water-cooled Schuler–Gollnow lithium hollow-cathode lamps, one containing lithium 6 the other containing lithium 7. A modified water-cooled Schuler–Gollnow hollow-cathode tube was used to vaporize the samples and standards and acted as the absorption cell. Both emission and absorption tubes were at low pressures in order to minimize collisional broadening.

Similarly, GOLEB has determined uranium 235 and 238[22] by a similar technique (using water-cooled Schuler–Gollnow tubes). In this case the 415·3 and 502·7 nm uranium lines were used, as these lines of uranium 235 have negligible nuclear spin splitting but exhibit an appreciable isotope shift between the 235 and 238 isotopes of 0·007 and 0·010 nm respectively. The sources contained the enriched isotopes, not the pure 235 and 238 isotopes. GOLEB also demonstrated a variation on the above method[22] whereby instead of placing samples and standards in the absorption tube, one demountable source lamp contained the sample and standards (one at a time) whilst the absorption tube contained the pure isotope (or the enriched isotope) to be determined. In this case a linear calibration curve was plotted between percentage of uranium 238 in the standards, versus the percentage transmission observed through the absorption cell. The advantages of this technique are that metal chips, oxides, and compounds can be used directly as samples and that the proportion of incident light absorbed is independent of the emission intensity (whilst when samples are placed in the absorption tube, the absorbance is very dependent on sputtering current flowing in the absorption tube).

Little overlap of the uranium isotopic component was observed. This was partially due to the large atomic mass of uranium, as the line widths are limited by Doppler broadening and the half-width of a line is inversely proportional to the square root of the atomic weight of an element. Thus, for a given Doppler temperature the uranium line half-width will be $\sqrt{(235/7)}$ times smaller than the lithium line half-width (ignoring the difference between the wavelengths of the lines).

Similarly KIRCHHOFF[58] has determined the lead 206 and lead 208 isotopic constitution of lead samples using the 283·3 nm line. In this case the samples and standards are excited in a demountable hollow-cathode lamp and the radiation is passed through a modified hollow-cathode lamp containing pure lead 208 (or pure lead 206). The calibration curve is then constructed by plotting the percentage transmission against the percentage of lead 208 (or lead 206) in the standards. The 283·3 nm lead line has four main hyperfine components (ignoring the small contribution of Pb204), see Fig. 9.13. It can be

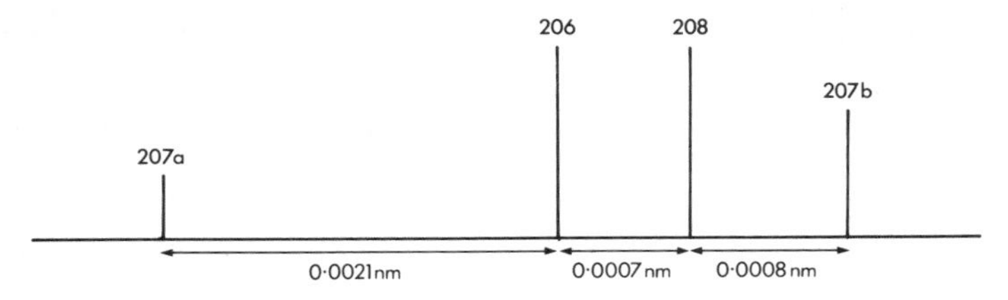

a & b represent the nuclear spin components of the 207 isotope.

Fig. 9.13 Hyperfine splitting of the 283·3 nm lead line

seen from the figure that isotopic splitting of the 206 and 208 component is much less than the nuclear spin splitting of the 207 component. (The 206 and 208 components exhibit no nuclear spin splitting.) Lead isotopes have also been determined using pure lead isotope hollow-cathode lamps and nebulizing the samples into a conventional air–acetylene flame.(59) Appreciable overlap was observed in this case.

Thus it is important to realize that for the determination of lead, lithium, uranium and certain other elements in flame cells, by atomic absorption, the element in the source lamp must have the same isotopic constitution as the element in the samples and standards.

CASE 3

The isotopic line absorption profiles completely overlap. (Fig. 9.11).

GOLEB(25) has observed this case when trying to determine boron 10 and 11 using an absorption tube. The isotopic shift of both lines of the 249·6/7 nm doublet is only 0·001 nm and, with a given boron isotopic lamp, the same absorption figures were obtained when either boron 10 or boron 11 was in the absorption tube.

Subsequently it was found that with krypton-filled boron hollow-cathode lamps (as GOLEB used) an unusual excitation mechanism occurs which results in a large Doppler half-width of the boron emission lines. HANNAFORD and LOWE(25a) successfully determined boron 10 and 11 with an absorption tube by using neon-filled boron 10 and 11 sources (these did not show excessive Doppler half-widths of the emission lines) and measuring at the 208·9/209·0 nm boron doublet (isotope separation 0·0025 nm) instead of the 249·6/7 nm doublet (isotope separation 0·001 nm).

It would appear strange that it is possible to determine lead 206 and lead 208 (isotope splitting 0·0007 nm) with good resolution but that under identical conditions, boron 10 and 11 are only very poorly resolved at the 249·6/7 nm doublet(25a) (isotopic splitting 0·001 nm). The line widths are controlled by Doppler broadening, and the Doppler half-width of an absorption or an emission line is proportional to the square root of the temperature, and inversely proportional to the square root of the atomic weight of the absorbing or emitting atoms. Boron is far more refractory than lead and therefore will require a much higher temperature, i.e., sputtering current, to produce an appreciable atom population. Also, from atomic weight considerations, the boron line, for a given temperature, will be $\sqrt{(207/11)}$ times broader than the lead line if the relatively small difference between the wavelengths of the two lines is ignored.

It can be seen that although isotopic analysis can be performed using atomic absorption techniques, it is more complicated procedure than conventional atomic absorption techniques. Many isotope shifts (Table 9.2) are somewhat less than average absorption line half-widths, in conventional flames. Thus, flames are not really suitable as absorption cells, although the use of low-pressure flames which will minimize collisional broadening should be more feasible. Most publications only report the determination of two isotopes for a given element. For an element with more than two major isotopes (assuming partial overlap) the procedure becomes more complicated. The determination of minor isotopic constituents (<2 per cent) is thought to be impractical. Thus, isotopic analysis using atomic absorption is thought to have limited application.

Chap. 9 References and further reading

1. WOLLASTON, W. H., *Phil. Trans.* 91 (1802) 365.
2. FRAUNHOFER, J., *Ann. d. Physik* 56 (1817) 264.
3. KIRCHHOFF, G. R., *Ann. d. Physik* 109 (1860) 148, 275. *Phil. Mag.* 20 (1860) 1.
4. MITCHELL, A. C. G., and ZEMANSKY, M. W., *Resonance Radiation and Excited Atoms* (University Press, Cambridge, 1961).
5. WALSH, A., The application of atomic absorption spectra to chemical analysis. Australian Patent No. 23041/53. Spectrochim. Acta 7 (1955) 108.
6. PARSONS, M. L., MCCARTHY, W. J., and WINEFORDNER, J. D., Approximate half-intensity widths of a number of atomic spectral lines used in atomic-emission and atomic-absorption flame spectroscopy. *Appl. Spectroscopy* 20 (1966) 223.
7. HINNOV, E., A method of determining optical cross sections. *J. Opt. Soc. Amer.* 47 (1957) 151.
8. HINNOV, E., and KOHN, H., Optical cross sections from intensity–density measurements. *J. Opt. Soc. Amer.* 47 (1957) 156.
9. HOLLANDER, TJ., JANSEN, B. J., PLAAT, J. J., and ALKEMADE, C. TH. J., Zeeman scanning of alkaline earth absorption line profiles in flames at atmospheric pressure. *J. Quant. Spectrosc. Radiat. Transfer* 10 (1970) 1301.
10. WILLIS, J. B., Atomisation problems in atomic absorption spectroscopy III. Absolute atomisation efficiencies of sodium, copper, silver and gold in a Meker-type air–acetylene flame. *Spectrochim. Acta* 26B (1971) 177.
11. YASUDA, K., Relationship between resonance line profile and absorbance in atomic absorption spectroscopy. *Anal. Chem.* 38 (1966) 592.
11. (a) WAGENAAR, H. C., and DE GALAN, L., Interferometric measurements of atomic line profiles emitted by hollow-cathode lamps and by an acetylene–nitrous oxide flame. *Spectrochim. Acta* 28B (1973) 157.
12. ALKEMADE, C. TH. J., Science vs. fiction in atomic absorption. *Applied Optics* 7 (1968) 1261.
13. CANDLER, C., *Atomic Spectra* (Hilger and Watts, 2nd Ed., London, 1964).
14. KUHN, H. G., *Atomic Spectra* (Longmans, London, 1964).
15. COX, A. H., Wavelengths in the spectrum of ^{86}Kr (1) between 6701 and 4185Å. *J. Opt. Soc. Amer.* 55 (1965) 780.
16. MEGGERS, W. F., and WESTFALL, F. O., Lamps and wavelengths of mercury 198. *J. Res. Nat. Bur. Stds.* 44 (1950) 447.

17. DAVISON, A., GIACCHETTI, A., and STANLEY, R. W., Interferometric wavelengths of thorium lines between 2650 Å and 3400 Å. *J. Opt. Soc. Amer.* 52 (1962) 447.
18. MANNING, D. C., and SLAVIN, W., Lithium isotope analysis by atomic absorption spectrophotometry. *Atomic Abs. Newsletter*, No. 1 (1962) 39.
19. OSBORN, K. R., and GUNNING, H. E., Determination of Hg^{202} and other mercury isotopes in samples of mercury vapour by mercury resonance radiation absorbiometry. *J. Opt. Soc. Amer.* 45 (1955) 552.
20. ATKINSON, R. J., CHAPMAN, G. D., and KRAUSE, L., Light sources for the excitation of atomic resonance fluorescence in potassium and rubidium. *J. Opt. Soc. Amer.* 55 (1965) 1269.
21. BURLING, D. H., CZAJKOWSKI, M., and KRAUSE, L., Light sources for the excitation of atomic resonance fluorescence in caesium and sodium. *J. Opt. Soc. Amer.* 57 (1967) 1162.
22. GOLEB, J. A., The determination of uranium isotopes by atomic absorption spectrophotometry. *Anal. Chim. Acta* 34 (1966) 135.
23. GOLEB, J. A., and YOKOYAMA, Y., The use of a discharge tube as an absorption source for the determination of Lithium-6 and Lithium-7 isotopes by atomic absorption spectrophotometry. *Anal. Chim. Acta* 30 (1964) 213.
24. ZAIDEL, A. N., and KORENNOI, E. P., Spectral determination of the isotopic composition and concentration of lithium in solution. *Optics and Spectroscopy* 10 (1961) 299.
25. GOLEB, J. A., An attempt to determine the boron natural abundance ratio B^{11}/B^{10} by atomic absorption spectrophotometry. *Anal. Chim. Acta* 36 (1966) 130.
25. (a) HANNAFORD, P., and LOWE, R. M., Determination of boron isotope ratios by A.A.S. *Anal. Chem.* 49 (1977) 1852.
26. SOBELEV, N. N., The shape and width of spectral lines emitted by a flame and a d.c. arc. *Spectrochim. Acta* 11 (1957) 310.
27. KIRKBRIGHT, G. F., TROCCOLI, O. E., and VETTER, S., The application of a piezo electric scanning Fabry–Perot Interferometer to the study of atomic line sources (II). Line-widths for calcium in air–acetylene and nitrous oxide–acetylene flames. *Spectrochim. Acta* 28B (1973) 1.
28. KIRKBRIGHT, G. F., and TROCCOLI, O. E., (III) Use of the channeled spectra produced with a continuum source for studies of the absorption line-width for calcium atoms in flames. *Spectrochim. Acta* 28B (1973) 33.
29. DAGNALL, R. M., THOMPSON, K. C., and WEST, T. S., The fluorescence characteristics and analytical determination of bismuth with an iodine electrodeless discharge tube as source. *Talanta* 14 (1967) 1467.
30. THOMPSON, K. C., and WILDY, P. C., Electrically modulated microwave excited discharge tubes in atomic spectroscopy. *Analyst*, 95 (1970) 562 and 776.
31. ALLAN, J. E., A spectral interference in atomic absorption spectroscopy. *Spectrochim. Acta* 24B (1969) 13.
32. BRUCE, C. F., and HANNAFORD, P., On the widths of atomic resonance lines from hollow-cathode lamps. *Spectrochim. Acta* 26B (1971) 207.
33. FASSEL, V. A., RAMUSON, J. O., and COWLEY, T. G., Spectral line interferences in atomic absorption spectroscopy. *Spectrochim. Acta* 23B (1968) 579.
34. MANNING, D. C., and FERNANDEZ, F., Cobalt spectral interference in the determination of mercury. *Atomic Abs. Newsletter* 7 (1968) 24.
35. MANNING, D. C., Chromium Absorption Using the Neon 3593·53 Å line. *Atomic Abs. Newsletter* 10 (1971) 97.
35. (a) NORRIS, J. D., and WEST, T. S., Some Applications of Spectral Overlap in AAS. *Anal. Chem.* 46 (1974) 1423.
36. KIRKBRIGHT, G. F., and SARGENT, M., The application of a piezoelectric scanning Fabry–Perot Interferometer to the study of atomic line sources (I). Assembly and general application of the instrumental system. *Spectrochim. Acta* 25B (1970) 577.
37. ZEEGERS, P. J. T., SMITH, R., and WINEFORDNER, J. D., Shapes of analytical curves in flame spectrometry. *Anal. Chem.* 40 (1968) 26A.
38. FASSEL, V. A., MOSSOTTI, V. G., GROSSMAN, W. E. L., and KNISELEY, R. N., Evaluation of spectral continua as primary sources in atomic absorption spectroscopy. *Spectrochim. Acta* 22 (1966) 347.
39. MCGEE, W. W., and WINEFORDNER, J. D., Use of a continuous source of excitation, and argon–hydrogen–air flame and an extended flame cell for atomic absorption flame spectrophotometry. *Anal. Chim. Acta* 37 (1967) 429.

40. ALLAN, J. E., Atomic absorption spectrophotometry absorption lines and detection limits in the air–acetylene flame. *Spectrochim. Acta* 18 (1962) 259.
41. ELSER, R. C., and WINEFORDNER, J. D., Double modulation—Optical scanning and mechanical chopping—in atomic absorption spectrometry using a continuum source. *Anal. Chem.* 44 (1972) 698.
41. (a) ZANDER, A. T., O'HAVER, T. C. and KELIHER, P. N. Continuum source A.A.S. with high resolution and wavelength modulation. *Anal. Chem.* 48 (1976) 1166.
42. SKOGERBOE, R. K., and WOODRIFF, R. A., Atomic-absorption spectra of europium, thulium and ytterbium using a flame as a line source. *Anal. Chem.* 35 (1963) 1977.
43. ALKEMADE, C. T. J., and MILATZ, J. M. W., A double beam method of spectral selection with flames. *J. Opt. Soc. Amer.* 45 (1955) 583.
44. RANN, C. S., Evaluation of a flame as the spectral source in atomic absorption spectroscopy. *Spectrochim. Acta* 23B (1968) 245.
45. CHEICK HUSSEIN, A. M., and NICKLESS, G., An investigation into the R. F. plasma as an excitation source in atomic absorption and fluorescence spectrometry. (International Atomic Absorption Spectroscopy Conference, Sheffield, July, 1969.)
46. FULLER, C. W., Electrothermal atomization for A.A.S. Chemical Society, London, 1977.
47. COWLEY, T. G., FASSEL, V. A., and KNISELEY, R. N., Free atom formation processes in premixed fuel-rich and stoichiometric oxygen–acetylene flames employed in atomic emission and absorption spectroscopy. *Spectrochim. Acta* 23B (1968) 771.
48. KIRKBRIGHT, G. F., PETERS, M. K., and WEST, T. S., Emission spectra of nitrous oxide supported acetylene flames at atmospheric pressure. *Talanta* 14 (1967) 789.
49. THOMPSON, K. C., Unpublished studies.
50. WILLIS, J. B., Atomisation problems in atomic absorption spectroscopy (I). A study of the operation of a typical nebuliser, spray chamber and burner system. *Spectrochim. Acta* 23A (1967) 811.
51. DE GALAN L., WINEFORDNER, J. D., Measurement of the free atom fraction of 22 elements in the air–acetylene flame. *J. Quant Spectrosc. Radiat. Transfer* 7 (1967) 251.
52. MAVRODINEANU, R., and BOITEUX, H., *Flame Spectroscopy* (Wiley, New York, 1965).
53. RANN, C. S., Absolute analysis by atomic absorption. *Spectrochim. Acta* 23B (1968) 827.
54. DE GALAN L., and SAMAEY, G. F., Measurement of degrees of atomisation in premixed, laminar flames. *Spectrochim. Acta* 25B (1970) 245.
55. WILLIS, J. B., Atomisation problems in atomic absorption spectroscopy (II). Determination of degree of atomisation in premixed flames. *Spectrochim. Acta* 25B (1970) 487.
56. KOIRTYOHANN, S. R., and PICKETT, E. E., Proc. of the 13th Colloquium Spectroscopicum Internationale, p. 270. Ottawa (1967).
57. WOODWARD, C. W., Ionisation of Metal Atoms in flames. *Spectroscopy Letters* 4 (1971) 191.
58. KIRCHHOFF, H., Bestimmung des Isotopenmischungsverhältnisses von Bleiproben mit der Atomabsorptionsmethode. [Determination of isotope ratios for lead samples by AA]. *Spectrochim. Acta* 24B (1969) 235.
59. BRIMHALL, W. H., Measurement of lead isotopes by differential A.A. *Anal. Chem.* 41 (1969) 1349.

Index